TEXTBOOK OF REMOTE SENSING AND GEOGRAPHICAL INFORMATION SYSTEMS

Kali Charan Sahu

ATLANTIC
PUBLISHERS & DISTRIBUTORS (P) LTD

Published by

ATLANTIC

PUBLISHERS & DISTRIBUTORS (P) LTD

B-2, Vishal Enclave, Opp. Rajouri Garden,
New Delhi-110027
Phones : +91-11-25413460, 25429987, 25466842

Sales Office
7/22, Ansari Road, Darya Ganj,
New Delhi-110002
Phones : +91-11-23273880, 23275880, 23280451
Fax : +91-11-23285873
web : www.atlanticbooks.com
e-mail : info@atlanticbooks.com

Chennai Branch
5, Nallathambi Street, Wallajah Road,
Chennai-600002
Phones : +91-44-64611085, 32413319
e-mail : chennai@atlanticbooks.com

Printed in India
at Printman (India), Delhi

Preface

Remote sensing technology in India started as early as in 1960s. During 1970s and 1980s, India emphasized space technology to develop satellites and sensors in the areas of communications, meteorological and Earth resources. Prof. U.R. Rao, who became the Chairman of the ISRO (Indian Space Research Organization) in October 1984, said that space technology had given India the opportunity to convert backwardness into an asset; like some other developing countries it could bypass the intermediate technology stage and leapfrog into the high technology area. India's IRS series started with the launch of IRS-IA on 17th March 1988 through Vostok, a Russian launch vehicle. Since then the Indian space technology has attained momentum and has made tremendous achievements by launching the Oceansat-1 for ocean resources monitoring; Resourcesat-1 for agricultural applications; and Cartosat-1 with a very high resolution panchromatic camera for cartographic applications. In India, remote sensing technology along with Geographic Information Systems is widely being used for more than two decades for inventorying, mapping and monitoring of Earth resources, and mitigation and management of natural disasters. Remote sensing and GIS education is extremely important in the present scenario to cover most of the global and regional phenomena. Geographical Information Systems will find increasing application as an Earth information technology to help boost agricultural productivity, predict climatic changes, preserve biodiversity and help disaster relief management. In days to come it would become the most powerful tool for management and distribution of information. GIS plays an important role in our day-to-day life. GIS is thriving on the web, providing city guides, driving directions, traffic reports, imagery and other online map based services. Mr. Raymond L. Orbach, Chancellor California University, Riverside says, GIS has come a long way from producing maps to three dimensional image. With advancement in computing technologies and the increasing availability of geospatial data and high resolution satellite images, remote sensing and GIS education will play a significant role in the information based economy of the twenty-first century.

There are many books on Remote Sensing and Geographic Information Systems which are mostly application-oriented and suited to those people who are primarily the users of remotely sensed data having sound knowledge on the basics of remote sensing and GIS. However, the subject grows in importance day by day. A significant number of people from various disciplines have been attracted to understand the subject to use in their field of research, apart from academic institutes. This book will serve as fundamental for them as well as textbook for introductory courses in universities and institutes.

This book is solely written to meet the requirements of UG courses in BE (Civil Engg.), B.Tech (Geoinformatics), the PG courses and M.Tech in Remote Sensing and ME (Geoinformatics) of various universities and institutions. Apart from the above, the active research workers in different institutions and remote sensing laboratories in the country and abroad from various disciplines like Geosciences, Ocean Science, Earth Science, Geomatics, Forestry, etc. can make use of the book. As a whole, it is designed to be used for students, scholars and professionals.

Efforts have been made to cover wide varieties of syllabus including the UGC/Department of Space designed course curriculum, universities and institutions and national and international training courses. In India, today, more than 60 universities, institutions and colleges impart education on remote sensing and GIS at Postgraduate, and Graduate levels, as part of the subjects such as Agriculture, Soils, Oceanography, Marine Sciences, Environmental Science, Geology, Geography, Forestry, Wildlife, Civil Engineering, Water Resources and Computer Applications, etc. Policymakers, natural resource management groups and working level resource scientists in most of the government departments are today aware of the potentials of remote sensing and GIS technology and many of them are able to independently work with the remote sensing data. This book can serve them as a reference book.

The book is written in a simple and lucid manner so that those who are inexperienced in the topic and have just started to study remote sensing can understand its various aspects more clearly, without depending on other resources and experts. The principles of remote sensing are explained through simple illustrations. Mathematics and statistics involved are nicely explained, particularly in digital image processing. The book covers a wide range of literatures globally and nationally.

The book is structured in eight chapters. First two chapters are fully devoted to the basics of remote sensing, highlighting what is remote sensing and how it performs, including electromagnetic principles, history and scope of remote sensing, interaction with atmosphere, matter, spectral response curves, etc. One will have clear idea of basics of remote sensing if one concentrates on its contents.

There are individual chapters based on the spectral signatures from different parts of the electromagnetic spectrum, viz. thermal remote sensing and interpretation of thermal imageries and microwave remote sensing. In microwave remote sensing, the radar terminologies and equations are clearly explained, which any one can understand and appreciate.

After a gradual entry into understanding of different parts of the EM spectrum and its functioning and interpretation, the reader will then be exposed to satellites and sensors. In this chapter the new developments with range of spectral configurations are included. The book includes recent developments in space technology like Quickbird, Ikonos, Modis, Oceansat, Resourcesat, Radarsat, etc.

Visual image interpretation is treated in an individual chapter containing the techniques of image interpretation and interpretation keys. All its elements are well explained in this chapter.

Digital image processing, a collection of techniques to process the raw satellite data is an important chapter of this book. For easy understanding the different techniques are explained under (i) image restoration; (ii) enhancement; and (iii) information extraction. Wherever necessary, figures and satellite images are provided to create interest among the readers. The statistics part is given due care, particularly in case of Gaussian stretching, Tassel cap transformation, PC analysis and similar other topics involving statistics.

Chapter 8 deals with the Geographical Information Systems. In this chapter the acquisition, retrieval, analysis and display of spatial data and their raster/vector representation are well conceptualized. The sections are designed keeping in mind the present trends of the diversified audience. This chapter as a whole provides the fundamental basis of GIS technology, spatial data modeling, attributes data management, GIS data analysis and modeling. References have been given at the end to provide access to extensive literature on the subject.

This book will be extremely useful to the students and teachers of Life Sciences, Bioinformatics, Geology, Remote Sensing and the researchers in these fields.

Kali Charan Sahu

Acknowledgements

I wish to extend my thanks to a number of persons for their support and assistance in the preparation of this book. My scholar B.N. Mishra, who is presently working in the Department of Forest, Government of Orissa took much pains for art and design work involved in the preparation of tables and figures using his GIS expertise. Unmesh Chandra Panda, a hard laborious scholar of mine, presently working in Chilika Development Authority, Government of Orissa, proofread the manuscript and assisted in the preparation of References. I express my indebtedness to Dr. P. Kumar, Scientist, Orissa Remote Sensing Application Center, Bhubaneswar for his help in designing the contents of the book and his continuous inspiration and encouragement during the course of preparation of the manuscript.

The hidden blessings showered on me while writing the script are from my teacher and guide B.K. Sahu, Professor of Marine Sciences, Berhampur University, who is presently the Vice Chancellor of Berhampur University. I wish to record his invariable support and constant encouragement.

I also expresses my thanks to my colleagues in the department for their inspiration and encouragement to write a textbook on Remote Sensing and GIS, which they felt was required for the students.

Gratitude is expressed to the personnel of Atlantic Publishers (P) Ltd., New Delhi for their patience, cooperation and assistance in the preparation and production of this book. Particularly, I extend my heartfelt indebtedness to Dr. K.R. Gupta, Chairman, Atlantic Publishers & Distributors (P) Ltd., New Delhi for his constant support and encouragement.

Finally, I am grateful for the patience of my wife Susama and daughter Swetleena who often saw a back in front of the computer during the several months of writing.

Kali Charan Sahu

Contents

Contents

1

Principles and Basic Concepts of Remote Sensing–I

1.1 INTRODUCTION

Remote sensing obviously implies sensing something from a distance. It is generally understood to imply the acquisition of information about an object or phenomenon on the Earth's surface by scientific means or devices called sensors without there being any physical contact between the object and the sensing device. This is done by sensing and recording reflected or emitted energy and processing, analyzing, and applying that information. We all are now familiar with the concept photography. It is a type of remote sensing (RS) technique. The camera is the sensor in this case. Let us examine our five natural senses of sight, hearing, smell, taste, and touch. The functioning of the first three can be compared to remote sensing. The only two fundamental forces of nature that human beings can directly experience through their five senses are gravity and light. Eyes, ears and nose serve as the sensors of our remote visible, audio and chemical environments. In fact, the phenomenon of seeing and recognition of objects around us provides a very close analogy with the modern remote sensing systems. The eye is the sensor in this case for the electromagnetic radiation emanating from or reflected by an object. The brain serves as an ultra high speed computer. It analyses the radiations making use of the stored information in the memory, which the remote sensing specialists call "signatures" and completes the recognition process (Sahai, 1991).

Let us first consider what is the connotation of the term 'remote sensing' as most widely used or understood today. We are familiar with force field sensors and instruments. Those are magnetometers, seismometers, gravity meters, and sensors detecting sound waves like sonars, sodars etc. used in mineral and geophysical exploration. In principle, they all are sensing, remotely but they are sensing fields or sound waves. The most widely used remote sensing techniques today are based on the detection and sensing of electromagnetic radiations (EMR) and hence in its current connotation "Remote Sensing" refers to the sensing or detection of electromagnetic radiations which are reflected/scattered or emitted by an object.

The remote sensing techniques thus based essentially on the use of sensors capable of detecting and/or measuring the reflected or emitted radiations. Remote sensing can now be defined as the acquisition of data for deriving information about objects or materials (targets) located at the Earth's surface or in its atmosphere by using sensors mounted on platforms located at a distance from the targets. Here, measurements are made in different spectral regions on interactions between the targets and electromagnetic radiation. Electromagnetic waves thus act as vehicles to carry information about objects/phenomena of interest. The field of remote sensing encompasses techniques that obtain precise information about Earth's surface from a distance.

The modern remote sensing technology is a spin off from the military reconnaissance activities and the space technology. It is not something entirely new although the term was coined only around 1960 by Evelyn L. Pruitt—a geographer. One of the first and still an extremely important remote sensing technique, within the meaning of the term as currently understood and used, is aerial photography. First aerial photograph in the history was taken by Felix Tournachon, nicknamed, Nadar, when he used a tethered or captive balloon to photograph a Paris suburb on a sunny autumn day in 1858 A.D. However, in spite of rapid strides made, aerial photography has, by and large, remained confined to the visible part of the electromagnetic spectrum which constitutes only a very small fraction of the electromagnetic spectrum. The present day remote sensing techniques have extended the scope of utilization of the electromagnetic spectrum to almost its entire range. The interaction of non-visible electromagnetic radiations (e.g. Infrared, microwaves etc.) with matter is converted into interpretable information and thus human "visibility" has been extended into the non-visible part of the electromagnetic spectrum. A verity of new sensors and platforms has enlarged the scope of remote sensing technology. In a close look for the scope of defining Remote Sensing, it is a space sensing technology to record data about the physical processes in the Earth's surface from far off distance. This technology is used when direct measurement is impractical (because of great distance) or when a greater breadth of data can be obtained (e.g., wider view from high above Earth).

1.2 SCOPE OF REMOTE SENSING

A basic idea of using remote sensing is to extend the human visibility to capture beyond reach information. Human brain fails to remember every minute detail of daily happenings/events surrounding him. Remote sensing can enhance our memory serving the best possible. With remote sensing images, we can measure and map spatial dimensions of objects. Furthermore, we use remotely sensed data to monitor the dynamics of the phenomena on the Earth surface. These include monitoring the vigor and stress of vegetation and environmental quality, measuring the temperature of various objects, detecting and identifying catastrophic sites caused by fire, flood, volcano, earthquakes etc., estimating the mass of various components, such as biogeochemical constituents of a forest, volume of fish schools in water, crop production of agricultural systems, water storage and runoff of watersheds, population in rural and urbanized areas, and quantity and living conditions of wildlife species. Remote sensing is one of a suite of tools available to land managers that provides up-to-date, detailed information about land condition. Remote sensing uses instruments mounted on satellites or in planes to produce images or 'scenes' of the Earth's surface.

Satellite retrieved images can be used in many applications, both in land and ocean resources inventorying, mapping, monitoring and management of the resources. The uniqueness of satellite remote sensing lies in its ability to show large land areas and to detect features at electromagnetic wavelengths which are not visible to the human eye. Data from satellite images can show larger areas than aerial survey data and, as a satellite regularly passes over the same plot of land capturing new data each time, changes in the land use and condition can be routinely monitored.

In the Land Monitor project, satellite images are being used to provide information on land condition and the changes in that condition through time, specifically salinity and the status of remnant vegetation, to help farmers, environmental managers and planners better manage the land. One of the outcomes of the Land Monitor project will be an archive of satellite images of the south-west agricultural region. To get additional information about land condition, the satellite images are combined with other data such as air photos; digital elevation maps (DEMs) and ground data.

Farmers, landcare workers and field officers, with their detailed knowledge of the vegetation and soils in their own paddocks or regions, can extract information on productivity from simple displays of the satellite images. The information from remotely sensed images can be used in a number of ways for a number of purposes. It is usually combined with information from other data sources (ancillary information), and with information from on-the-ground observations (*in situ* measurements), called 'ground truth', to get a more complete picture of what is happening and to check suspected features or changes.

1.3 REMOTE SENSING SYSTEM

Principally remote sensing is used for inventorying, mapping and monitoring of the Earth resources and the health of the environment. Earth Observation Satellites continuously observe the Earth from space, and the acquired data are provided as satellite images and are used to study environmental problems, to monitor disasters and to explore resources. The information, so generated, is extremely important for the resources managers and policy makers.

The major elements of such a system described by Lillesand and Kiefer (1979) are: (a) Data acquisition and (b) Data processing and analysis (**Figure 1.1**).

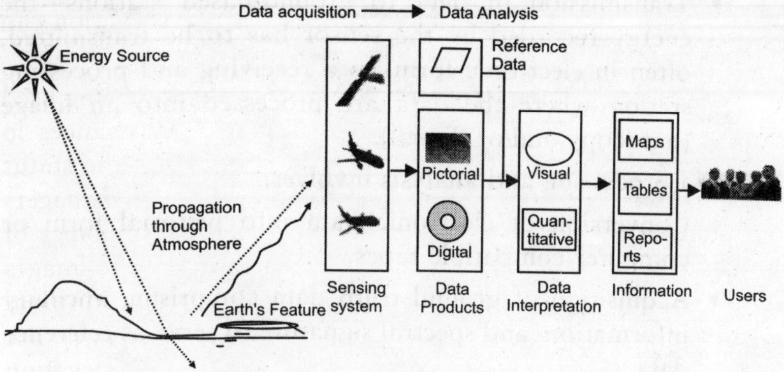

Figure 1.1

Data acquisition process involves the following:

- A source of energy—the first requirement for remote sensing is to have an energy source which illuminates or provides electromagnetic energy to the target of interest.
- Energy in the form of electromagnetic radiation—solar radiation and terrestrial radiation that carries information about the target of interest.
- Propagation of radiation through the Earth's atmosphere—as the energy travels from its source to the target; it will come in contact with and interact with the atmosphere it passes through. This interaction may take place a second time as the energy travels from the target to the sensor.
- Interaction of radiation with matter—once the energy makes its way to the target through the atmosphere, it interacts with the target depending on the properties of both the target and the radiation.
- Sensors—after the energy has been scattered by, or emitted from the target, we require a sensor (remote—not in contact with the target) to collect and record the electromagnetic radiation—active or passive.

- Platforms—airborne or space borne.
- Recording of sensor signals either in pictorial form or electronically in numerical form or magnetic tapes.
- Transmission of data to ground-based stations—the energy recorded by the sensor has to be transmitted, often in electronic form, to a receiving and processing station where the data are processed into an image (hardcopy and/or digital).

Data processing and analysis involves:

- Conversion of electronic data into pictorial form or computer compatible tapes.
- Acquisition of ground truth data comprising ancillary information and spectral signature to serve as reference data.
- Interpretation of the data using interpretation devices, aids or computers, making using of the reference data to extract the information about the target which was illuminated.
- Ground checks or evaluation of the data.
- Generating information products for the users in the form of maps, tables, pictures and reports to aid them in their decision-making process in the management of various Earth resources. Examples are land use maps, crop inventories, etc.

1.4 ELECTROMAGNETIC WAVES

There are different kinds of waves we come across in our real life. The familiar examples are ocean waves and wind waves, which we can see. There are other kinds of waves, which we cannot see but experience. These are electromagnetic waves. Electromagnetic waves get their name from a physical interaction between an electric field and a magnetic field that creates the wave. Electromagnetic waves are produced by the motion of electrically charged particles and hence these are also called electromagnetic radiation. Unlike other waves, the electromagnetic waves do not need medium to travel through. They can travel through empty space as well as through air and other substances. Radio waves are a type of electromagnetic wave. Electricity and magnetism can be static but together when they change or move, they make electromagnetic wave. When an electric field combines with the magnetic field, an Electromagnetic wave is formed. The electromagnetic radiation has two properties. It propagates as a wave and carries a stream of particles (called "photons") that have no mass. High energy photons correspond towards shorter wavelengths of the electromagnetic spectrum and *vice versa*. Perhaps the greatest theoretical achievement of physics in the 19th century was the discovery of electromagnetic waves. An English scientist, James Clerk Maxwell discovered that when there is a change in the magnetic field, automatically there will be a change in electric field and *vice versa*. This results in an electromagnetic wave to exist. Once created, the electromagnetic wave will continue unless it is absorbed by matter.

The radio, TV, microwave oven, etc. make use of the electromagnetic waves. Radio waves, microwaves, visible light, X-rays all are examples of electromagnetic waves that differ from each other in wavelength.

Electromagnetic theory of radiation as explained by Maxwell through his famous equations, describes electromagnetic radiations as rapid variations in the electromagnetic field being caused by the vibrations of the particle (Slater, 1980). Electromagnetic radiation is, classically speaking, a wave of electric and magnetic fields propagating at the speed of light c through empty space. Two characteristics of electromagnetic radiation are particularly important for understanding the phenomenon of remote sensing. These are the wavelength and frequency. The wavelength is the length of one wave cycle, which can be measured as the distance between successive wave crests (**Figure 1.2**). Wavelength is usually represented by the Greek letter lambda (λ). Wavelength is measured in metres (m) or some factor of meters such as nanometers (nm, 10^{-9} metres), micrometers (μm, 10^{-6} metres) (μm, 10^{-6} metres) or centimeters (cm, 10^{-2} metres). The electric and magnetic fields in an electromagnetic wave change their magnitude and direction each second. This rate of change is the frequency υ measured in cycles per second—namely, in hertz. Frequency refers to the number of cycles of a wave passing a fixed point per unit of time. Frequency is normally measured in hertz (Hz), equivalent to one cycle per second, and various multiples of hertz.

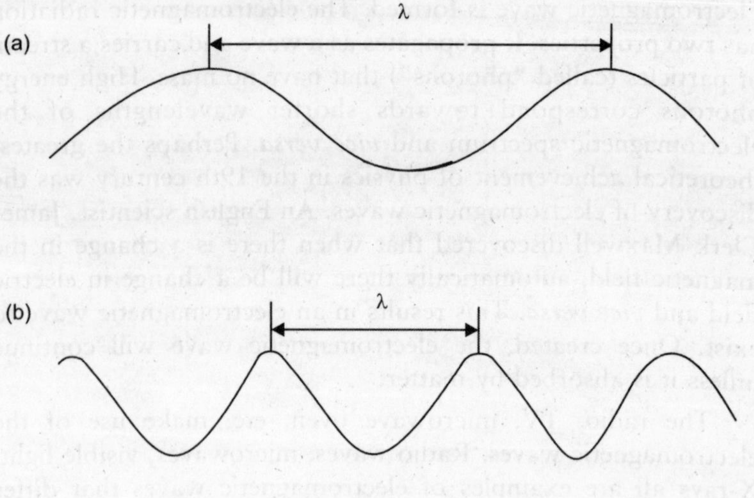

Figure 1.2

The wavelength λ and frequency υ of the radiation are related as: $\lambda \upsilon = c$.

Where, 'c' is the velocity of light—a universal constant. Longer wavelength radiations (Figure 1.2a) like microwaves are of lesser frequencies while shorter wavelength radiations (Figure 1.2b) like visible are of higher frequencies. The velocity of radiations in a medium is reduced by $1/\eta$ times where η is the refractive index of that medium. **Figure 1.3** illustrates oscillating electric and magnetic components of an electromagnetic wave that carries electromagnetic energy (EMR) and momentum.

An electromagnetic wave consists of two oscillating fields, one electric field E and the other magnetic field B. Electromagnetic waves are formed when an electric field combines with a magnetic field. Such waves have the following properties:

(i) Electric field E and the magnetic field B are perpendicular to each other.

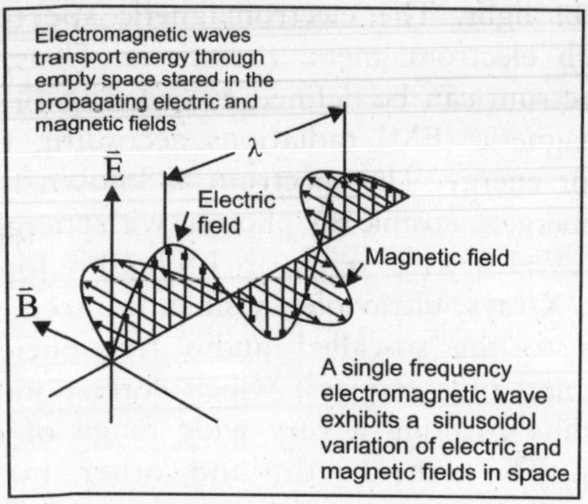

Figure 1.3

(ii) Both **E** and **B** are perpendicular to the direction of motion of the wave front (i.e. direction of propagation). Thus the electromagnetic waves are said to be transverse.

(iii) The velocity of the waves is the same as velocity of light *c*.

(iv) The velocity *c* is reduced by a factor $1/\eta$ while traveling in a medium with refractive index η.

(v) There is as much energy carried by the electric component of the wave as by the magnetic component, and the energy is proportional to the square of the field strength.

The electromagnetic radiation (EMR) is the source of all signals collected by most remote sensing instruments.

1.4.1 Electromagnetic Spectrum

The light that we see through our eyes is the electromagnetic radiation, a part of the electromagnetic spectrum. Electromagnetic radiation (EMR) is energy in the form of electromagnetic waves. The most familiar form of EMR is visible light. However, electromagnetic radiation consists of much more than just visible light. It also includes: X-rays, Gamma rays, Microwaves and Radio waves. The characteristic property that distinguishes different electromagnetic waves is the wavelength. This visible part contains spectrum of colors from red and orange, through blue and purple. Each of these colors corresponds to a different wavelength of light. The electromagnetic spectrum contains array of such electromagnetic radiations. Thus, the electromagnetic spectrum can be defined as ordering or arranging of the electromagnetic (EM) radiations according to frequency, wavelength or energy. The spectrum as known today extends from highly energetic cosmic ray photons with energies equivalent to 10^{14} eV (eV=1.59×10^{-12} ergs or 1.59×10^{-19} Joules) through gamma rays, X-rays, ultraviolet, visible, infrared, microwaves, radio waves to the so-called audio frequencies (used for navigational and time services) which correspond to energies ~ 10^{-13} eV, thus covering a very wide range of energies and wavelengths. The sun, Earth, and other bodies radiate electromagnetic energy of varying wavelengths. The similarity of these radiations lies in the fact that all transport energy and travel with the same speed. All electromagnetic radiations pass through space at the speed of light (300 million meters per second) in the form of sinusoidal waves. They mainly differ in the way they are generated or detected and the way they interact with matter. Radiation can be defined by its frequency and wavelength. If a wave is passing by a point the wavelength is the distance between peaks in the wave, i.e. the distance from crest to crest. The frequency is the time interval between passing peaks. Since they all move at the same speed, the frequency of oscillation is related to the wavelength. Therefore, the higher the frequency, the shorter the wavelength is and *vice versa*. In the visible range, the red has the longest wavelength (lowest frequency and energy) while the blue has the shortest wavelength (highest frequency and energy). We can write:

Frequency = Speed of EM radiation / Wavelength

(Basic unit of Frequency is the Hz or Cycles per Second)

Wavelength = Speed of EM radiation / Frequency

(Measuring unit of wavelength is the Micron)

Figure 1.4 shows the electromagnetic spectrum and its functional division. The spectrum of waves is divided into sections based on methodology of generation, sensing or use of the radiations. There are no natural breaks in the EM spectrum. These separations are made by us for our convenience. The shortest waves are gamma rays, which have wavelengths of 10^{-12} microns (trillionth of a meter) or less. The longest waves are radio waves, which have wavelengths of many kilometers. Visible light is a particular band of electromagnetic radiation that can be seen and sensed by the human eye. This energy consists of the narrow portion of the spectrum, from 0.4 micron (blue) to 0.7 micron (red). The infrared range starts at the end of the red spectrum with wavelengths greater than 0.7 micron. One should note that the electromagnetic spectrum of light contains wide range of energies, wavelengths and frequencies. Remote sensing technology makes use of the wide range electromagnetic spectrum from very shortwave Gamma rays to a very long Radio wave. The two regions which are currently being exploited the most in remote sensing are the optical region (wavelengths 300 nm to 1500 nm) and the microwave region (wavelength 1 mm to about 1 m). The Electro-Magnetic Radiation (EMR), which is reflected or emitted from an object, is the usual source of Remote Sensing data. **Table 1.1** below explains the portions of the electromagnetic spectrum and their spectral ranges.

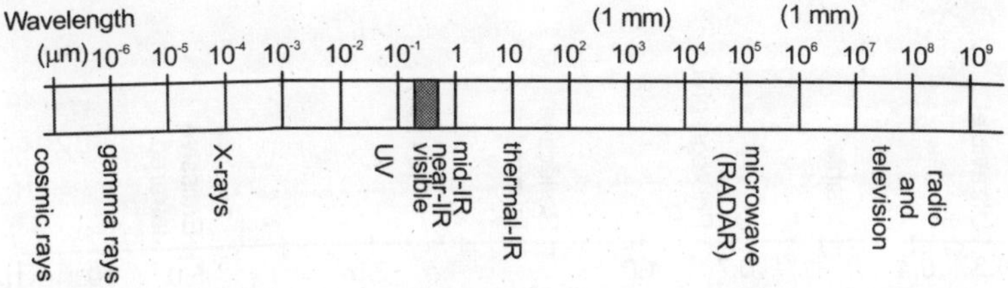

Figure 1.4

TABLE 1.1: Electromagnetic spectrum and their spectral ranges

Type of EM Wave	Typical Unit of Measure
Radio	Meter (m) Centimeter (cm) = 0. 01 m
Microwave (radar)	Millimeter (mm) = 0.001 m
Infrared	Micrometer (μm) = 10^{-6} m
Visible	Nanometer (nm) = 10^{-9} m; 10^{-3} μm
Ultraviolet	Angstrom (A^0) = 10^{-10} m

Optical region of the Electromagnetic Spectrum

The portion of the electromagnetic spectrum extending in wavelength from 0.3 to 15 micron is known as the Optical wavelength region. This is the region in which all objects on the Earth's surface, under natural conditions, either reflect incident solar radiations or emit radiations as a function of their temperature and molecular structure. **Figure 1.5** depicts optical wavelength regions used in remote sensing and their divisions. Although solar radiation is spread over the entire electromagnetic spectrum, major fraction of the solar radiation that reaches the Earth surface is concentrated within the wavelength region extending from 0.3 to 3 micron. Thus the radiation sensed in

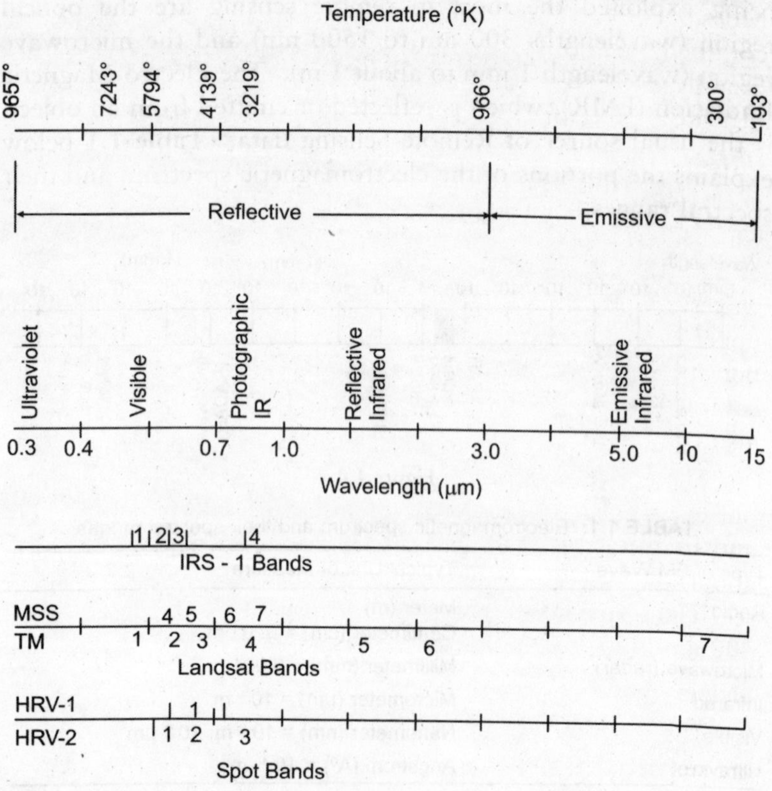

Figure 1.5

this region being predominantly the solar radiations reflected off the objects, the region is called 'reflective region'. It contains Ultraviolet (0.3 – 0.4 μm), Visible (0.4 – 0.7 μm) and Reflective or solar infrared (0.7 – 3.0 μm).

The wavelength band from 0.3 to 0.9 μm is often termed 'photographic band' as it is amenable to detection by photography. The portion of the infrared band from 0.7 to 0.9 μm is called photographic infrared because photographic films sensitive to this wavelength region are available. Energy recorded in the 3 to 15 μm portion of the spectrum is mostly due to emissions from an object as a result of its thermal activity which is a function of the true temperature and the emissivity of the object. The region often referred to as 'emissive infrared'. Most of the remote sensing studies in this region are, however, confined to 8-15 μm portion as the portion from 3 to 8 μm (middle infrared) has many atmospheric bands and, therefore, the interpretation of the spectra becomes rather difficult.

Ultraviolet

The region of the electromagnetic spectrum that extends in wavelength from 400 billionths of a meter to about 10 billionths of a meter is the ultraviolet region. Sunlight contains ultraviolet waves. Ozone prevents most of these radiations in the Earth's upper atmosphere. A small fraction of it is beneficial to humans, but at the same time, larger amount can cause skin cancer and cataracts. Some remote sensing observations of the Earth are also concerned with the measurement of ozone.

There are many millions of colours we can distinguish with our eyes. Each color of light we see has a different wavelength and frequency. The longest wavelength light we see is red, at about 700 nm (700×10^{-9} m). Any longer than this, we cannot see. The shortest wavelength light we see is the violet, at about 400 nm (400×10^{-9} m). Any shorter than this, we see nothing.

Visible light

This is the small portion of the EM spectrum that humans are susceptible to sensation. It is the part of the electromagnetic spectrum that we see. It ranges between wavelengths 400 and 700 billionths of a meter (400 to 700 nanometers or 0.4 to 0.7 µm). It resolves into major components viz. Blue (0.4-0.5 µm), Green (0.5-0.6 µm) and Red (0.6-0.7 µm). Visible electromagnetic radiations play a vital role in remote sensing for inventorying, mapping and monitoring of the Earth resources.

Infrared

The region of the electromagnetic spectrum that extends from the visible region to about one millimeter (in wavelength) i.e. between the visible and microwave portions of the electromagnetic spectrum (**Figure 1.6**). Thus infrared radiation follows microwave radiation on the electromagnetic spectrum. Infrared region that familiar in remote sensing contains a range of wavelengths from 0.7 to 15 µm. Infrared radiation can be emitted from an object or reflected off a surface. Emitted infrared is detected as heat energy, and is therefore known as thermal infrared. The entire region is considered under three divisions: Near infrared (reflected) which can be recorded in film emulsions, Middle infrared (reflected), which can be detected by electro optical sensors and Far infrared (thermal, emissive), which can only be detected using electro optical sensors. Near infrared band (NIR) is nearest to visible light and thermal infrared is nearer to the microwave region of the electromagnetic spectrum. Sensors operating in infrared region are especially important in constructing temperature profiles of the Earth's land and water surfaces and to observe the temperature of cloud tops in the atmosphere, which provide us with clues about weather patterns.

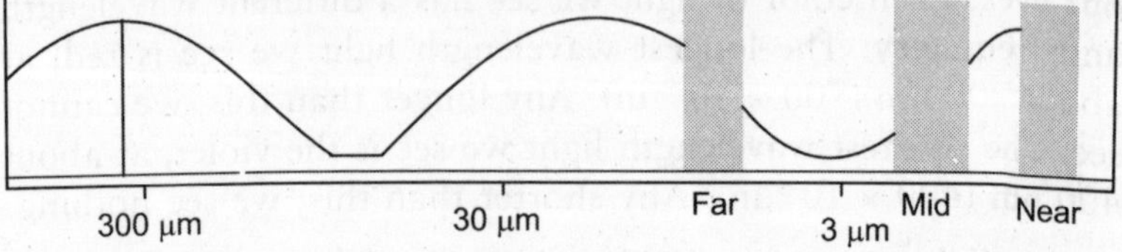

Figure 1.6

Vegetation on the Earth's surface reflects near infrared radiation in varying amounts, depending on many factors including the pigment composition of the plant species, the amount of water available to the plants, the presence of pest or disease stress on the vegetation, and the time of year (within the growing season). Thus, near infrared sensors can be used to quantify vegetative cover on the surface and construct vegetation profiles of the planet. Infrared imageries are very important in detecting disease affected crops and forest firing.

Microwave region

The portion of the electromagnetic spectrum extending in wavelength from one millimeter (the thickness of a pencil lead) to one meter is termed as the microwave region. The upper and lower limits are rather arbitrary. In this region the natural radiations reflected off the objects are a result of the atomic or molecular vibrations and are thus characteristic of the molecular structure and temperature of the object. Microwaves can penetrate solids and unconsolidated Earth materials. That is why the microwaves are good transmitter of information from one place to another. This property makes it possible to detect radiations emanating from below the soil, snow and ice covers etc. and thus open up possibilities for many new applications. It provides an all weather remote sensing capability as the radiations are not appreciably affected by haze, clouds, rain, smoke, hail or falling snow. Earth emits microwave radiation. Such radiation can be detected to retrieve information about the objects of interest. Radars can detect objects by sending microwave energy to illuminate the object and intercepting its backscattered energy. They are considered active remote sensing systems, as they use their own source of energy. Passive remote sensing refers to the sensing of electromagnetic waves which did not originate from the satellite or sensor itself rather natural like sun's energy. The microwave intensities being low, passive microwave remote sensing results in comparatively poorer temperature resolution as well as spatial resolution. There is another very important feature associated with microwaves. The surface which appears rough to us in visible light may be quite smooth in the microwave region, as the microwaves have much longer wavelengths.

Passive microwave radiometers and active sensors like SLAR (Side Looking Airborne Radar), SAAR (Synthetic Aperture

Airborne Radar) are some of the important sensors used in this region of the EM spectrum. The active sensors send out their own EM radiation from an antenna towards the Earth at angles less than 90°. The same is scattered, absorbed or reflected back to a receiving antenna. The return for a given λ depends upon the angle of incidence, orientation of topography with respect to the transmitting antenna, surface roughness and the electrical properties of the object.

Units of wavelength

The types of electromagnetic waves and their units of measurements are given in **Table 1.2.** In basic science, the unit of wavelength is the meter (m). However, in remote sensing, the wavelengths in the visible and infrared regions of the electromagnetic spectrum are measured in terms of micrometer or microns (10^{-6} m). On the other hand for longer wavelengths like microwaves in remote sensing, the unit micrometer is not used. This is due to the fact that the microwave radiations are too long as compared to the shorter visible radiations (0.4-0.7 micrometers) including ultraviolet and near infrared. Astronomers express the unit of wavelengths for shorter radiations as angstrom (10^{-10} m).

TABLE 1.2: Electromagnetic waves and their units

Regions	Spectral ranges
Gamma ray	< 0.03 nanometers
X-ray	0.03 - 3 nanometers
Ultraviolet	3 nanometers - 0.4 micrometers
Visible	0.4 - 0.7 micrometers
Near infrared	0.7 - 1.3 micrometers
Mid-infrared	1.3 - 3.0 micrometers
Thermal (far) infrared	3.0 - 5.0 micrometers AND 8 - 14 micrometers
Microwave	0.3 - 300 centimeters

$$1\ \text{A}^\circ = 10^{-10}\ \text{m and } 1\ \mu\text{m} = 10^{-6}\ \text{m}$$

$$1000\ \text{A}^\circ = 1\ \mu\text{m}$$

i.e., One micrometer (micron) equals 10,000 angstroms.

The remote sensing scientists some times use the term micron in stead of micrometer, which is equal to 10^{-6} m.

1.5 ENERGY SOURCES

The principal source of energy for passive remote sensing systems is the Sun. All passive remote sensing systems depend directly or indirectly on the Sun as the source of energy.

The Sun

Sun emits radiant energy or electromagnetic energy called solar radiation. The primary source of energy for our solar system is the Sun. It is the sustainer of life on Earth. Our Sun has an average diameter of 1.3914×10^6 km. Its mean distance from the Earth is 149.6×10^6 km which is called one astronomical unit (A.U.). Its circumference is 4.373×10^6 km, i.e. 342 times the diameter of the Earth. It is closest to the Earth on January 3 and farthest on July 5. The respective distances on these two dates are 0.983 A.U. and 1.0167 A.U. Average angular subtends of the sun's diameter on the Earth is 9.3 mill radian or 0.5329° or 31′58″. Thus, the Earth receives only a fraction of energy radiated away by the sun.

The energy generates at sun is due to thermonuclear reaction in the core of the sun where the temperatures as high as 14 million °C are present to convert 584 million tons of hydrogen to 580 million tons of helium per second. The balance of 4 million tons of hydrogen is converted into high energy gamma rays. These gamma rays get softened up by a blanket of gas of progressively decreasing thickness as they move outwards from the core. The outer layer of the sun—the photosphere—becomes the apparent source of energy for the solar system. It has a black body temperature of about 6000 K. The spectrum is continuous.

1.5.1 Solar Radiation

Most remote sensing instruments are designed to detect solar radiation and terrestrial radiation. Solar radiation is being characterized by electromagnetic radiation emitted from sun which passes through the atmosphere and is reflected in varying degrees by Earth's surface and atmosphere. This is only detectable during daylight. Visible portion of the sun called photosphere has temperature—6000K. It radiates energy from gamma to radio waves.

Solar radiation (short-wave) is a term used to describe visible and near-visible (ultraviolet and near-infrared) radiation emitted from the sun. These radiations are within wavelength range of 0.20 - 4.0 μm whereas terrestrial radiation (long-wave) is a term used to describe infrared radiation emitted from the atmosphere. The solar and terrestrial radiations and their approximate wavelength ranges are as follows:

- Ultraviolet: 0.20 - 0.39 μm
- Visible: 0.39 - 0.78 μm
- Near-Infrared: 0.78 - 4.00 μm
- Infrared: 4.00 - 100.00 μm

Approximately 99% of solar, or short-wave, radiation at the Earth's surface is contained in the region from 0.3 to 3.0 μm while most of terrestrial, or long-wave, radiation is contained in the region from 3.5 to 50 μm.

Major fraction of the energy received by the Earth's surface is the solar radiation. Around 99.97% of the heat energy required for the physical processes taking place in the Earth—atmosphere system comes from the sun. The other energy sources such as heat from the interior of the Earth, radioactivity, radiations from other celestial bodies, tidal forces, etc. contribute a negligible fraction. The Earth's surface on a clear day, at noon receives the direct solar beam radiation of approximately 1000 watts/meter^2 for many locations. The radiation spectrum from the sun outside the Earth atmosphere is very similar to that of a blackbody at 6,000K. The solar energy received by the Earth is expressed in terms of solar constant (S) which may be defined as the flux of solar radiation incident at the outer boundary of the Earth's atmosphere per unit area (1 cm^2) held perpendicular to the Sun's direction at the mean distance between the Sun and the Earth. The value of S turns out to be 2 cal.cm^{-2} min^{-1}, 2 langleys min^{-1}, or 1400 Wm^{-2} . Based on more accurate recent measurements, S=1375 ± 21Wm^{-2} (Slater, 1980). Thus, the total solar energy intercepted by Earth is $\pi a^2 S$, where a = 6377 km, the mean radius of the Earth.

$$W = \pi a^2 S = 2.555 \times 10^{18} \text{ cal.min}^{-1} = 3.679 \times 10^{21} \text{ cal.day}^{-1}$$

The amount of electromagnetic energy incident on a surface per unit time per unit area is called irradiance. In the past this quantity has often been referred to as "flux". The amount of energy received at the top of the Earth's atmosphere on a surface oriented perpendicular to the Sun's rays (at the mean distance of the Earth from the Sun) is the Solar Constant. The generally accepted solar constant of 1368 W/m^2 is a satellite measured yearly average. In general, solar radiation is received at the Earth's surface. The rate at which direct solar radiation is incident upon a unit horizontal surface at any point on or above the surface of Earth is called Insolation.

The amount of solar radiation incident at the top of the atmosphere depends upon the latitude, longitude, time of the year and time of the day. Not this entire radiation incident on the top of the atmosphere reaches the surface of the Earth. On a global basis of the above radiation (Sellers, 1965),

- 6% is reflected and scattered by atmospheric constituents to space
- 24% is reflected and scattered to space by clouds
- 3% is absorbed by clouds
- 14% is absorbed by atmospheric constituents
- 22% reaches surface of the Earth as diffuse sky radiation
- 31% reaches as direct beam solar radiation on the surface of the Earth.

Thus, the radiation reaching the surface of the Earth is composed of direct beam solar radiation and the diffuse sky radiation.

1.5.2 Terrestrial Radiation

Earth as well as atmosphere emits long wave radiation in the form of electromagnetic waves. Outgoing long wave radiation (OLR) or thermal infrared radiation is called terrestrial radiation. It is determined by the temperature and composition of the Earth's atmosphere and surface. The Earth primarily because it absorbs solar radiation has an average temperature of about 285 K (this is often taken as 300 K for ease calculation) i.e. the ambient temperature of the Earth is 300 K. It radiates almost like a blackbody (or more strictly a gray body) with a power peak (maximum emission) at around 10 μm. The Earth surface materials also attain more or less the same temperature. Earth radiates 160,000 times less than the sun. It radiates essentially all energy at thermal infrared wavelengths between 4–25 μm. **Figure 1.7** shows the electromagnetic spectra for both the solar and terrestrial radiations (after Sellers, 1965). The solar radiation spectrum is similar to the one shown in **Figure 1.8** which, however, shows the spectrum only in the reflective optical region. The energy units used as stated before is langleys (LY) (1 LY = cal cm^{-2}). The terrestrial radiation spectrum is at the outer limits of the atmosphere. Again, water vapor, CO_2, etc. present in the atmosphere is responsible for absorption of the outgoing radiation. Thus only the radiations which are not absorbed by the atmosphere will be sensed by a space borne sensor.

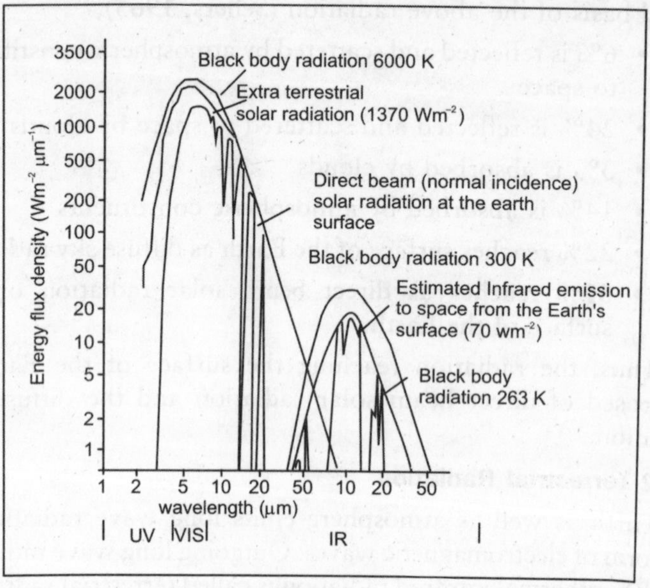

Figure 1.7

1.6 PROPAGATION OF RADIATIONS IN THE ATMOSPHERE

The electromagnetic radiation emanating from any object on the Earth's surface carries information about the physical and chemical structure of the object. During its passage through

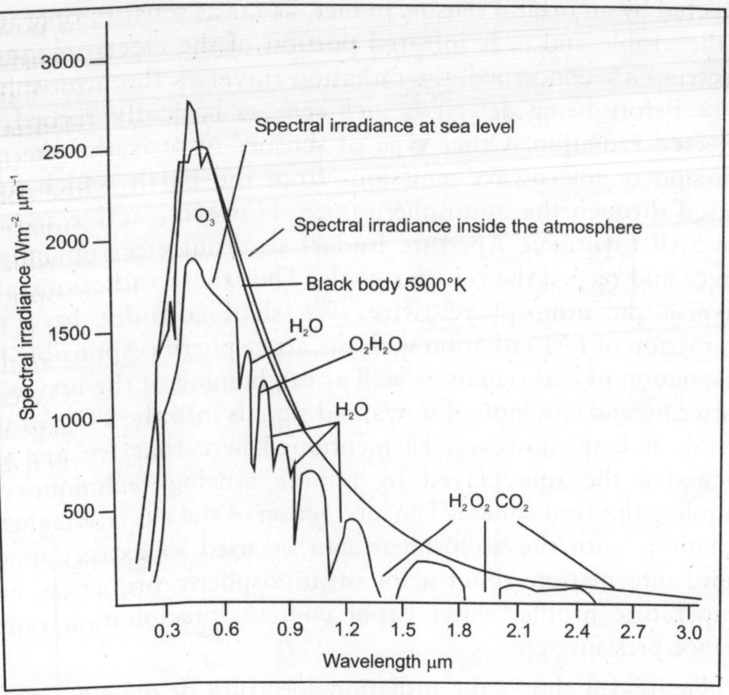

Figure 1.8

the atmosphere before reaching a remote sensor, mounted on a space borne platform, EM radiation undergoes interaction with the constituents of the atmosphere. This interaction could result in either scattering or absorption of the radiation in the atmosphere. Besides, the atmospheric constituents and pollutants present in the atmosphere also emit energy. Thus the atmosphere acts both as an attenuator and a source of radiant energy. This affects the electromagnetic radiation reflected or emitted by an object on the surface of the Earth into the field of view (FOV) of a sensor. The atmospheric contribution i.e. extraneous radiation emitted or scattered by the atmosphere into the sensor's FOV acts as a background noise and is a hindrance in extracting accurate and reliable information about the target on the surface of the Earth.

The radiation emanating from the Earth's surface has to traverse full thickness of the Earth's atmosphere before being detected by an orbital sensor. In fact, as far as sensors operating in the visible and near infrared portion of the electromagnetic spectrum are concerned, the radiation traverses the atmosphere twice before being detected. Such sensors basically record the reflected radiation. Other type of sensors records the thermal emission or microwave emissions from the Earth which have passed through the atmosphere once. However, active sensors like SAR (Synthetic Aperture Radar) send out electromagnetic waves and record the return signals. Thus these radiations also traverse the atmosphere twice. We shall consider here the interaction of EM radiation with the atmosphere responsible for attenuation of real signals as well as brightening of the image by scattering and emission of unwanted signals into the FOV of the sensor. It may, however, be mentioned here that we are not discussing the role played by remote sensing techniques in sounding the atmosphere. The interaction of the electromagnetic radiations with the atmosphere can be used to extract much useful information about a lot of atmospheric properties like temperature profile, water vapor content, precipitation rates, surface pressure, etc.

Figure 1.8 shows the radiation spectrum of the sun at the top of the Earth's atmosphere as well as at the sea level. The lines superposed on the radiation spectrum on the top of the atmosphere are Fraunhofer lines which result from absorption in the outer layer of the sun's atmosphere. The smooth curve is that for a blackbody at 5,900K. The solar radiation curve at sea level shows many absorption lines/bands attributed to water vapor, carbon dioxide, oxygen (O_2) and ozone (O_3). The curve is for the sun zenith angle of 0°. At any other zenith angle, this will depend upon the optical air mass traversed, vertical air mass being taken as 1.

1.6.1 Composition of the Earth's Atmosphere

Earth's atmosphere is a layer of gases surrounding the planet Earth and retained by the Earth's gravity. The Earth's atmosphere is composed of two components. One with a nearly permanent concentration extends to a height of about 200 km are nitrogen (N_2), oxygen (O_2) molecules, argon, carbon dioxide

(CO_2), methane (CH_4), nitrous oxide (N_2O), carbon monoxide (CO) water vapor, etc. and another with variable concentrations of other gases. Nitrogen and oxygen are the main components of the atmosphere by volume. Nitrogen contains 78%, oxygen 21% and argon dominates over 1% mixture of minor gases (**Figure 1.9**). The next most abundant gas in the atmosphere is water vapor. Water (H_2O) vapor is one of the principal variable gases which take primary role in evaporation, cloud formation, and precipitation. Water vapor concentration decreases rapidly with latitude, almost following an exponential function. The highest concentrations of water vapor are found near the equator over the oceans and tropics whereas the volume of the water vapor can approach zero per cent in subtropical and higher latitudes. The fifth most abundant gas in the atmosphere is carbon dioxide. The volume of this gas has increased by over 25% in the last three hundred years. This increase is primarily due to human induced burning for fossil fuels, deforestation, and other forms of land-use change. Methane is a very strong greenhouse gas. Since 1750, methane concentrations in the atmosphere have increased by more than 150%. The average concentration of the greenhouse gas nitrous oxide is now increasing at a rate of 0.2 to 0.3% per year. Its part in the enhancement of the greenhouse effect is minor relative to the other greenhouse gases. Ozone (O_3) concentration also varies

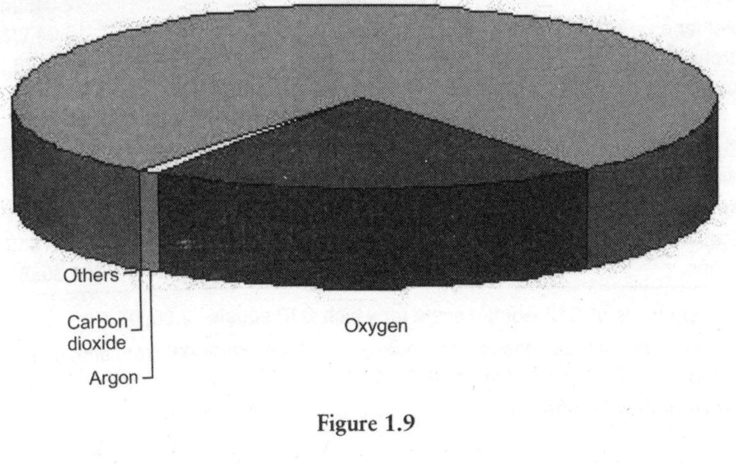

Figure 1.9

significantly with space and time, and occurs principally at altitudes of about 15-30 km. A significant variable gas is a mixture of chlorofluorocarbons (CFCs) produced by industrial activities. **Table 1.3** gives the normal composition of clean, dry atmospheric air near sea level. Water vapor although constituting on the average only 0.3% of the total atmospheric mass, has a high variable concentration; ranging from 30 g/m^3 (hot humid climate) to 10^{-2} g/m^3 (very cold dry climate). Average surface concentration is 10 g/m^3 and average total mass of water vapor in a column of the atmosphere is 2.5 g/cm^2. The atmospheric composition, except for water vapor, up to an altitude of 90 km. remains constant. Hardly 10% of the atmospheric mass lies above 90 km. Ninety-nine per cent of the atmospheric mass lies below a height of 30 km.

TABLE 1.3: Normal composition of clean, dry atmospheric air near sea level (from US Standard Atmospheres, 1962)

Constituent Gas	Gas Symbol	Content (% by volume)	Molecular weight
Nitrogen	N_2	78.084	28.0134
Oxygen	O_2	20.9476	31.9988
Argon	Ar	0.934	39.948
+Carbon dioxide	CO_2	0.0314	44.00995
Neon	Ne	0.001818	20.183
Helium	He	0.000524	4.0026
Krypton	Kr	0.000114	83.80
Xenon	Xe	0.0000087	131.30
Hydrogen	H	0.0000.5	2.01594
+Methane	CH_4	0.0002	16.04303
Nitrous Oxide	N_2O	0.00005	44.0128
+Ozone	O_3	Summer: 0 to 0.000007	47.9982
		Winter: 0 to 0.000002	47.9982
+Sulphur dioxide	SO_2	0 to 0.0001	64.0628
+Nitrogen dioxide	NO_2	0 to 0.000002	46.0055
+Ammonia	NH_3	0 to trace	17.03061
+Carbon monoxide	CO	0 to trace	28.01055
+Iodine	I_2	0 to 0.000001	253.8088

Note: On basis of C12 isotope scale for which C12 equals 12.0000.

\+ The content of these gases may undergo significant variations from time to time or from place to place relative to the normal.

Source: **Valley (1965).**

Apart from the atmospheric constituents listed in Table 1.3 there is particulate matter of various sizes originating from smoke, duct and rock debris, etc. present in the atmosphere. The particulate matter with sizes much larger than the molecules is of two types (Chahine, 1983):

(a) Haze, fog, smoke, smog, pollen, dust particles, typically 0.1 – 10 μm in radius, often referred to as aerosols or particulates.

(b) Clouds of liquid or solid water with radii of 1 – 100 μm.

The particulates are primarily concentrated in the troposphere, usually up to a height of 5 km near industrial areas. Typically lifetime for various particulates is 7-10 days. The relative composition of Earth's atmosphere varies from place to place due to the presence of the aerosols [see above at (a)] and water vapor both of which vary widely in concentration. Low, thick clouds primarily reflect solar radiation and cool the surface of the Earth. High, thin clouds primarily transmit incoming solar radiation; at the same time, they trap some of the outgoing infrared radiation emitted by the Earth and radiate it back downward, thereby warming the surface of the Earth. Both the gaseous molecules as well as the particulate matter play a major role in attenuating the EM radiation and contributing path radiance to the signal.

1.6.2 Layers of the Atmosphere

There is no definite boundary between the Earth's atmosphere and the outer space. The Earth's atmosphere thins upward slowly and fades away to the outer space. There are three physical quantities that determine the vertical structure of the atmosphere. Those are: (i) density (mass per unit volume), (ii) atmospheric pressure (the force the atmosphere exerts per unit area), and (iii) temperature. The Earth's atmosphere contains discontinuities in the vertical structure. According to the discontinuities, the parcel of the atmosphere is divided and further subdivided into various atmospheric layers. Those layers in terms of meteorology are the Troposphere, Stratosphere, Mesosphere, Thermosphere, Exosphere, and Ionosphere.

Troposphere

The lowermost layer of the atmosphere is named as the troposphere. It extends vertically upward from the surface to about 11 km (0 – 7 mi) with some variation due to weather factors. It contains approximately 80% of the total air mass. Clouds and all weather related phenomena (wind, storms, rain, and snow), occur within the troposphere. The temperature decreases with the increasing height. The parcel of air in the troposphere is continually mixed vertically due to solar heating at the surface. The air masses due to heating warm up and rise vertically upward to release the latent heat as sensible heat that further uplifts the air mass. This process continues until all water vapor is removed.

Above the troposphere lies a region known as the tropopause. The tropopause separates the troposphere from the stratosphere. The temperature above the tropopause increases slowly with height up to about 50 km. The troposphere is directly below the stratosphere.

Stratosphere

The stratosphere lies directly above the troposphere and directly below the mesosphere. It extends from about 11 km to 49 km (7 – 30 mi). In this layer the temperature increases with the increasing altitude. In the lower part of the stratosphere, in a region known as the isothermal region, the lapse rate is zero. Higher up, near a height of 20 km, the air temperature begins to increase resulting in a temperature inversion. At the top of the stratosphere the temperature is about 270K. This top layer is called the stratopause. Temperature again decreases with altitude above the stratopause. Ozone (O_3) is an important constituent in the stratosphere. The stratosphere is warmer than the upper troposphere, primarily because of a stratospheric layer that absorbs solar ultraviolet radiation.

Stratopause

The stratopause is the boundary between the stratosphere and the mesosphere. The stratopause is generally considered to be located at an altitude of 50 km. In the stratosphere the temperature increases with altitude, and the stratopause is the section where a maximum in the temperature occurs.

Mesosphere

The atmospheric layer, which is directly above the stratosphere and directly below the thermosphere, is called the mesosphere. The mesosphere is located 50-80/85 km above Earth's surface. Within this layer, temperature decreases with increasing altitude.

Mesopause

The mesopause is the transition between the mesosphere and the thermosphere. It is at an altitude of about 80 km from the Earth's surface.

Thermosphere

The thermosphere begins about 80 km above the Earth to 500 km. Thermospheric temperatures increase with altitude due to absorption of highly energetic solar radiation. Great variations in temperature are found in the thermosphere due to the low density of this layer.

1.7 INTERACTION OF RADIATION WITH ATMOSPHERE

As mentioned in the foregoing discussion, quite an appreciable fraction of the radiation incident on the top of the atmosphere does not reach the surface of the Earth. Interaction of electromagnetic radiations with the atmosphere can affect their speed, direction, intensity, wavelength and spectral distribution. Understanding all of these phenomena is very important in remote sensing for the proper choice of sensors, selection of spectral bands and also for the proper interpretation of the images. When sun's energy penetrates through the parcel of the atmosphere there are three optical events taking place. First it can be reflected (known as backscattering), second it can be absorbed and third it can be scattered. At this stage the energy is said to be depleted and attenuated by the atmospheric constituents and pollutants. This prevents a substantial part of it from reaching the Earth's surface due to the presence of thin layer of ozone in the upper atmosphere (Brooks and Miller, 1963; McVeigh, 1977; Sabins, 1978; Michel's, 1979;

WHO, 1979). This is called diffuse solar radiation. The solar radiation that reaches the Earth's surface without being diffused is called direct beam solar radiation. The sum of the diffuse and direct solar radiation is called global solar radiation. Atmospheric conditions can reduce direct beam radiation by 10% on clear, dry days and by 100% during thick, cloudy days.

The most important interactions which the electromagnetic radiation undergoes in the atmosphere are absorption and scattering. Absorption and scattering together result in the attenuation of the radiant flux. After attenuation by the atmosphere almost all the radiant flux of solar radiation which reaches the Earth's surface is of wavelength shorter than 3 μm whereas that emitted by the Earth is principally in the broad waveband from 4 μm to 40 μm. There is thus practically no overlap between the type of radiation reaching the Earth from the sun and the radiation emitted by the surface of the Earth. The atmosphere not only absorbs a part of the solar radiation reaching the Earth, but also part of the long wave (terrestrial) emitted radiation. The type of interaction depends on the frequency/wavelength of the incident radiation. In general, scattering dominates in the ultraviolet and visible radiations, whereas absorption dominates in the infrared and higher wavelengths.

1.7.1 Absorption

Absorption is the process by which the energy of a photon (electromagnetic energy) is taken up by the atmospheric constituents. The photon is destroyed in the process. The absorbed energy is re-emitted as radiant energy. The absorption of energy during wave propagation is often called attenuation. The absorbing medium will only absorb a portion of the total energy. The other energy will be reflected, refracted, or scattered. The energy absorbed can also be transmitted back into other parts of the atmosphere.

The atmospheric constituents and trace constituents present in the atmosphere are responsible for the absorption of the incident radiation. Most important constituents responsible for absorption are water vapor, carbon dioxide, ozone and molecular oxygen. Absorption of visible and infrared radiations takes place at many wavelengths which correspond to their absorption lines/bands. The absorption may result in thermal agitation and re-emission of the radiation at longer wavelengths.

The absorption of the radiation in the atmosphere is primarily by the gaseous molecules. Apart from the kinetic energy associated with translation of molecules, the energy of a gaseous molecule can exist in three types of internal energy states: rotational, vibrational and electronic. Translational energy is due to translational motion of the centre of mass of the molecule. The average translational kinetic energy of a molecule is equal to kT/2 where k is the Boltzmann's constant and T is the absolute temperature of the gas. Rotational energy can be defined as the energy due to rotation of the molecule about an axis through its centre of mass. Vibrational energy is the energy due to vibration of the component atoms of a molecule about their equilibrium positions. This vibration is associated with stretching of chemical bonds between the atoms. Energy due to the energy levels of the electrons of the molecule is Electronic energy. The sum of these energies is equal to the energy of the system. Electronic energy levels are separated by 2-10 eV, pure vibrational states by 0.2-2 eV, and pure rotational states by $10^{-3} - 10^{-5}$ eV. The three forms of energy are quantized and interact with the incident radiations. Incident radiation can get absorbed by a molecule if its energy corresponds to the energy difference between two states i and j i.e. $h\upsilon_{ij} = |E_i - E_j|$.

Pure rotational band has energies ~ 10^{-3} to 10^{-5} eV and hence the emissions are in the far infrared or microwave region. Oxygen and water vapor have pure rotational lines in the microwave region. Rotation – vibration energy levels are 0.2 – 2 eV and radiations fall in the far infrared, thermal infrared, infrared or red regions of the electromagnetic spectrum (Lintz and Simonett, 1976). For example, CO_2 shows rotation-vibration bands in the 1.46-15 μm region and water vapor in the region 0.69-6.26 μm. Ozone has strong absorption bands at 9.6 and 14 μm. The electronic energy levels are ≈ 2-10 eV and the corresponding radiations fall in the ultraviolet or visible part of the electromagnetic spectrum. Thus absorption of radiation takes place in all the regions of the electromagnetic spectrum.

The UV radiation < 0.12 μm are absorbed by oxygen and nitrogen in the upper atmosphere above 100 kilometers, UV radiation between 0.18 and 0.34 μm is absorbed between 50 kilometers and 10 kilometers by ozone and the infrared part is partially absorbed by water vapor in the troposphere below 10 kilometers. Absorption of ultraviolet (UV) in the atmosphere is chiefly due to electronic transitions of the atomic and molecular oxygen and nitrogen. Due to the ultraviolet absorption, some of the oxygen and nitrogen molecules in the upper atmosphere undergo photochemical dissociation to become atomic oxygen and nitrogen. These atoms play an important role in the absorption of solar ultraviolet radiation in the thermosphere. Ozone in the stratosphere absorbs about 99% of the UV radiation shorter than 0.32 μm. It has been observed by measurement from space platforms that the ozone layers are depleting over time, causing a small increase in solar ultraviolet radiation reaching the Earth. Infrared (IR) and near infrared (NIR) radiations are absorbed by rotational transitions associated with the vibrational modes in water vapor. These water bands are the cause of the intense and extensive atmospheric absorption; they are by far the largest contributors for the greenhouse effect (Ludlam, 1980; Ramanathan and Vogelmann, 1997; Tyndall, 1861). The water and carbon dioxide molecules have absorption bands centered at the wavelengths from near to long wave infrared (0.7 to 15 μm). Water vapor absorbs strongly in wavelengths between 5-8 μm and 20 μm^{-1}, 1,000 μm (1 cm). However, they (water vapor and carbon dioxide) absorb in narrow bands 0.9-2.7 μm. In the far infrared region, most of the radiation is absorbed by the atmosphere. There is little absorption of the electromagnetic radiation in the visible part of the spectrum. The atmosphere is practically transparent to the microwave radiation. Absorption of atmospheric gases has maximum influence in wavelengths < 0.3 μm and minimum impact on wavelengths greater than 0.6 cm. The atmospheric effects of various electromagnetic radiations are mentioned in **Table 1.4.** The portions of the electromagnetic spectrum which are exploited the most in remote sensing are shown in **Table 1.5.**

TABLE 1.4: Electromagnetic radiations and their atmospheric responses

EMR	Wavelength (nm)	Atmospheric Effects
Gamma ray	< 0.03	Completely absorbed in the upper atmosphere by Oxygen and Nitrogen.
X-Ray	0.03-3	Completely absorbed in the upper atmosphere by Oxygen and Nitrogen.
Ultraviolet (B)	3-300	Completely absorbed by oxygen, nitrogen, and ozone in the upper atmosphere.
Ultraviolet (A)	300-400	Transmitted through the atmosphere, but atmospheric scattering is severe.
Visible	400-700	Transmitted through the atmosphere, with moderate scattering of the shorter waves.
Reflected Infrared	700-3000	Mostly reflected radiation, water vapor and carbon dioxide absorb in narrow bands.
Thermal Infrared	3000-14000	Strong absorption by water vapor between 5-8 μm and 20 μm-1,000 μm (1 cm), carbon dioxide absorbs 14-20 μm, and ozone 9-10 μm, with two major atmospheric windows.

TABLE 1.5: Important atmospheric windows exploited in remote sensing

Band width		EMR
1.5 - 1.8 μm		Mid infrared
2.0 - 2.4 μm		Mid infrared
3.0 - 5.0 μm		Mid infrared
8.0 - 14.0 μm	(Below ozone layer)	Thermal Infrared
10.5 - 12.5	(Above ozone layer)	Thermal Infrared
> 0.6 cm		Microwave

Source: www.udel.edu/Geography/DeLiberty/Geog474/geog474_energy_interact.html

1.7.2 Scattering

Scattering is an optical phenomenon whereby the radiant energy while interacting with the atmospheric constituents is forced to deviate in all direction from its unidirectional original path. When radiation is scattered only by one localized scattering center, this is called single scattering. It is very common that scattering centers are grouped together, and in those cases the radiation may scatter many times, which is known as multiple scattering. The main difference between the effects of single and multiple scattering is that single scattering can usually be treated as a random phenomenon and multiple scattering is usually more deterministic.

The atmosphere is considered to be an attenuating medium. The incident solar radiation is reflected and scattered primarily by clouds (moisture and ice particles), particulate matter (dust, smoke, haze, and smog), and various gases. The physical phenomenon, atmospheric scattering disperses radiation in all directions. The important scattering agents are gaseous molecules; suspended particulates (aerosols) and clouds. There are three types of atmospheric scattering which the electromagnetic radiations can undergo during their passage through the atmosphere. Such scatterings are important while performing remote sensing events. The atmosphere modifies the frequency, intensity, the spectral distribution and the direction of the incident radiation passing through it. Scattering of EMR by the atmospheric constituents leads to adverse effect in remote sensing. It reduces the image contrast, and it changes the reflectance characteristics (spectral signature) of ground objects as seen by the sensor. Scattering of light is important particularly in Radar. Scattering is divided into two distinct forms (i) elastic, and (ii) inelastic. Elastic scattering of EM radiation includes Rayleigh scattering and Mie scattering. Inelastic EM scattering effects include Brillouin scattering, Raman scattering, inelastic X-ray scattering and Compton scattering. Light scattering is a physical phenomenon that contributes to the visible appearance of most objects around us. An object looks white due to almost completely scattering of light by the surface of the object itself. The absence of surface scattering leads to a shiny or glossy appearance. Light scattering can also give color to some objects.

Rayleigh scattering

As the name indicates, Lord Rayleigh modeled first the effect of scattering successfully. As per his model to apply, the sphere must be much smaller in diameter than the wavelength (λ) of the scattered wave; typically the upper limit is taken to be about 1/10 the wavelength. Accordingly, Rayleigh scattering is caused by very small particles and molecules with radii far less than the electromagnetic radiation of interest. This type of scattering is important for molecules of the atmospheric constituents, mainly O_2 and N_2 molecules for which

$$a << \lambda$$

where 'a' is the diameter (~ 0.1 – 1 nm) of the scattering particle. The intensity of scattering is inversely proportional to the fourth power of the wavelength. σ_R is proportional to λ^{-4}. Therefore, shorter wavelengths, such as blue light (λ ~ 0.45 μm) in the visible spectrum gets scattered much more than the red radiation, thus are affected the greatest. Blue color of the sky in a clear day or the red sun set can be explained by Rayleigh scattering. Rayleigh scattering is dominant for near ultraviolet and visible radiations and becomes negligible for wavelengths longer than 1 micrometer.

Mie scattering

This is very important when the particle size is comparable to the wavelength of the incident radiation. This is caused by particles with radii between 0.1 and 10 μm such as dust, smoke and aerosols. For larger diameters, the problem of electromagnetic scattering by spheres was first solved by Gustav Mie, and scattering by spheres larger than the Rayleigh range is therefore usually known as Mie scattering. In this case, the shape of the scattering center becomes much more significant and the theory only applies well to spheres and, with some modification, spheroids and ellipsoids. Water vapor, dust and many other particles have sizes of the same order of magnitude as the wavelength of radiation and hence Mie scattering takes place. Mie scattering may or may not depend upon the wavelength of the incident radiation. Scattering coefficient of aerosols varies as λ^{-1} to λ^{-3}. In both Mie and Rayleigh scattering there is no substantial change of the energy (and thus wavelength and frequency) of the light. However, the scattered radiation undergoes a Doppler shift, which can be detected and used to measure the velocity of the scattering center/s in forms of techniques such as LIDAR and RADAR. This shift involves a slight change in energy.

Non-selective scattering

The last type of scattering is non-selective scattering. It occurs in the lower portion of the atmosphere when the size of the particles is much larger than the incident radiation. It causes all interacting wavelengths to scatter equally. The particles taking part in such an event are of radii exceed 10 μm (e.g. water droplets, ice fragments or large dust particles). Such particles directly reflect any incident radiation. When the particles are sufficiently dense they act as an opaque layer between the sensor and the Earth's surface. Cloudy and dust-laden atmospheres cause non-selective scattering in the visible and infrared wavelengths and thus reduce the effectiveness of data in these spectral regions for remote sensing of the Earth's surface. Fog can have a similar effect on thermal infrared wavelengths. Microwaves have much longer wavelengths so are usually only affected by very heavy rain where the size and density of water droplets is much higher.

To summarize, the electromagnetic radiations emitted by the Earth surface objects as a function of their temperature and molecular structure, and the radiations reflected off the objects, undergo absorption and scattering within the Earth's atmosphere before being sensed by the remote sensors. There are certain atmospheric windows within which radiations are transmitted without any appreciable losses in intensity due to absorption. Scattering of radiations is most important for visible and near IR radiations. Remote sensing studies of the atmosphere itself, however, use absorption and scattering characteristics to measure atmospheric state and composition. In these studies the absorption bands, rather than windows, are preferred and scattering effects indicate atmospheric condition.

1.7.3 Atmospheric Transmission Windows

Each atmospheric molecule has its own set of absorption bands in various parts of the electromagnetic spectrum. The radiations which reach the space borne or airborne sensors are the ones which have wavelengths not falling within the absorption bands of such atmospheric constituents. The wavelength bands which are free of absorption bands and hence of least attenuation are known as Atmospheric Transmission Windows. A range of wavelengths over which there is relatively little absorption of radiation by atmospheric gases is known as atmospheric window. The infrared window loses much of its transparency under very humid conditions due to continuum absorption by water vapor, and can become completely opaque when clouds are present. **Figure 1.10** shows the atmospheric windows which are important for remote sensing in the optical region and microwave region. The major windows (**Table 1.6**) are the visible window (including

TABLE 1.6: Atmospheric windows

Wavelengths	Window
< 300 nm	Absorbed by ozone
0.3 - 0.9 μm	Major UV-visible-near IR window
1 - 5 μm	IR window between H_2O and CO_2
8 - 15 μm	Major IR windows
1.3 cm - 1.9 mm	Centimeter/millimeter windows
1.8 - 1.1 mm	Millimeter window
0.8, 0.45, 0.35 mm	Submillimeter windows
2 cm - 10 m	Radio window
> 10 m	Ionospheric absorption

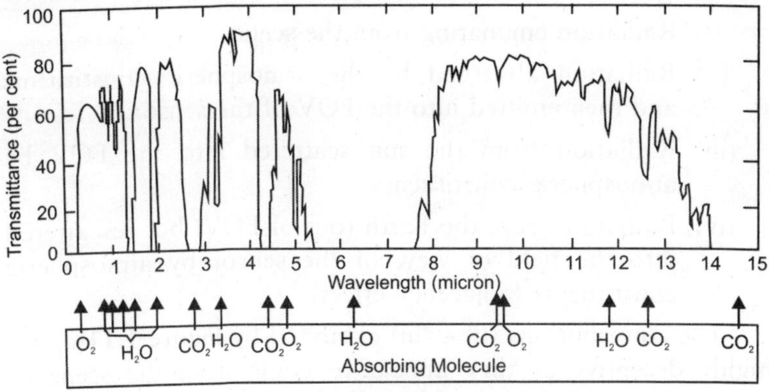

Figure 1.10

UV and NIR), from 0.3 to 0.9 μm; the infrared window, from 8 to 15 μm; and the microwave window, at wavelengths longer than 1 mm. The wavelength bands used in remote sensing systems are usually designed to fall within these windows to minimize the atmospheric absorption effects. The sensors in the orbiting space borne systems must be sensitive to the radiations which fall within the atmospheric windows to be able to sense the radiations emanating or being reflected by the Earth surface materials.

1.7.4 Atmospheric Effect on Remotely Sensed Data

While sensing the Earth from the space, the sensors are looking through a parcel of atmosphere in between the sensor and the target on the Earth's surface. Electro-magnetic energy (solar as well as diffuse sky radiation) when incident on the Earth's surface, the target on the Earth's surface gets irradiated. The irradiance is then reflected, transmitted or emitted by the target and recorded by the sensor. The intervening atmosphere significantly influences target leaving radiance measured by space borne sensors. Because of this, the intensity of the EM radiation recorded at the sensor gets modified. Hence, it is felt important to calibrate and validate the remotely sensed data in order to get the real signal reproducible to remote sensing data products. The radiant energy that reaches the field of view (FOV) of the sensor from a target of interest may contain the following components:

(i) Radiation emanating from the scene.

(ii) Radiation absorbed by the atmospheric constituents and then emitted into the FOV of the sensor.

(iii) Radiation from the sun scattered into the FOV by atmospheric constituents.

(iv) Radiation from the Earth (out of FOV) but rescattered into the field of view of the sensor by atmospheric constituents (adjacency effect).

The contributions (ii) – (iii) exhibited by **Figure 1.11**(a) are highly deceptive as they appear to come from the scene of interest at O. The object N is really speaking not in the field of view of the sensor. This flux contains no spatial or spectral modulation from the scene. "Its effect is to reduce of the contrast of the scene and thus make fine detail invisible or at least harder to detect. Furthermore, in adding a uniform flux level to that from the ground scene, it confuses the interpretation of the spectral signatures of the scene" (Slatter, 1980). On the other hand, as shown by **Figure 1.11(b)**:

(i) A photon reflected from the object of interest at O is absorbed in the atmosphere.

(ii) A photon from O is scattered out of FOV.

(iii) A photon before reaching O is scattered out of FOV.

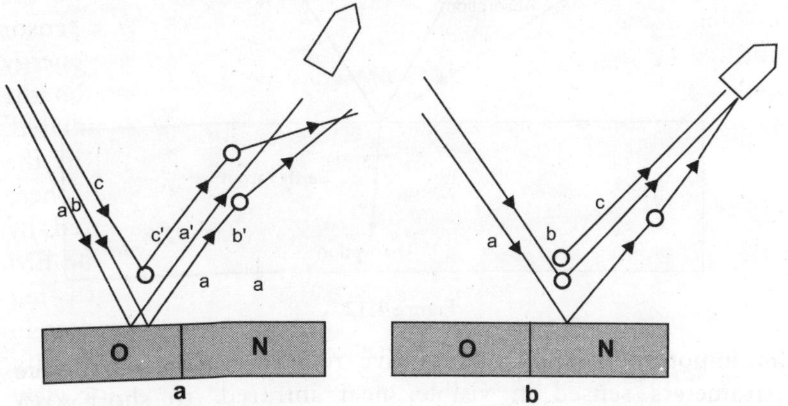

Figure 1.11(a) **Figure 1.11(b)**

This leads to attenuation of the signal. Some of the effects of the atmosphere on remote sensing processes are illustrated schematically in **Figure 1.12.**

1.7.5 Corrections for Atmospheric Effects

Remotely sensed data have been used for developing and validating various land and ocean studies. However, because of the atmospheric interference, the large volume of data collected from the remote sensors get corrupted. In order to retrieve the reflectance from the target in an ideal remote sensing condition, removal of the effects caused by Rayleigh (due to air molecules) and Aerosol (mainly Mie scattering) scattering i.e. atmospheric contribution is highly essential. Atmospheric correction plays

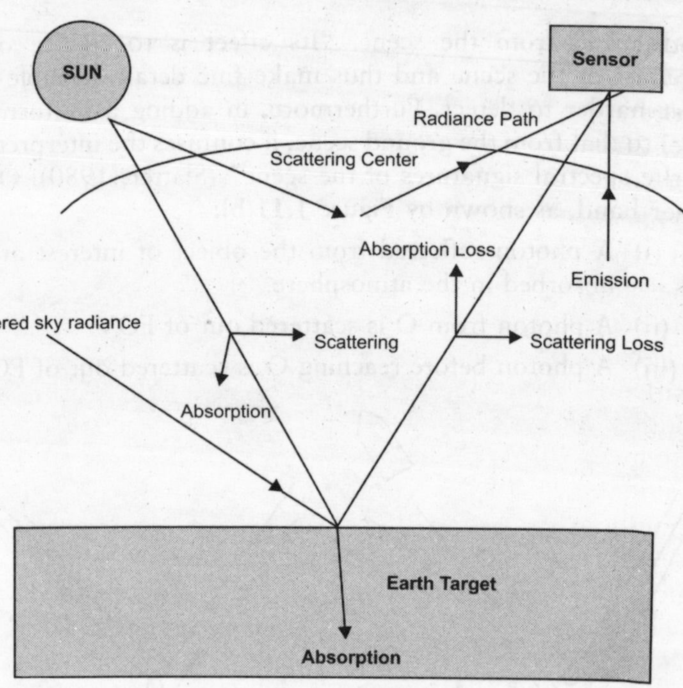

Figure 1.12

an important role in quantitative remote sensing of surface parameters sensed in visible, near infrared, to short-wave infrared waveband. For surfaces with low reflectance such as ocean surface, atmospheric contribution to the reflected solar radiance reaching the sensor dominates over the radiance reflected by the ocean surface. The atmospheric contribution due to combined Rayleigh and Mie scattering is termed as the atmospheric path radiance. It is quite strong and constitutes more than 85% of the radiance at the top of the atmosphere (TOA). Therefore, to estimate the oceanic constituents correctly, the very vital step is to remove the atmospheric contribution from the detected radiances. Thus atmospheric correction is considered as the major step for ocean color remote sensing. Different atmospheric correction models in ocean color remote sensing were suggested by different authors (Gordon, 1977; Sturm, 1980; Gordon and Clark, 1980; and Gordon et al.,

1980). The details of atmospheric correction procedures adopted by different ocean color researchers can be found in literature (Moore et al., 1999; Wang, 1999; Hu et al., 2000; Chauhan et al., 2001; Nayak et al., 2001; Sathyendranath et al., 2001; Chomko et al., 2003).

The effect of atmospheric scattering is called adjacency effect, which is the atmospheric contribution reflected or emitted by the target on the Earth surface into the Field of View (FOV) of the sensor. This fraction creates background noise for the real signal, analysis of which the object or phenomenon can be identified. This adjacency effect plays an important role in high resolution remote sensing of inhomogeneous surfaces. Extensive research has been focused towards appropriate atmospheric correction. In principle, if one knows the atmospheric condition, particularly, atmospheric aerosol information such as aerosol optical depth (AOD), aerosol particle size distribution (number of particles per unit volume), and reflective indices, surface can be retrieved by radiative transfer model calculation for horizontal homogeneous atmosphere with homogeneous surface.

Radiative transfer is a phenomenon of energy transfer by radiation in media which can absorb, emit, and scatter electromagnetic energy when they interact with it. The concept of radiative transfer modeling is a fundamental and key aspect in remote sensing. Radiative transfer theory has been recognized as the principal modeling method that accounts for the solar radiation in the atmosphere. Within the last decade, however, the development of efficient radiative transfer algorithms and freely available gaseous transmission codes, coupled with the steady improvements in computer technology have made detailed atmospheric radiative transfer modeling accessible to a much larger audience. The atmospheric measurements can be utilized to test whether the radiative transfer procedures treat the existing conditions in a realistic manner, and to establish the errors introduced when certain atmospheric properties are derived in the atmospheric correction algorithms (IOCCG Report, 1999).

For inhomogeneous surfaces, one needs to reduce the adjacency effect, which is related to radiative transfer model for horizontally homogeneous atmospheric and inhomogeneous surfaces (Minzheng, 1999). In fact, in many cases, information for atmospheric condition is not available. On the contrary, evaluation of atmospheric aerosol optical depth and related parameters is an important task for the optical remote sensing researchers as on the other hand it plays significant role in the present day global warming. In general, dynamic monitoring of both the atmospheric scattering (path radiance) as well as Earth leaving radiance becomes an urgent task in Earth Science research and various applications. In this sense, we need to develop retrieval algorithms for simultaneous remote sensing of atmospheric aerosol optical depth (AOD) and surface reflectance. This can be possibly done by using known empirical (by some other research) spectral dependence of atmospheric scattering and surface reflectance. For example, Doerffer (1992) has derived an empirical relationship by equating the radiance detected by a space borne sensor at the Top of the Atmosphere (TOA) with the aerosol path radiance, Rayleigh path radiance (molecular scattering) and water leaving radiance in ocean color remote sensing. Some workers use dark pixel (in certain wavelength) to derive atmospheric aerosol optical depth, then by using empirical relationship to obtain aerosol optical depth in other wavelengths, which may be used as the basis of atmospheric correction for other non-dark surface pixels. In this procedure, there should be specific provision for existence of few additional channels to record dark pixel radiance. IRS-P4 OCM payload, for example, has 765 and 865 nm channels in NIR region in which ocean surface will act as a dark back ground due to high infrared absorption by water. In these wavelengths all the detected radiances are due to scattering in the atmosphere. From the radiances measured in the long wavelengths, the atmospheric path radiances in the ocean color wavelengths ($\lambda < 700$ nm) are determined and removed from the sensor radiances through an atmospheric procedure.

As for the correction of adjacency effect, there have been detailed discussions with preliminary success (Tenner et al.,

1987; Zagolski and Gastellu-Etchegorry, 1995). Recent development of various methods includes various empirical methods and radiative transfer models that attempt to account for wavelength specific absorption and scattering. Empirical methods include ratio, subtraction and empirical line corrections. In recent years, new developments in atmospheric correction algorithms have come up. An atmospheric correction algorithm based on inverse modeling by a neural network is applied to spectral measurements, which is essentially used by the scientists in multispectral remote sensing of data correction. The accuracy of image classification has significantly improved after the application of different atmospheric correction algorithms. Atmospheric correction algorithms basically consist of two major steps. First, the optical characteristics of the atmosphere are estimated either by using special features of the ground surface or by direct measurements of the atmospheric constituents or by using theoretical models. Various quantities related to the atmospheric correction can then be computed by the radiative transfer algorithms given the atmospheric optical properties. Second, the remotely sensed imagery can be corrected by inversion procedures that derive the surface reflectance.

In order to correct for atmospheric effects, the knowledge of the following, apart from the characteristics of the sensor system, is essential:

(i) Irradiance at the ground feature of interest with reflectance R.

(ii) Atmospheric spectral transmittance between the ground and the sensor.

(iii) The upwelling atmospheric radiance between ground and sensor, also called the atmospheric path radiance.

If we are considering radiance emitted by the object, then of course, irradiance is not needed. However, the following two methods are essentially used:

(i) Image redyeing

(ii) Dark object subtraction.

In this case the atmosphere is essentially acting as an attenuator of the signal. In the image redyeing method, corresponding pixels from different spectral bands are ratioed. This removes not only atmospheric effects but also effect of the sun angle variation or topographic effect.

In the dark object subtraction method, the signatures are corrected by assuming that the signal produced by the least reflective object in the scene is primarily due to atmospheric path radiance and hence this can be subtracted from the signals produced by other features in the scene.

2

Principles and Basic Concepts of Remote Sensing–II

2.1 INTERACTION OF ELECTROMAGNETIC RADIATIONS WITH MATTER

When electromagnetic radiation is incident on matter, the energy gets transferred into the matter. The induced energy in the matter may re-emit back the way it came (i.e. reflected), scattered in a different direction or transmitted forward into the material. During the process of reflection and transmission of electromagnetic radiation, the electromagnetic waves transfer energy to the electrons of the material. This causes the electrons of the material to vibrate. If the material is transparent, then the vibrations of the electrons are passed on to neighboring atoms through the bulk of the material and re-emitted on the opposite side of the object. If the material is opaque, then the vibrations of the electrons are not passed from atom to atom through the bulk of the material; but rather the electrons vibrate for short periods of time and then re-emit the energy as a reflected light wave. The light may be re-emitted from the surface of the material at a different wavelength which would change its color.

Einstein got the Noble prize in physics for explaining the photoelectric effect based on quantum property of electromagnetic radiation. All electromagnetic radiation "carries" energy. When electromagnetic energy is absorbed by any substance the absorbed radiation delivers its energy in discrete indivisible bundles called quanta or photons. Each photon carries energy equal to the product of frequency and the Planck's constant.

$$Q = h\upsilon$$

where $h = 6.625 \times 10^{-34}$ J.s and υ = frequency of the radiation.

The frequency of an EM wave in above is given by $\upsilon = c/\lambda$, where 'c' is the speed of light.

In remote sensing, the intensity of electromagnetic radiation is called the brightness. The electromagnetic intensity is related to the number of photons (or quanta) passing through a unit area in space per unit time; or equivalently, the number of photons incident on a unit area (surface) per unit time. Few of the radiometric quantities are described below to understand the principles in a more clear way.

Radiant flux = rate of energy transfer by electromagnetic radiation. Units: joules per second or watts. For example, the sun's radiant flux is 3.9×10^{26} W.

Irradiance (E) = radiant flux divided by area through which it passes e.g. for the sun, with visible radius about 7×10^8 m.

$$E = \frac{3.9 \times 10^{26}}{4\left(7 \times 10^3\right)^2} = 6.34 \times 10^7 \text{ Wm}^{-2}$$

Monochromatic irradiance E(λ) = irradiance covering an infinitesimal wavelength interval of the electromagnetic spectrum. Units: $\text{Wm}^{-2}\ \mu\text{m}^{-1}$. It is defined by:

$$E = \int_0^\infty E_\lambda \, d\lambda$$

Sometimes when referring to the Planck function, $B(\lambda, T)$ may be used in place of $E(\lambda)$. In general, the irradiance upon an element of surface may consist of contributions which come from infinity of different directions.

Identifying that part of the irradiance that comes from some specified infinitesimal arc of solid angle $d(\Omega)$, allows a definition of radiance.

Radiance L = irradiance per unit solid angle. Units: $\text{Wm}^{-2}\ \text{ster}^{-1}$. L is also often referred to as intensity (or B when referring to the Planck function). L is the quantity measured by a passive sensor on board a spacecraft.

In order to express the relationship between irradiance and radiance it is necessary to define the zenith angle theta, which is the angle between the direction of the radiation and the normal to the surface in question. The component of the radiance normal to the surface is given by L cos θ.

The irradiance then represents the combined effects of the normal component of the radiation coming from the whole hemisphere, i.e.:

$$E = \int_{\Omega} L \cos \theta \, d\Omega$$

For radiation emitted uniformly in all directions, called isotropic radiation, then E = (pi)L. **Figure 2.1** below shows the geometry in which an infinitesimal area is emitting or receiving radiation and is located at the centre of a sphere of unit radius.

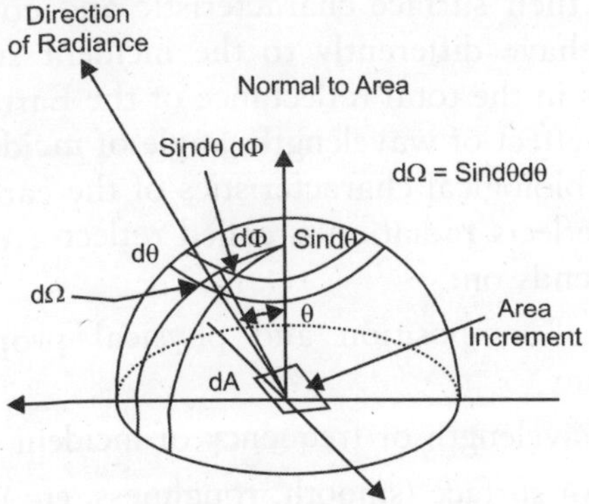

Figure 2.1

It may be pointed out here, that the wavelength or frequency of electromagnetic radiation is independent of penetrating the EM wave but it is the interaction of EMR with matter that penetrates. For instance, microwaves penetrate glass very easily, but they are strongly absorbed by water. When we move up to slightly higher frequency, infrared is strongly absorbed by both glass and water, but both substances transmit visible light. Ultraviolet is stopped by glass, but not so readily by water. Therefore, interaction matters but not the frequency or wavelength of the electromagnetic radiation.

Remote sensing techniques are thus concerned with the detection and measurement of reflectance and/or emittance by the objects of interest. The reflectance is primarily because of the incident solar radiation (except in case of active sensors). The emittance measured is characteristic of the temperature and molecular structure of the object.

2.1.1 Reflection of Electromagnetic Radiations

Electromagnetic radiation incident on any object on the Earth's surface can either through back (reflect) by a surface upon which the radiation is incident or penetrate (transmit) the radiation through the object. In general, a reflecting surface is the boundary between two materials of different electromagnetic properties, such as the boundary between air and glass, air and water, or air and metal. All matters on the Earth's surface differ principally in their surface characteristic and composition and hence they behave differently to the incident solar radiation. The variations in the total reflectance of the Earth's surface are the combined effect of wavelength, angle of incidence, physical, chemical and biological characteristics of the earth cover. The surface that reflects radiation is called reflector or mirror. The reflection depends on:

(i) Surface composition and physical properties of the medium
(ii) The wavelength or frequency of incident radiation
(iii) Type of surface (smooth, roughness, etc.) and
(iv) The angle of incidence.

When electromagnetic radiation strikes the matter, three types of interaction of EM radiation viz. transmission, reflection and absorption with matter are taking place (**Figure 2.2**).

Transmission

It is a process of interaction whereby the interacting incident solar radiation penetrates through the matter. The attenuated energy cause change in velocity and wavelength but not frequency.

Specular reflection

Process whereby incident radiation "bounces off" the surface of substance in a single, predictable direction. It is caused by smooth surfaces. The reflected energy causes no change in

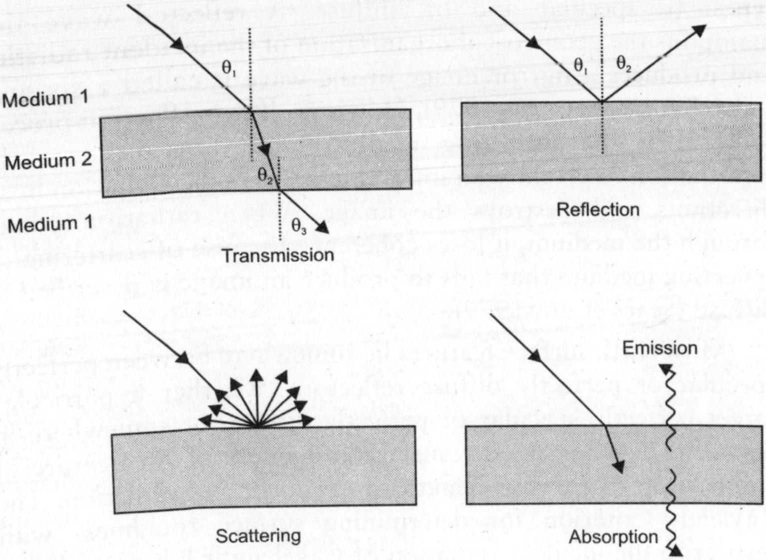

Figure 2.2

velocity or wavelength. When a surface is smooth, we get specular or mirror-like reflection where all (or almost all) of the energy is directed away from the surface in a single direction.

Scattering (diffuse reflection)

Occurs when incident radiation is dispersed or spread out unpredictably in many different directions. It is caused by rough surfaces. The scattered energy causes no change in velocity or wavelength. Diffuse reflection occurs when the surface is rough and the energy is reflected almost uniformly in all directions.

We are familiar with laws of reflection that concern the incident, reflected rays and the normal to the reflecting surface all lie in one plane and that the angle of incidence equals the angle of reflection. Reflection depends on the smoothness of the material's surface relative to the wavelength of the radiation. A rough surface will affect both the relative direction and the phase coherency of the reflected wave. Thus, this characteristic determines both the amount of radiation that is reflected back to the first medium and the purity of the information that is preserved in the reflected wave. The reflection can be of two types: (i) specular and (ii) diffuse. A reflected wave that maintains the geometrical organisation of the incident radiation and produces a mirror image of the wave is called a specular reflection (Sears et al., 1979). However, if the reflection process fails, then the reflecting medium does not preserve the information to produce an image but scatters the radiation in all directions and destroys the image. When radiation passes through the medium, it loses coherence because of scattering. A reflecting medium that fails to produce an image is described as diffuse (Sears et al., 1979).

Most earth surface features lie somewhere between perfectly specular or perfectly diffuse reflectors. Whether a particular target perfectly specular or perfectly diffuse, or somewhere in between, depends on the surface roughness of the feature in comparison to the wavelength of the incoming radiation. The Rayleigh Criterion for determining surface roughness with respect to the incident radiation of wavelength λ is given by

$$H \leq \frac{\lambda}{8 \cos \theta}$$

where H is the average height of irregularities above the mean sea level and θ is the angle of incidence with respect to the normal to the surface. For $\lambda < h$, the surface acts as a diffuse reflector whereas for $\lambda >> h$, the surface is considered to be smooth or specular (Short, 1982). For example, fine-grained sand would appear fairly smooth to long wavelength microwaves but would appear quite rough to the visible wavelengths. Thus a surface can be treated as smooth as long as the roughness is of the order of a fraction of the wavelength of the incident radiation. In specular reflection the angle of incidence equals the angle of reflection and almost all the radiations incident at angles less than 90° are reflected away. In diffuse reflectance which occurs at non-smooth surface the angle of reflection is not the same as angle of incidence. The radiations are scattered randomly in all possible directions. Some are even scattered back along the path of the incident radiation. The relative contribution of specular and diffuse reflection will depend on the roughness of the surface. Reflection of radiation from a very rough surface will be almost entirely diffuse. Reflectance is confined to the reflective region (0.3–3 μm) and sensors operating in this region will essentially record the reflected radiations.

2.1.2 Emission of Electromagnetic Radiation

All objects on the earth's surface at temperature above absolute zero (-273.15° Celsius) radiate EM energy to their surrounding environment by virtue of their atomic or molecular vibrations. This energy, or radiation, is emitted in the form of electromagnetic waves that travel at the speed of light. In a sense, each matter is a radiation source. The radiation laws apply when the radiating body is a black body. Generally, black body conditions apply when the radiator has very weak interaction with the surrounding environment and can be considered to be in a state of equilibrium. The spectral radiant emittance W_λ from a black body is governed by Planck's Distribution Law, as given below:

Planck Radiation Law

Planck's Law of black body radiation predicts the spectral intensity of electromagnetic radiation at all wavelengths from a black body at a given temperature T. The law that we are familiar with is given by:

$$W_\lambda = a\,\lambda^{-5}\,(e^{h\upsilon/kT} - 1)^{-1}\ \text{wm}^{-2}\ \mu\text{m}^{-2}$$

where, a is a constant = $2\pi hc^2 = 3.74151 \times 10^{16}$ wm^{-2}

h = Planck's constant = 6.6256×10^{-34}

c = velocity of light = 2.9979×10^{8} ms^{-1}

k = Boltzmann's constant = 1.38054×10^{-23} JK^{-1}

T = absolute temperature in degrees (K)

λ = wavelength in meters (m)

υ = frequency

The terms we come across in different laws of radiation are:

I = spectral radiance = energy per unit time per unit surface area per unit solid angle per unit frequency ($J\cdot s^{-1}\ .\ m^{-2}.\ sr^{-1}\ .\ Hz^{-1}$)

υ = frequency (the measurement of the number of times that a repeated event occurs per unit of time. It is also defined as the rate of change of phase of a sinusoidal waveform (hertz)

T = temperature (Temperature is a measure of the average energy contained in the microscopic degrees of freedom of a system) of the Black body (Kelvin)

h = Planck's constant (is a physical constant that is used to describe the sizes of quanta) (Joule per hertz)

c = speed of light (meter per second)

e = base of the natural logarithm

k = Boltzmann's constant (physical constant relating temperature to energy) (joule per Kelvin).

The expression makes it clear that spectral radiant emittance holds a definite relationship with temperature of the body and the wavelength (hence on frequency). The basis of Max Planck's theory of Black body Radiation postulated in 1900 is "All bodies on the earth's surface radiate electromagnetic energy continuously". A black body is defined as the one which absorbs the entire radiation incident upon it. Such a body emits maximum possible thermal radiation at any given wavelength for any body at that temperature. A perfect black body at a constant temperature absorbs and re-emits all the electromagnetic radiation that falls on it. A black body is, therefore, a perfect absorber as well as a perfect radiator. If we look at the spectral radiance from a black body at different temperatures (**Figure 2.3**) following observations can be inferred from a study of these radiance curves:

(a) The radiant power peak i.e. the wavelength at which the radiant emittance is maximum shifts to higher wavelength as the temperature is lowered. This is why the dominant colour of a glowing heated object shifts from a red towards shorter wavelength as the temperature is raised.

(b) The curves corresponding to black bodies at different temperatures never cross each other.

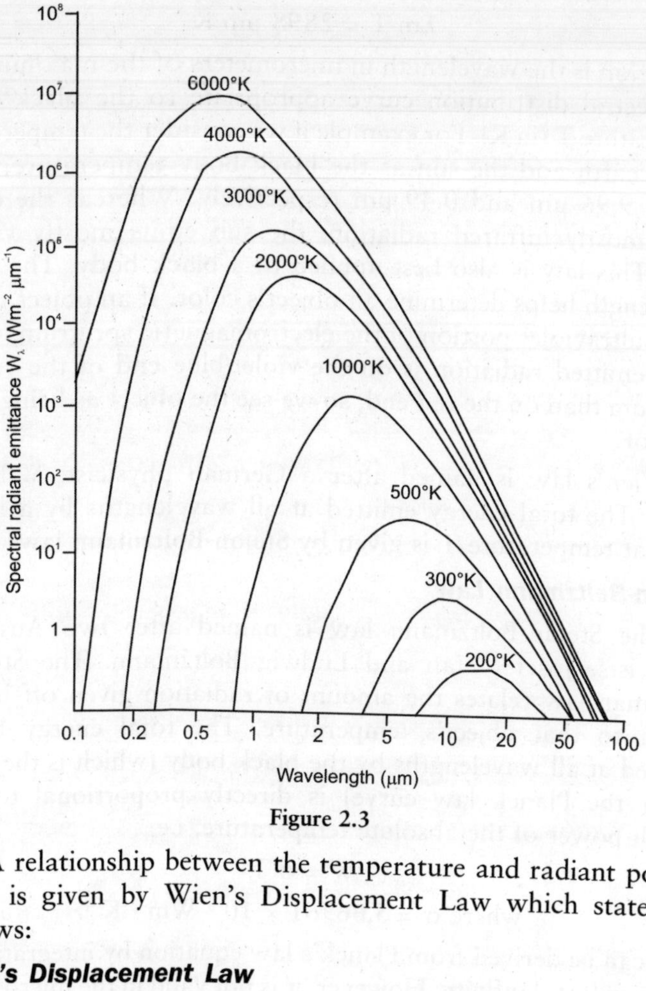

Figure 2.3

A relationship between the temperature and radiant power peak is given by Wien's Displacement Law which states as follows:

Wien's Displacement Law

An object emits radiation at several wavelengths. However, the peak wavelength emitted depends on the object's temperature. Wien's law states that the peak wavelength (λm) given off by a black body radiator depends on the temperature of that object. The cooler the object, the longer the wavelength at which most of the radiation is emitted. The wavelength at which the radiant emittance is maximum is inversely proportional to the absolute temperature:

$$\lambda m\ T = 2898\ \mu m\ K$$

where, λm is the wavelength in micrometers of the maximum of the spectral distribution curve appropriate to the black body temperature T (in K). For example if we consider the temperature of the earth and the sun as the black body temperature, then λ_{max} is 9.96 μm and 0.49 μm respectively. Whereas the earth emits mostly infrared radiation, the sun emits mostly visible light. This law is also best applied to a black body. The peak wavelength helps determine an object's color. If an object peaks in the ultraviolet portion of the electromagnetic spectrum, more of its emitted radiation is on the violet/blue end of the visible spectrum than on the red end, so we see the object as being blue in color.

Wien's law is named after a German physicist, Wilhelm Wien. The total energy emitted at all wavelengths by a black body at temperature T is given by Stefan-Boltzmann law.

Stefan-Boltzmann Law

The Stefan-Boltzmann law is named after two Austrian physicists, Josef Stefan and Ludwig Boltzmann. The Stefan-Boltzman law relates the amount of radiation given off by an object to that object's temperature. The total energy being emitted at all wavelengths by the black body (which is the area under the Planck law curve) is directly proportional to the fourth power of the absolute temperature, i.e.

$$W = \sigma T^4$$

where $\sigma = 5.66961 \times 10^{-8}\ Wm^{-2}\ K^{-4}$.

This can be derived from Planck's law equation by integrating it between 0 and infinity. However, it is not valid in the microwave region as $h\upsilon << kT$ and W becomes proportional to T. This is known as Rayleigh-Jeans law.

Physical matter, however, emits radiation in accordance with the black body radiation except for a proportionality factor å called emissivity whose value is always less than 1. Sometimes physical matter is described in terms of a gray body. A gray body is one that has a spectral radiant emittance of the same functional form as an ideal black body but the intensity is a constant fraction of the ideal curve at every wavelength (see **Figure 2.4**). Black body curves are used to illustrate the Stefan-Boltzmann's law and Wien's law. A hot object has a high curve (more total energy, Stefan-Boltzmann's law), which peaks at a short wavelength (Wien's law). A cooler object has a low curve (less total energy), which peaks at a longer wavelength. Thus Stefan-Boltzmann's law for a general case is written in the form of

$$W = \varepsilon \sigma T^4$$

for a Black body $\varepsilon_\lambda = \varepsilon = 1$

for a gray body, $\varepsilon_\lambda = \varepsilon =$ constant but $0 < \varepsilon < 1$

The Stefan-Boltzmann law explains the growth in the height of the curve as the temperature increases.

For a perfect radiator or reflector, $\varepsilon = 0$

The temperature of an object measured by a remote sensor is thus its black body temperature reduced by a factor ε for that object. The observed temperature is known as apparent, radiometric or brightness temperature.

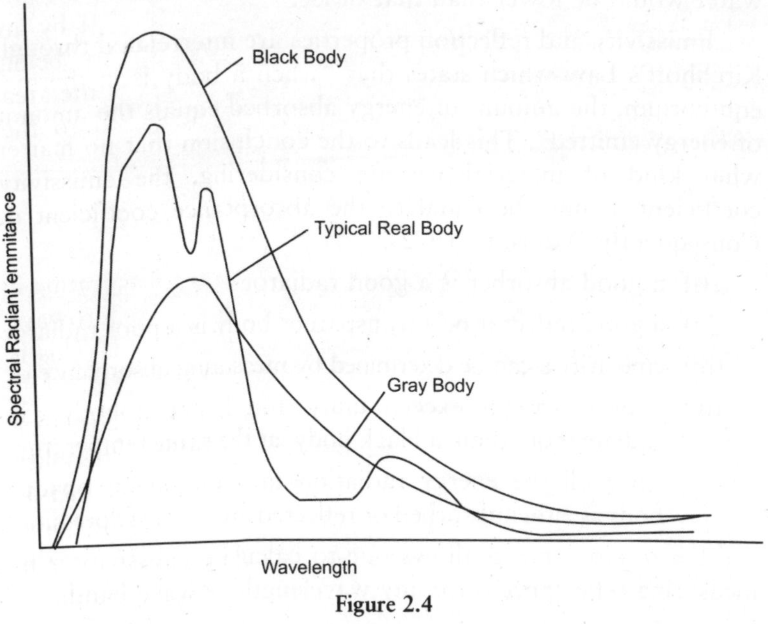

Figure 2.4

The Earth is essentially a gray body. For other objects whose spectral radiant emittance has a functional form different than Planck's law, one can still apply the concept of emissivity but the emissivity will be a function of wavelength, i.e. $\varepsilon_\lambda = f(\lambda)$. For example the atmosphere shows a highly variable emissivity which varied with wavelength. Similarly, white paint has a high reflectance in the visible and hence ε is low. It behaves as a gray body around λ = 3 μm where ε = 0.5 and as a black body beyond λ = 3 μm where ε = 0.9. **Figure 2.4** illustrates the spectral radiant emittance of a black body, a gray body and a typical real object. In general, emissivity is nearly independent of temperature for most common terrestrial objects.

The emissivity of a material is a very important property. If there are two objects at the same temperature in the same environment with one having a higher emissivity than the other, the former will radiate more strongly.

For example, $\varepsilon_{ice} \sim 2\varepsilon_{water}$ in microwave region. Therefore, if ice and water are imaged together by a microwave radiometer, the apparent or brightness (or radiometric) temperature of water would be lower than that of ice.

Emissivity and reflection properties are interrelated through Kirchhoff's Law which states that "when a body is in thermal equilibrium, the amount of energy absorbed equals the amount of energy emitted". This leads to the conclusion that no matter what kind of material we are considering, the emissivity coefficient å must be equal to the absorptance coefficient ñ. Consequently (Vanzetti, 1972):

(i) a good absorber is a good radiator

(ii) a good reflector or a transparent body is a poor radiator

(iii) emissivity ε can be determined by measuring absorptance α

(iv) ε or α cannot exceed unity and hence nobody can radiate more than a black body at the same temperature

(v) since all the energy radiation incident on an opaque body is either absorbed or reflected away, the expression.

$1 = \alpha + \rho = \varepsilon = \rho$ allows one to calculate emissivity α by measuring reflectance ρ for any wavelength or wave band.

Kirchhoff's law

An object radiates unique spectral radiant flux depending on the temperature and emissivity of the object. This radiation is called thermal radiation because it mainly depends on temperature. According to Kirchhoff's law the ratio of the radiated energy from an object in thermal static equilibrium, to the absorbed energy is constant called emissivity and only dependent on the wavelength and the temperature T. A black body shows the maximum radiation as compared with other matter. Therefore, a black body is called a perfect radiator. In remote sensing, a correction for emissivity should be made because normal observed objects are not black bodies. Emissivity can be defined by the following formula:

$$\text{Emissivity} = \frac{\text{Radiant energy of an object}}{\text{Radiant energy of a black body with the same Temperature as the object}}$$

Emissivity ranges between 0 and 1 depending on the dielectric constant of the object, surface roughness, temperature, wavelength, look angle etc. The temperature of the black body which radiates the same radiant energy as an observed object is called the brightness temperature of the object.

The ratio of emitted radiation to absorbed radiation is the same for all black bodies at the same temperature. This law forms the basis for the definition of emissivity:

$$\varepsilon = \frac{M}{M_b}$$

where ε = emissivity, M = emittance of a given object, M_b = emittance of a black body at same temperature.

2.2 SIGNATURES

When solar radiation hits a target surface, it may be transmitted, absorbed or reflected. Different materials reflect and absorb differently at different wavelengths. The reflectance spectrum of a material is a plot of the fraction of radiation reflected as a function of the incident wavelength and serves as a unique signature for the material. In principle, a material can be identified from its spectral reflectance signature if the sensing system has sufficient spectral resolution to distinguish its spectrum from those of other materials. This premise provides the basis for multispectral remote sensing.

Sensors collect and store data about the spectral reflectance of natural features and objects, both of which reflect radiation. This radiation can be quantified on an electromagnetic spectrum.

Remote sensing techniques which are exploited for surveying the earth resources seek to recognize and classify objects and/or their characteristic electromagnetic signatures. The whole edifice of remote sensing techniques is built on the premise that all earth surface objects have characteristic spectral signatures. For any given material, the amount of solar radiation that is reflected (absorbed, transmitted) will vary with wavelength. This important property of matter allows us to separate distinct cover types based on their response values for a given wavelength. The response of earth surface materials to incident radiation (i.e. reflectance) and or energy emitted by all objects as a function of their temperature and molecular structure (i.e. emittance) essentially determines the signatures. Broadly speaking the determination of signatures implies establishing a relationship between the spectral response pattern and the inherent characteristics of an object. A more formal definition is "A spectral signature of a ground surface feature is a set of values for the reflectance or radiance of the feature, each value corresponding to the reflectance or radiance averaged over a different, well defined wavelength interval. In general, the set of reflectance or radiance values is peculiar to the irradiance and viewing geometry existing when the area was sensed" (Slatter, 1980). Spectral signatures are the specific combination of reflected and absorbed electromagnetic (EM) radiation at varying wavelengths which can uniquely identify an object. Every material in nature has a spectral signature. It is a function of (i) incidental EM wavelength and (ii) interaction of the matter with that portion of the electromagnetic spectrum. A spectral signature is a unique reflectance value in a specific part of the spectrum. Spectral signatures (spectral response patterns) change over time and space.

However, we must clearly understand that signatures though quite often distinctive, are not absolute or unique in the real world. What we observe in the real world is the spectral response pattern rather than signature. The spectral response patterns observed by the remote sensors may be quantitative but are not absolute. They may be distinctive but they are not necessarily unique (Lillisand and Kiefer, 1979). Degree to which an object reflects incident electromagnetic energy in different regions of the electromagnetic spectrum is called spectral response pattern.

Objects at the Earth's surface emit and reflect many wavelengths of radiation. In principle, each object has a unique spectral signature, which could be used for identification of the object, much like a finger print. Remotely sensed spectral signatures could be utilized for recognizing and mapping for all manner of features.

2.2.1 Spectral Response of some Earth Surface Features

How the typical Earth surface materials respond to the incident EM radiation are understandable by critically examining the plot of reflectance versus wavelength, what we call spectral reflectance curves. Examples of spectral curves of typical materials such as vegetation, soil and water are shown in **Figure 2.5**.

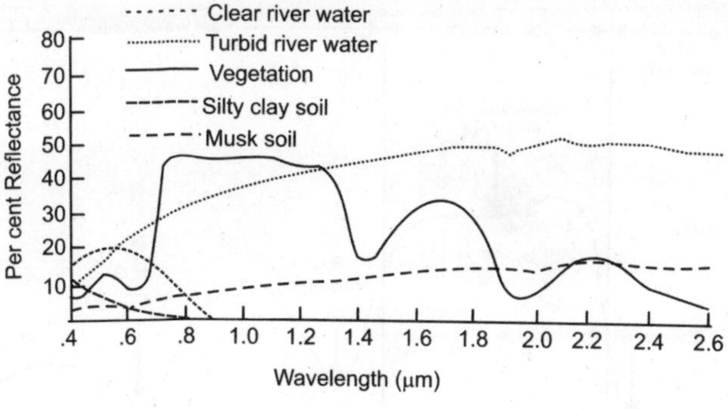

Figure 2.5

Vegetation

Remotely sensed data has shown its capability for recognition, discrimination, classification and growth potential of different types of vegetation cover at several levels. The spectral reflectance of vegetation (**Figure 2.6**) is quite distinct from other materials due to its strong absorption bands at blue and red, high reflectance at green, and very strong reflectance in the near IR region of the EM spectrum. Its spectral response curve manifests the "peak and valley" configuration. At maximum chlorophyll content, vegetation looks the "greenest". However, at less chlorophyll content vegetation appears red or yellow owing to less absorption and proportionately more reflection of the red wavelengths. Plant pigments, leaf structure and total water content are the three important factors that influence the spectrum in the visible, near IR, and middle IR wavelength regions respectively. The unique signature built up in such wavelengths help distinguish readily from other land cover types. Different types of vegetation cover can also be discriminated in visible, near IR and middle IR regions. If we look at the spectral response pattern of typical vegetation, we observe minima around 0.45 and 0.65 μm. These minima in the blue and red regions correspond to two chlorophyll absorption

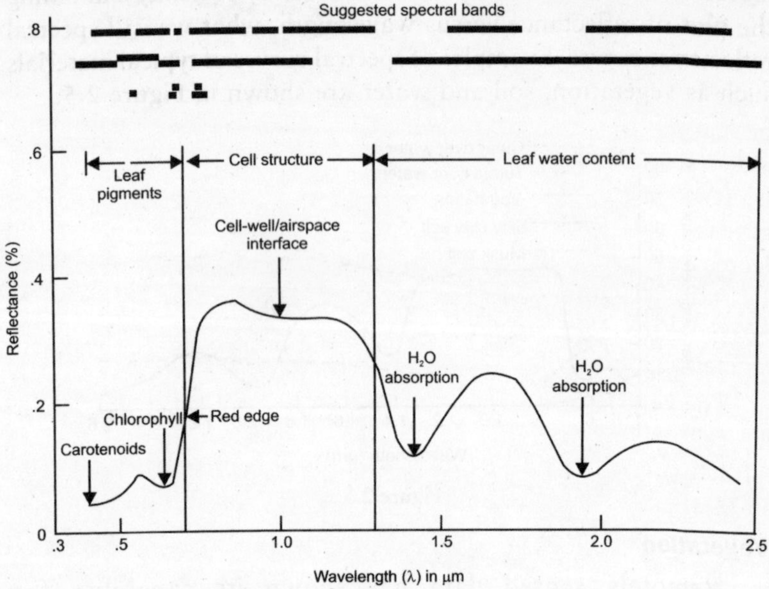

Figure 2.6

bands. The first reflectance peak and relative lack of absorption at 0.55 μm is familiar green maximum. In the near infrared region, the reflectance is fairly high of nearly 45%, transmittance of similar magnitude and absorptance of only about 5% with only minor absorption bands at 0.97 μm and 1.21 μm. The almost flat topped region from 0.7 to 1.3 μm is of primary concern in biological sciences. The middle IR has strong water vapor absorption bands just towards left of 1.6 and 2.0 μm. It has been shown that total incident solar radiation absorbed in this region is directly proportional to the total leaf water content. The peak reflectance in middle IR occurs at 1.6 and 2.2 μm. Thus remote sensing measurements in 1.55 – 1.75 μm and 2.08 – 2.35 μm can give information about moisture content of the plant canopy.

A small amount of light is reflected from the leaf cuticle. Remaining large amount of light transmitted into the spongy mesophyll gets reflected/critically reflected from the leaf cell walls. The discontinuities in the refractive indices within a leaf structure determine its near IR reflectance. For thicker leaves, the reflectance is substantially higher than the transmittance. The low absorption of incident solar energy in the near IR helps the sunlit leaves to keep their temperatures lower. The reflectance in the near IR is controlled by the internal cellular structure of the leaves. As the leaves grow, intercellular air spaces increases and so the reflectance increases. As vegetations become stressed or senescent, chlorophyll absorption decreases, reflectance increases and also there is decrease in intercellular airspaces, decreasing the reflectance in the near IR. This is the reason why the ratio of the reflectance in the near IR to red or any of the derived indices from theses data are sensitive indicators of vegetation growth/vigor (Navalgund, 1993).

The near IR reflectance from 0.7 to 1.3 μm is dominated by the cell wall/intercellular air spaces, and to a lesser extent, by refractive index discontinuities of cellular constituents (Gausman, 1974). Beyond 1.3 μm reflectance is primarily controlled by leaf water content.

During the growth cycle of vegetation the reflectance decreases in the visible wavelength and increases in the near IR wavelengths until maximum canopy development is reached. Then with senescence (senescence is the combination of processes of deterioration which follow the period of development of an organism), the visible reflectance increases while the near-infrared reflectance decreases, although relatively less than the visible increases. Thus vegetation reflectance usually progresses from a background, such as soil to full greenness and then returns to the background again.

The broad absorption areas near 1.4 and 1.95 μm are atmospheric water vapor bands and should be avoided in remote sensing. However, the 1.6 and 2.2 μm regions are useful for distinguishing succulent (average leaf water content of 92%) from non-succulent (average leaf content of 72%) plants (Gausman et al., 1978). According to Collins (1978) the sharp spectral reflectance rise between the chlorophyll maximum absorption and the cellular reflectance maximum, the red edge, can be very useful in detecting phonologic changes and geochemical changes. In a study of maturation changes of crop plants such as corn, wheat and sorghum, Collins detected a red shift (of 0.007 to 0.010 μm) of the red edge to longer wavelengths (0.690 to 0.70 μm) associated with the conversion from the vegetative growth to reproductive growth (heading and flowering). The red shift was useful in separating some crop types, particularly the non-grains from the grain crops (the shift is not as pronounced in the non-grain crops). When plants become stressed, a decrease in chlorophyll productivity causes a shift of the red edge towards shorter wavelengths. This kind of blue-shift has been detected in the reflectance spectrum from a forest canopy growing over copper-lead-zinc sulfide mineralization (Collins et al., 1977).

Some of the external factors which affect the vegetation reflectance are Moisture stress, Soil nutrients, Salinity, Diseases, Seasonal variations and Climatic factors. Leaf temperature increases when there is reduction in transpiration due to inadequate availability of water. Hence measurement of crop canopy, temperatures and their difference with ambient temperature has been used as an indicator to diagnose moisture stress.

A number of studies have linked responses in leaf spectral reflectance, transmittance, or absorptance to physiological stress. A variety of stressors including dehydration, flooding, freezing, ozone, herbicides, competition, disease, insects, and deficiencies in *ectomycorrhizal* development and N fertilization have been imposed on species ranging from grasses to conifers and deciduous trees. In all cases, the maximum difference in reflectance within the 400–850 nm wavelength range between control and stressed states occurred as a reflectance increase at wavelengths near 700 nm. In studies that included transmittance and absorptance as well as reflectance, maximum differences occurred as increases and decreases, respectively, near 700 nm. This common optical response to stress could be simulated closely by varying the chlorophyll concentration of model leaves (fiberglass filter pads) and by the natural variability in leaf chlorophyll concentrations in senescent leaves of five species (Carter and Knapp, 2001). Changes in leaf chlorophyll content can serve as relative indicators of plant vigor and environmental quality.

Soil

Typically soil reflectance curve shows there is an approximate monotonic increase in reflectance with increasing wavelength in the visible and near IR regions. The present day high resolution satellite imagery provides the visible boundaries of soil types. The ground truth equipments such as radiometers help in collecting spectral signatures for the soil surfaces. Then possible soil classification is made interpreting the remote sensing images with the help of the spectral signatures.

As we know, the reflectance and emittance of an object depend upon its physical and molecular state. The spectral reflectance of soil is controlled by six parameters: moisture content, organic matter content, particle size distribution, iron oxide content, soil mineralogy, and soil structure. Surface dimension also affects the reflectance spectrum. Reflectance percentage with respect to wavelength increases from the shorter visible band towards the shortwave infrared band. In visible spectrum, soils usually show high reflectance as compared to vegetation. However, the reverse is the case in near IR band i.e. vegetation gives higher reflectance than soil. Remote sensing allows for classification of soils, providing information on the productivity of forests, meadows, wildlife habitat conditions, land-use and recreational suitability.

Water is a relatively strong absorber of all wavelengths, particularly those longer than the red part of the visible spectrum. As the moisture content of the soil increases, the overall reflectance of the soil tend to decrease and as such two significant water absorption bands are conspicuous at around 1.4 µm and 1.9 µm (**Figure 2.7**). The 0.4 to 1.0 µm region is not useful for monitoring soil moisture content (Reginato et al., 1977), although the entire reflectance curve is generally suppressed with increasing moisture. The region centered at 2.2 µm has the highest correlation with soil moisture; this region was similarly important with vegetation. Radar waves may not be able to penetrate the moist soil. On the soil reflectance spectra, this soil moisture will develop parallel curves. Spectral bands of red and near-infrared bands are independent from the soil moisture. It is also shown that degree of polarization (per cent) is more pronounced for soil with higher moisture content. In a thermal IR image moist soils look

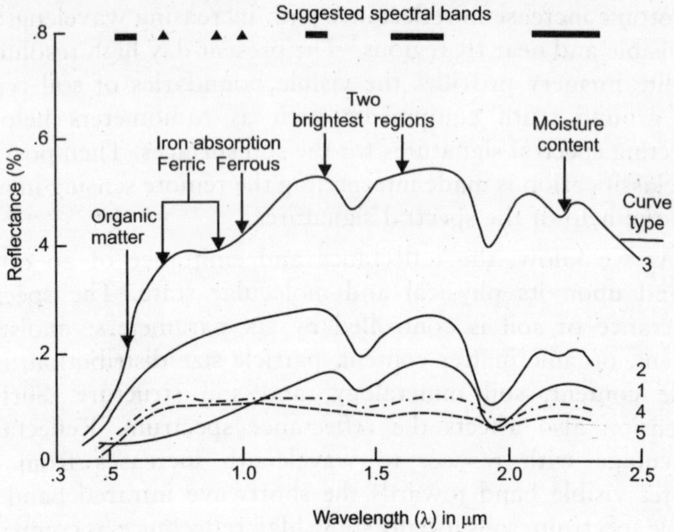

Figure 2.7

darker compared to the dry soils. In view of the large differences in dielectric constant of water and soil at microwave frequencies, quantification of soil moisture becomes possible.

Organic matter and roughness of the soil surface will cause a decrease in reflectance. Organic matter present in the soil has a strong influence on the soil reflectance. As the organic matter content increases, soil reflectance decreases. Such influence of organic matter is based on the soil structure and water retention capacity. High organic matter in soil may produce spectral interferences for band characteristics of minerals like Mn and Fe. Soils rich in iron oxide reflect proportionally more of the red than other visible wavelengths and therefore appear red (rust color) to the human eye. A sandy soil on the other hand tends to appear bright white in imagery because visible wavelengths are more or less equally reflected; when slightly less blue wavelengths are reflected this results in a yellow color. Soil texture (roughness) also affects soil optical properties. Light is trapped in the rough surfaces of the coarse soil particles. For example, if iron and lime are present, a stronger reflectance is received than if the soil material was fine textured and dry. Variations in soil reflectance occur where there is a change in distribution of light and shadow areas with surface roughness areas. This factor is important in the thermal infrared and microwave spectral domains. Soil size and shape influence the reflectance properties. On aerial photographs finely textured soil materials such as clay usually display darker tones than coarse soils. This is the result of either higher water retention or particle size or both. If the size of a soil aggregate expands in diameter, a decrease in reflection will result. The shape is related to the surface (texture); a smooth, even surface will probably result from a more spherical soil aggregate, instead of a jagged soil aggregate. In dry soil, inter-particle spaces are occupied by air. As the moisture content of the soil increases, these air-spaces get occupied by water. The water-soil particle refractive index is lower than the air-particle refractive index. Soil having more moisture content has lower reflectance. Increasing particle diameter results in decreased soil reflectance. Structure of the soil is another parameter, which has an important influence on its reflectance. Soils without much of the structure reflect 15 to 20% more than the soils which have well defined structure. Fine textured soils have low reflectance compared to those soils having sandy texture. The two regions of highest soil reflectance, centered at approximately 1.27 and 1.65 μm, correlate with many soil properties (Stoner and Baumgartner, 1980). With sandy textured soils, a decrease in particle size increases reflectance. However, with medium to fine textured soils, a decreases in particle size decreased reflectance.

Minerals

The solar spectral range, 0.4 to 2.5 micrometers, provides abundant information about many important Earth-surface minerals and research has proved the ability of hyperspectral systems to uniquely identify and map many of these minerals. The reflectance properties of minerals vary considerably with the chemical composition, weathering, types of rock and rock outcrop. Natural surfaces are rarely composed of pure minerals. They are composed of several minerals with weathered product. Weathered rocks generally, give low reflectance. Homogeneous rock types give similar spectra at various points on the sample surface while inhomogeneous rocks show significant variations. True spectrum of a mineral is obscured due to spectral interference of water, organic material in soils vegetation cover. Major constituents of most of the minerals namely silicon, aluminium, magnesium and oxygen do not provide absorption features in the visible and near infrared range. The most commonly absorbed feature in the VNIR region is due to the presence of iron oxide and water or OH group. Rock forming minerals show absorption lines in the 7-12 μm regions.

The reflectance spectra of the minerals are governed by two processes: (i) electronic, and (ii) atomic/molecular vibration. Diffuse reflectance spectra of 15 mineral species commonly associated with sulfide-bearing mine wastes show diagnostic absorption bands related to electronic processes involving ferric and/or ferrous iron, and to vibrational processes involving water and hydroxyl. Many of these absorption bands are relatively broad and overlapping; however, spectral analysis methods, including continuum removal and derivative analysis, permit most of the minerals to be distinguished (Crowley et al., 2003). Because secondary iron minerals are sensitive indicators of pH, Eh, relative humidity, and other environmental conditions, spectral mapping of mineral distributions promises to have important application to mine waste remediation studies. Electronic processes in the rock-forming minerals produce broadband in the ultraviolet, visible, and in the near IR up to 1 μm. Spectral features in such materials are generally due to iron. An increase in iron oxide can cause the decrease in reflectance. Iron oxide (Fe_2O_3) is one of the primary causes of the red colors in many soils. Iron oxide content and organic matter content are the two most important soil properties affecting the spectral reflectance characteristics of eroded soils, particularly in the 0.5 to 1.2 μm region. Chemically removing the extractable iron oxides from a soil sample results in increase in reflectance especially at wavelengths less than 1.1 μm. A broad absorption feature attributed to iron oxide is centered at 0.9 μm. Spectral features due to vibrational process are sharp. No fundamental vibrational mode is seen below 2.5 μm. But overtone and combination tone bands do appear in the near IR and visible range. Water molecules are invariably present with rocks and minerals and have influence on their reflectance spectra. The two characteristic bands at 1.4 μm and 1.9 μm are due to water present in the minerals/rocks. Overtone or combinations of the internal vibrations of the carbonate ions show distinct features in the 1.1-2.5 μm. Laboratory measurement of reflectance spectra of a few rock specimen are shown in **Figure 2.8** (see Navalgund, 1993).

All silicate rocks show broad minima in their spectral emissivity in the 8-12 micrometer region. This is due to the Si-O stretch mode. It is observed that the reflectance maximum shifts to longer wavelengths as the SiO_2 content of the rocks decreases. The change in the diurnal temperature of the rock surfaces is another important parameter amenable to remote sensing. Degree of change is related to the heat capacity of the rock which is a function of its thermal conductivity, porosity, spectral, chemical composition, texture, structure and moisture content. In order to get sharp difference, measurements have to be made long before dawn and around noon.

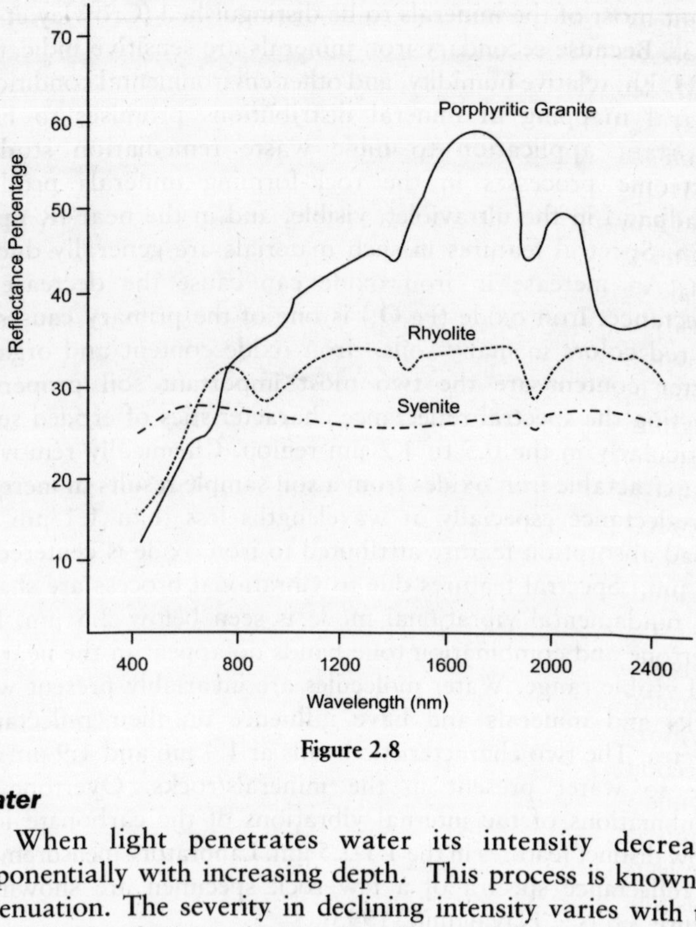

Figure 2.8

Water

When light penetrates water its intensity decreases exponentially with increasing depth. This process is known as attenuation. The severity in declining intensity varies with the wavelength of EM radiation. Irradiance on water surfaces is mostly absorbed or transmitted rather than reflected. Absorption takes place mostly in the longer visible wavelength, near IR and middle IR regions than shorter visible wavelengths. Thus water typically looks blue or blue-green due to stronger reflectance at these shorter wavelengths, and darker if viewed at red or near infrared wavelengths. This property enables easy delineation of even small water bodies. The strong absorption of near infrared energy, in contrast to the high reflectance of soil and vegetation at this wavelength, clearly delineates land/water boundaries. Variations in reflectance from water surfaces may be due to changes in water depth, turbidity, water color, substrate type (if shallow) and/or surface texture. Generally reflectance increases with decreasing water depth or particulate content. Water color may be due to suspended particulate matter, organic materials such as chlorophyll or colored dissolved organic matter (CDOM). Turbidity in water generally tends to better reflectivity and a bright appearance of the water. The apparent color of the water due to turbidity shows a slight shift of the reflectance peak towards longer wavelength. Suspended sediment can be easily confused with shallow (but clear) water, since these two phenomena appear very similar. Increase in chlorophyll concentration leads to greater absorption in the blue and red regions and more reflection in green band, making the water appear greener in color. Dissolved gases and many inorganic salts do not manifest any changes in the spectral response of water. However, the water surface roughness and bottom reflectance lead to complex problems while making water related interpretation.

The two spectral bands in thermal infrared (TIR) region which are currently being used in various applications of remote sensing are 3-5 and 8-14 μm, correspond to atmospheric windows. The shorter wavelength region (3-5 μm) is predominantly used for sensing very hot objects, such as fires, while the 8-14 μm region coincides with the peak energy emission for the Earth, so is used for most applications. Unlike reflected energy, emitted energy such as thermal infrared may also be detected at night. For example, AVHRR imagery is available for day and night overpasses. Thus day/night pairs of thermal imagery may be obtained for an area and these are frequently used to study properties such as thermal inertia. Thermal infrared data may also be calibrated to indicate apparent surface temperature. This technique is commonly used for sea surface temperature mapping (where actual sea surface temperature can be measured to within 2 degrees) and in meteorological applications. The relative temperature differences indicated by the thermal infrared channel may also be used in land areas to indicate differences in soil moisture (such as with irrigated crops) and/or vegetative cover.

Snow and clouds

Snow has very high reflectance, up to 0.8 μm and then decreases rapidly afterwards. Impurities and an increase in grain size reduce reflectance and consequently as snow grows, reflectance decreases. In case of clouds, there is non-selective scattering and they appear uniformly bright throughout the range 0.3-3 μm. The cloud tops and snow generally have same temperature and hence it is not easily possible to separate these in the thermal IR region. Hence, the two atmospheric windows in the middle IR wavelength regions 1.55 to 1.75 and 2.11 to 2.35 μm are important for snow cloud discrimination.

At 1.6 μm wavelength, snow has very low reflectance while the reflectance of cloud remains high (**Figure 2.9**). This allows a significant discrimination between the snow and clouds at 1.6 μm. Therefore, both cirrus and optically thick clouds can be directly classified and distinguished from snow at the 1.6 μm wavelength (Warren, 1982). This has been clearly demonstrated using the operational Landsat Thematic Mapper satellite, which has a channel centered near 1.6 μm (channel 5; 1.57-1.78 μm) (Dozier, 1987; Baglio and Holroyd, 1989).

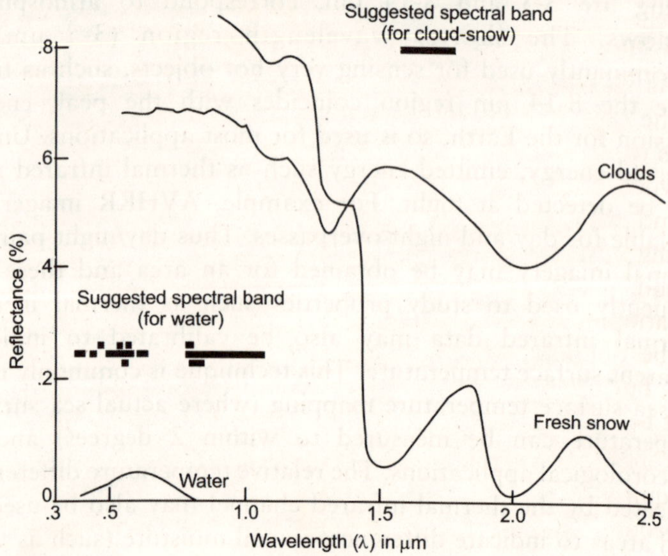

Figure 2.9

Man-made materials

Some of the man-made materials we come across in satellite imagery are concrete, bricks of various colors, painted materials, wood and roofing materials such as sheets and asbestos. They do not change their reflectance properties overtime but for ageing / degradation. The red wood has high reflectance at 1 μm and falls off at both shorter and longer wavelengths. Woods covered with paints show reflectance properties of paints generally. Concrete has lower reflectance in the shorter wavelengths of ultraviolet and blue regions and its reflectance is much less if it is wet or has accumulation of oil on the surface. Road surface materials have high reflectance in the shorter wavelengths. Metal surfaces do not show any fine structure in the visible and near IR reflectance spectra. However, rusted porous metals have higher reflectance in the red region. Galvanized iron, a common roofing material shows a dip at 1 μm due to ferrous iron oxide present. Road surfaces, buildings reach higher temperatures than the surroundings because of the absence of any evaporative cooling and low thermal masses. The metal roof tops generally show low brightness temperatures because of lower value of emissivity.

2.2.2 Spectral Signature Measurements

As we know, spectral signature serves as the ground truth for the objects about which accurate and reliable information needs to be extracted. Measurements of ground truth are made in synchronous with the satellite overpass. The purpose of ground investigation in remote sensing is to give the remote sensing investigator a realistic portrait of the target. These ground measurements are necessary in both research and operational applications. Sometimes the ground observations can be obtained from regularly collected or already available data, but often a ground investigation data collection program must be specifically designed for the remote sensing mission.

Careful analysis is necessary to determine what information is required to meet the need of particular task. Proper selection of ground investigation equipment is dependent on proper selection of the variables to be measured. When correctly used, the combined remotely sensed data and ground truth data reduce considerably the time for analysis of the scene. Some Earth surface processes and conditions are dynamic and transient, that may not persist for even a minute (e.g. internal waves). Under such conditions ground truth collection at the time of satellite overpass is the only way to verify, what was actually present and what the imagery may subsequently reveal. The ground investigation involves two kinds of observations with respect to remote sensing. These are (a) visual and (b) spectral measurements. The visual ground truth investigation involves observations such as color, size, shape, texture, pattern, association, and temperature of the object of interest. The spectral investigation involves measurements of spectral reflectance of a particular object, so called *in situ* observation, using specialized equipments such as field spectroradiometer, underwater spectroradiometer, etc.

The aim of field and laboratory study is to establish the capability for detection and identification of various Earth resources. These are, therefore, essential towards establishing spectral signatures and the parameters that influence them. This involves the measurement of reflectance and emittance of various materials in laboratory and at natural field conditions. All radiation measuring instruments can be broadly divided into two categories: (i) Spectrometer and (ii) Radiometer. The spectrometer measures the discrete energy content of the radiation emitted or reflected by a target at every wavelength of the spectral wave band. The radiometer on the other hand measures the total content of radiation within a predetermined spectral band. The two most common forms of dispersing elements are the prism and the diffraction grating. They can both be used in similar optical arrangements to produce a spectrum from a particular source. If the spectrum is observed by eye, the instrument is called a spectroscope. If the spectrum is recorded on photography film, the instrument is called a spectrograph. If the instrument produces a spectrum that can be scanned in some way with a detector to determine the wavelength position of emission or absorption lines in the source, the instrument is called a spectrometer. If in addition, the instrument can measure the radiant exitance of these lines, then it is called a spectroradiometer. These entire instruments are used in spectral analysis of a source of radiation.

Spectral signature measurements on various objects can be carried out from instruments placed on tripods, cherry picker, balloons, airborne and space borne platforms. In case of space borne instruments and to certain extent balloon/aircraft instrument the effect of the atmosphere on the spectral response of the objects need to be understood and corrected for before drawing meaningful conclusions. Spectrophotometer with provision for reflectance studies has been a standard instrument for making measurements in the laboratory. They generally operate in the 0.3 to 3.0 μm range and provide very fine spectral resolution (~ nm). A large number of studies have been carried out on single leaf of different crops grown under different conditions, or leaves stacked together to understand/simulate thus reflectance and transmission properties of crop canopies on soil samples with varying degree of moisture content and different compositions, water samples with different constituents, etc. Measurements on rock specimen belonging to different types have been done to understand the basic absorption properties of rocks/minerals. Laboratory measurements suffer from the fact that they do not correspond to real life situations. For example, soil sample taken for a laboratory measurement may not have the same packing structure, texture, etc. Spectrum from a single leaf or stacked leaves can be quite different from the reflectance of a crop canopy in view of its geometry itself. So, *in situ* measurements are required to understand the spectral response of objects in their natural environment. Portable hand held battery operated ground truth radiometers operating in discrete spectral bands have been used for spectral measurements under *in situ* conditions. Spectral bands of such instruments have been chosen so as to match those of space borne instruments. $BaSO_4$ coated reflectance panel having uniform high reflectance in the VNIR range is used for calibration. Spectroradiometers with provision for making measurements in narrower spectral regions are also available. Data-logging facility is available with these instruments to facilitate measurements in short line.

Infrared thermometer is another instrument widely used for making crop, canopy and ambient temperature measurements and their difference.

Measurement from ground based instruments has one important limitation. The measurements represent a small area of the ground corresponding to field of view, the instrument from 2 m altitude covers approximately an area of 25 cm radius. This is much too small compared to the ground resolution of a few tens of meters provided/seen by the space borne sensor. Airborne spectrometers/multispectral scanners provide data necessary to simulate "signature" from space.

2.2.3 Signature and Feature Space

How spectral signature measurements in different wavelengths regions are useful in discriminating objects can be illustrated using feature space concept. **Figure 2.10** shows a typical feature space involving measurement in the red and

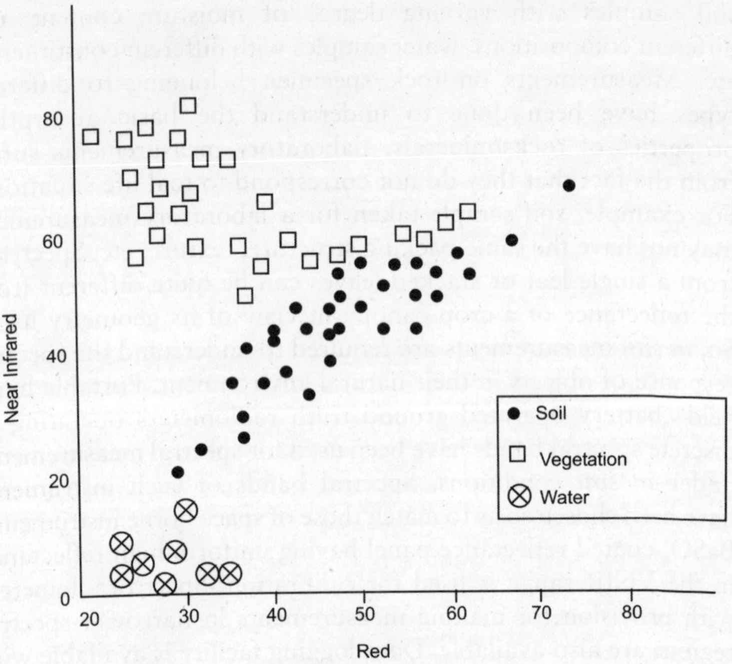

Figure 2.10

infrared regions. For simplicity, measurement of only three classes of objects is plotted. All points belonging to soil class lie predominately along a line, points corresponding to water near the origin and the vegetation points above the "soil line". The data show three natural clusters indicating an obvious approach for distinguishing vegetation, soil and water bodies. Vigor of vegetation is found to be proportional to the perpendicular distance of the vegetation point under consideration from the soil line. This concept extended to measurements in large number of spectral bands from the basis of multivariate analysis.

3

Thermal Remote Sensing

3.1 INTRODUCTION

An object on the Earth's surface, under natural conditions, at above absolute zero (-273° or 0°K), emits electromagnetic energy both day and night. This is due to its thermal activity which is a function of the true temperature and the emissivity of the object. All the features we encounter in the landscape on a typical day (vegetation, soil, water, and even people) emit thermal infrared electromagnetic radiation. Humans sense thermal energy primarily through the sense of touch. Our eyes cannot detect the differences in thermal IR energy because they are sensitive to short wave solar radiation from 0.4 – 0.7 µm. Our eyes are not sensitive to the reflected IR (0.7 – 3.0 µm) or thermal IR (3.0 – 14 µm). However, detectors are being developed to sense the thermal radiation. These thermal IR sensors extend human sensibility into the invisible world of information as they monitor the thermal characteristics of the landscape. Infrared sensing refers to the detection of remote objects by recording the amount of infrared energy emitted from various surfaces as a continuous tone image on photographic film. Thermal infrared imageries are usually obtained in the wavelength regions from 3 to 5.5 µm and from 8 to 14 µm because of atmospheric absorption at other wavelengths. Emission from solids and liquids occurs when they are heated and results in a continuous spectrum. This is called thermal emission and it is an important source of EMR from the viewpoint of remote sensing. All matter is composed of atoms and molecules with particular compositions. Therefore, matter will emit or absorb electro-magnetic radiation on a particular wavelength with respect to the inner state. All matter reflects, absorbs, penetrates and emits Electro-magnetic radiation in a unique way.

Although remote sensing is recognized as a powerful tool in the collection, analysis, and modeling of environmental data, less attention has been given to the use of thermal infrared (TIR) remote sensing. Thermal Remote sensing techniques have proven to be very useful in the field of archaeology. The individual buried site features are identified by recording the heat difference in the ground with the help of modern airborne thermal scanners. The heat difference is due to the fact that the buried stone structures during the day time they absorb the incident solar radiation which they release in the afternoon and evening. Thermal scanners can also bring about small variations in vegetation and soils. Airborne thermal infrared imageries are not widely available for the public use. This is due to (1) the relatively high cost of the thermal sensor (2) the cost of mobilizing an aircraft to acquire the remotely sensed data (3) the difficulty of calibrating and correctly interpreting the imagery (Jensen, 2004). The capability of thermal infrared imagery in allocating firefight resources during a forest fire (Lytle, 1996) or targeting enemy facilities as in the Gulf War in 1991 (McDaid, and Oliver, 1997) is well established. In agriculture, surface temperature can be used as criteria to determine crop water stress for irrigation scheduling or to detect ground features. Thermal remote sensing in this regard is useful for tracking and monitoring those ground features. The use of thermal imagery to study the fluxes and redistribution of materials is a key aspect of land surface processes and land-atmosphere interrelationships.

3.2 HISTORY

The development of thermal IR imaging technology dates back to 1950s. The discovery of infrared portion of the electromagnetic spectrum was made in the early 19th century by Sir Frederick William Herschel, the astronomer. Herschel used a prism to refract light from a hot source and detected the infrared, beyond the red part of the spectrum, through an increase in the temperature recorded on a thermometer, described in his famous paper "Investigations of the powers of the Prismatic Colors to Heat and Illuminate Objects: with Remarks."

Melloni in 1835 has come up with the first thermopile IR detector. In 1879, S.P. Langley began a research program to find a superior radiation detector. One year later he invented the bolometer, an exceedingly sensitive instrument that was able to obtain measurable temperature variations of 1/10,000°C. In World War I, S.O. Hoffman could detect men at 120 m and aircraft. In the 1930, Germany developed the Kiel system for discriminating between bombers and night fighters. The single most important development in infrared technology was the development of the detector element by nations at war during World War II. Early infrared detectors were lead salt photo detectors (Fischer, 1983). However, presently very fast detectors are available, consisting of mercury-doped germanium (Ge: Hg), indium antimonide (InSb) and other substances that are very responsive to infrared radiation. We also have computers to rapidly process and display the thermal radiometric measurements.

The imageries were initially restricted for military uses and classified for security reasons. In due course of time, the civilian geology community realised the importance and use of thermal IR imageries. In 1960s, manufacturers were allowed to acquire images for civilian use. In 1968, US government declassified thermal infrared imaging systems that did not exceed certain standards for spatial resolution and temperature sensitivity. Typical thermal IR images record broad spectral bands, 8-14 μm for aircrafts and 10.5 – 12.5 μm for spacecrafts.

The first classified satellite remote sensor data were collected by the U.S. Television IR Operational Satellite (TIROS) launched in 1960. The coarse resolution thermal infrared data were ideal for monitoring regional cloud patterns and frontal movements. NASA launched the Heat Capacity Mapping Mission (HCMM) on April 26, 1978 that obtained 600 × 600 m spatial resolution thermal IR data (10.5-12.6 μm) both day (1.30 pm) and night (2.30 am). This was one of the first scientifically oriented (geology) thermal infrared systems. NASA's Nimbus 7 launched on October 23, 1978 had a Coastal Zone Color Scanner (CZCS) that included a thermal infrared sensor for monitoring sea surface temperature. In 1980, NASA and the Jet Propulsion Laboratory developed the thermal infrared multispectral scanner (TIMS) that acquires thermal infrared energy in six bands at wavelength intervals of < 1.0 μm. Landsat Thematic Mapper 4 and 5 sensors were launched on July 16, 1982 and March 1, 1984 respectively, and collected 120 × 120 m thermal infrared data (10.4-12.5 μm) along with two bands of middle infrared data. Today, the NOAA Geostationary Operational Environmental Satellite (GOES) collects thermal infrared data at a spatial resolution of 8 × 8 km for weather prediction. Full-disk images of the Earth are obtained every 30 minutes both day and night by the thermal infrared sensor. The NOAA Advanced Very High Resolution Radiometer (AVHRR) collects thermal infrared location area coverage (LAC) data at 1.1 × 1.1 km and global area coverage (GAC) at 4 × 4 km. The routine collection of thermal infrared data is a part of each person's daily life as we watch the nightly weathered report. Now, Landsat 7, measures in two bands, high and low gain, at 60 m by 60 m spatial resolution.

3.3 IR REGION OF THE ELECTROMAGNETIC SPECTRUM

The infrared region of the electromagnetic spectrum covers wavelengths from 0.7 to 300 μm. The reflected infrared region refers to electromagnetic waves with a wavelength of between 0.7 to 3 μm. It includes the photographic IR band between 0.7 – 0.9 μm, that may be detected from IR film. The Infrared region ranging in wavelength from 3.5 to 20 micrometers is called the thermal infrared region. Most remote sensing applications make use of the 8 to 14 micrometer range. The thermal infrared differs from the color infrared mainly by the fact that thermal infrared is the emitted energy that is detected and recorded digitally by special optical mechanical scanners whereas the near infrared or "photographic infrared" is reflected infrared energy that is detected by photographic film. Photographic film does not detect thermal IR. Special quantum or photo detectors and optical mechanical scanners detect and record images in the thermal IR. Infrared radiation at wavelengths larger than 14 micrometer is not utilized in remote sensing as the radiation is absorbed by the Earth's atmosphere. Since thermal infrared radiation is recorded digitally, the products are not referred to as thermal photos, but thermal imageries. However, all photos can be called imageries. The different wavelength regions in infrared portion of the electromagnetic spectrum are summarised below in **Table 3.1.**

TABLE 3.1: Spectral regions in IR portion of the electromagnetic spectrum

Spectral regions	Wavelength (λ) in μm
Infrared (IR)	0.7 – 300
Near Infrared (NIR)	0.7 – 1.5
Shortwave Infrared (SWIR)	1.5 – 3.0
Middlewave Infrared (MWIR)	3 – 8
Longwave Infrared (LWIR)	8 – 14
Far Infrared (FIR)	> 14

The NIR and SWIR are also known as the Reflected Infrared, referring to the main infrared component of the solar radiation reflected from the Earth's surface. The MWIR and LWIR are the Thermal Infrared.

3.4 SATELLITE THERMAL IR SYSTEMS

The thermal sensors acquire thermal infrared rays, which is radiated from land surface heated by sunlight. It can observe the high temperature areas, such as volcanic activities and fires. By examining the strength of radiation, we can understand surface temperatures of land and sea, and status of volcanic activities and forest fires. This method can observe at night when there is no cloud.

The following extract is a typical report of various thermal sensors in operation (**Table 3.2**).

TABLE 3.2: Satellites and Thermal bands

Satellite/sensor	Thermal bands
NOAA/AVHRR (1978)	4 (10.5-11.5 μm), 5 (11.5-12.5 μm)
Landsat-3 (1978)	8, failed shortly after launch
Landsat 4 & 5/TM: (1982, 1984)	6 (10.4-12.5 μm), (120 m resolution)
Landsat-/ETM+7 (1999)	6 (10.4-12.5 μm), 60 m resolution
MODIS 2	(10.78-11.28 μm), (11.77-12.27 μm) (1 km resolution)
ASTER	(8.125-8.475 μm), (8.475-8.825 μm), (8.925-9.275μm), (10.25-10.95 μm), (10.95-11.65 μm) (90 m resolution)

3.5 LIMITATIONS OF THERMAL IMAGERY

The acquisition of satellite imagery is not always straightforward. The followings are considered to be the limitations while using thermal imagery.

1. Imagery acquisition and processing is very expensive.
2. Operational/technical parameters in thermal imaging scanners are very sensitive.
3. Calibration of the electro-optical system in thermal remote sensing is very difficult.
4. Since the emitted long wave radiation is very weak, detector materials must be kept extremely cold during use.
5. Instrument calibration is very difficult.
6. Imagery acquisition is computationally expensive because of the iterative nature of filtering software.
7. Comparatively difficult for interpretation of thermal imageries. It needs some FCCs to make use of while interpreting.

In case of water bodies, since the longer IR wavelengths are rapidly absorbed in the top few meters, thermal imageries can have limitations of measurements only very top layer of the water surface.

3.6 THERMAL PROPERTIES OF MATERIAL

3.6.1 Emissivity

The emissivity of a real body means the emitting ability of a real body as compared to that of a black body. Emissivity (ε) of a material is defined as the ratio of radiant flux emitted from the real body to radiant flux emitted by a reference black body at the same temperature. It is a measure of a material's ability to absorb and radiate energy. Emissivity describes how efficiently an object radiates energy compared to a black body (a theoretical object that absorbs and then emits all incident energy at all wavelengths). Water is the only object whose emissivity is constant under different conditions. The emissivity depends upon the factors such as temperature, emission angle, and wavelength. A true black body would have an emissivity, $\varepsilon = 1$ while any real object would have $\varepsilon < 1$.

The planet Earth does not contain radiating black bodies but contains selectively radiating bodies. They are such as rocks, soil, and water that emit only a fraction of the energy emitted from a black body at the same temperature. Emissivity for all selectively radiating bodies is from 0 to less than 1 that fluctuates according to the wavelengths of energy being considered. A gray body has an output of constant emissivity that is less than one at all wavelengths. It is defined as a body with constant emissivity over all wavelengths and temperatures. Such an ideal body does not exist in practice, but exist under certain assumptions, called, the gray body assumption i.e. the spectral emissivity and the absorptivity do not depend upon wavelength, so that the emissivity is a constant. Non-black body surfaces, however, deviate from ideal black body behavior and their emissivity depends on surface geometry and the chemical composition. They follow the Kirchhoff's law of thermal radiation, which states emissivity equals the absorptivity (for an object in thermal equilibrium), so that an object that does not absorb all incident light will also emit less radiation than an ideal black body. A material's surface emissivity is a measure of the energy emitted when a surface is directly viewed. Surface emissivity is generally measured indirectly by assuming that, emissivity = 1 – reflectivity. A Typical surface emissivity is given in **Figure 3.1**.

A perfect specular surface may reflect 98% of the incident energy absorbing the remaining 2%; however a black body does the reverse. Therefore, effective emissivity is the ratio of the total amount of energy emitted by a black body to that which is predicted by Planck's law. **Table 3.3** gives the emissivity of representative samples in the 8 to 12 μm. The emissivity of most of the substances falls within the relatively narrow range of 0.81 to 0.96 μm.

3.6.2 Kinetic Heat, Temperature, Radiant Energy and Radiant Flux

A body in a thermal state can be expressed by two temperatures: (i) internal temperature or kinetic temperature

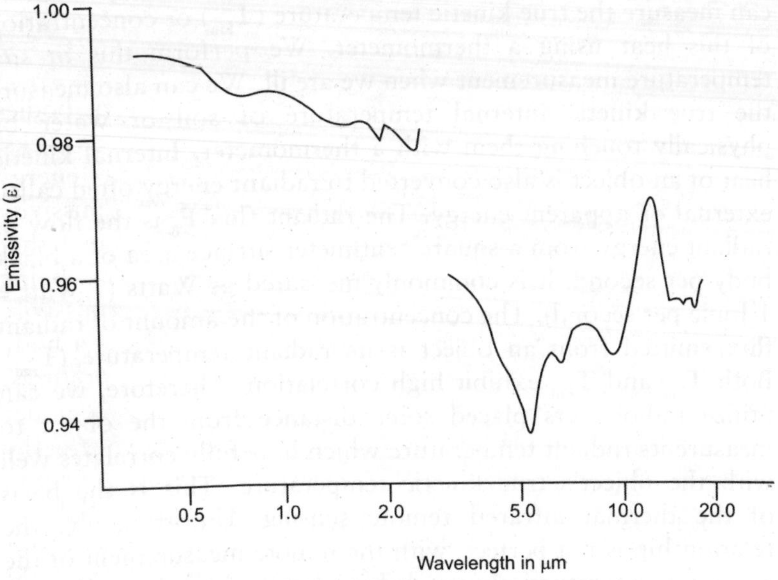

Figure 3.1

TABLE 3.3: Emissivity of selective materials

Material	Emissivity (μm)
Granite	0.815
Quartz Sand, large grains	0.914
Asphalt paving	0.959
Concrete walkway	0.966
Water with thin film of oil	0.972
Pure water	0.993
Polished metal surfaces	0.060

(T_{kin}) due to the kinetic motion of its atoms. It is measured by a thermometer in contact with the body, and (ii) external temperature or radiant temperature (T_{rad}). It is measured remotely by a radiometer. Kinetic heat (so-called internal, real, or true heat) refers to energy generated due to random motion of particles of matter. All objects having a temperature above absolute zero exhibit this random motion. When these particles collide they change their energy state and emit electromagnetic radiation. The amount of heat can be measured in calories. We can measure the true kinetic temperature (T_{kin}) or concentration of this heat using a thermometer. We perform this *in situ* temperature measurement when we are ill. We can also measure the true kinetic internal temperature of soil or water by physically touching them with a thermometer. Internal kinetic heat of an object is also converted to radiant energy often called external or apparent energy. The radiant flux F_B is the flow of radiant energy from a square centimeter surface area of a black body per second. It is commonly measured as Watts (1 Watt = 1 Joule per second). The concentration of the amount of radiant flux emitted from an object is its radiant temperature (T_{rad}). Both T_{kin} and T_{rad} exhibit high correlation. Therefore, we can utilize radiometers placed some distance from the object to measure its radiant temperature which hopefully correlates well with the object's true kinetic temperature. This is the basis of the thermal infrared remote sensing. Unfortunately, the relationship is not perfect, with the remote measurement of the radiant temperature always being slightly less than the true kinetic temperature of the object. This is due to a thermal property called emissivity, as discussed earlier.

Beyond the visible region, we encounter the reflective IR region from 0.7 – 3.0 μm and the thermal IR region from 3.0-14 μm, the only region we can use remote sensing devices to detect infrared energy to be transmitted from the terrain to the detectors. Regions that pass energy are called atmospheric windows. Regions that absorb most of the infrared energy are called absorption bands. Water vapor, carbon dioxide and ozone are responsible for most of the absorption. For example, atmospheric water vapor absorbs most of the energy exiting the terrain in the region from 5 to 7 μm making it almost useless for remote sensing.

The radiant flux of a black body F_B and its kinetic energy T_{kin} (in Kelvin units) are related by the Stefan-Boltzmann Law, which in simplified form is given as:

$$F_B = \sigma T_{kin}^4$$

The above equation holds only for perfect black bodies. However, for other bodies, called real or gray bodies the radiant flux will always be less than the black body flux. The radiant flux of a real material (gray body) F_r can be expressed as:

$$F_r = \varepsilon \sigma T_{kin}^4$$

where σ is the Stefan-Boltzmann constant,

$$\sigma = 5.67 \times 10^{-12} \text{ W cm}^{-2} \text{ K}^{-4}.$$

T is the absolute kinetic temperature (°K = °C + 273)

Infrared radiometers record radiant temperatures, T_{rad} expressed as:

$$T_{rad} = \varepsilon^{1/4} . T_{kin}$$

The final product of a thermal mapping system is a map showing apparent surface temperature. The system is sensitive not only to changes in surface temperature but also to change in emissivity of the scanned surfaces. Common surface materials like rocks, soil, vegetation etc. have emissivity values falling between 0.75 and 0.95 μm and hence surface temperature differences resulting from emissivity changes alone are quite small when compared to the temperature differences produced by other factors. Therefore, interpretation of only the most subtle anomalies requires consideration of emissivity.

Spectral distribution of energy radiated from black bodies at various temperatures are given by Planck's curve, according to which, the hotter the radiating body, the greater is its radiance over its range of wavelengths, and the shorter is its peak emission wavelength. The Sun with a radiant temperature of about 6000°K, radiates enormous amount of energy at wave lengths all across the UV, Visible, and IR bands with the peak emission is at around 0.5 μm. A forest fire peaks around 5.0 μm. The Earth, on the other hand, as observed from space, having radiant temperature of about 273°K, distributed in abroad low curve peaks at 9.7 μm. Thus, sun radiates maximum short wave solar radiation, while the Earth emits maximum long wave radiation.

Two rocks lying next to one another on the ground could have the same true kinetic temperature but have different apparent temperatures when sensed by a thermal radiometer simply because their emissivities are different. The emissivity of an object may be influenced by a number of factors, including:

(i) Colour: Darker coloured objects are usually better absorbers and emitters (i.e. they have a higher emissivity) than lighter colored objects which tend to reflect more of the incident energy.

(ii) Surface roughness: The greater the surface roughness of an object relative to the size of the incident wavelength, the greater the surface area of the object and potential for absorption and re-emission of energy.

(iii) Moisture content: The more moisture an object contains, the greater its ability to absorb energy and become a good emitter. Wet soil particles have a high emissivity similar to water.

(iv) Compaction: The degree of soil compaction can affect emissivity.

(v) Field of view: The emissivity of a single leaf measured with a very high resolution thermal radiometer will have a different emissivity than an entire tree crown viewed using a coarse spatial resolution radiometer.

3.6.3 Thermal Conductivity

Thermal conductivity (k) is the intrinsic property of a material which indicates its ability to conduct heat. Conduction of heat refers to transfer of energy within a material without any motion of the material as a whole. Conduction takes place when a temperature gradient exists in a solid (or stationary fluid) medium. The rate of heat transfer depends upon the temperature gradient and the thermal conductivity of the material. It is measured as the calories delivered in 1 second across a 1 cm^2 area through a thickness of 1 cm at a temperature gradient of 1°C. Heat flows in the material in the direction of decreasing temperature. Energy is transferred from the more energetic to the less energetic molecules when neighboring molecules collide.

If Q = quantity of heat transmitted, L = thickness of the material,

A = surface area of the material and, ΔT = temperature gradient of the material under steady state conditions, then thermal conductivity within the material is given by the following equation:

Thermal conductivity = heat flow rate × distance / (area × temperature gradient), i.e.

$$k = \frac{Q}{t} \times \frac{L}{A \times \Delta T}$$

Thermal conductivity is not a simple property, and depends intimately on structure and temperature.

Thermal conductivity in terms of thermal diffusivity can be defined as $k = \rho\ c_p\ K$, where c_p is the heat capacity, ρ is the density of the material and, K is the thermal diffusivity.

$$\text{For air, } k_{air} = 0.03\ \text{Wm}^{-1}\ \text{K}^{-1}\text{(in MKS)}$$

3.6.4 Thermal Capacity

Thermal capacity c is the number of calories required to raise the temperature of 1 gm of material by 1°C and is expressed in cal.gm^{-1}. It also represents the ability of a material to store heat. Water has the highest thermal capacity of 1.01. The temperature of a lake usually varies very little between night and day. However, rocks do not store heat well and exhibit significantly different temperatures in the night and day. The .hermal capacity of most geologic materials is quite similar. If ρ is the density of the material then an additional parameter called Thermal storage C can be written as the product of the density and the thermal capacity i.e. $C = \rho c$. Water has a very high thermal capacity compared to other materials.

3.6.5 Thermal Inertia

Heating of a material leads to raise its temperature. Heat capacity as we discussed, is meant to resist the change in temperature. Thermal inertia, therefore, can be said of, how much something resists having its temperature raised. The bigger is the heat capacity, more the difficult to raise the temperature. Thermal inertia P is a measure of the thermal response of a material to temperature changes. It is given as:

$$P = (k\rho c)^{1/2}\ [\text{J m}^{-2}\ \text{K}^{-1}\ \text{s}^{-1/2}]$$

where: k = thermal conductivity ($Wm^{-1} K^{-1}$)

ρ = density ($kg\ m^{-3}$)

c = Thermal capacity ($J\ kg^{-1} K^{-1}$)

The higher the density, higher is the thermal inertia. Materials having high thermal inertia (sandstone, basalt) are cooler in day time and warmer at night. However, on the contrary, materials having low thermal inertia are warmer in the day time and cooler at night. As the name indicates, thermal inertia represents the ability of a material to conduct and store heat. It is a measure of the subsurface's ability to store heat during the day and re-radiate or emit it during the night. The thermal inertia 'P' depends predominantly on the physical properties of the near surface materials such as particle size, degree of indurations (i.e., cementation of grains), rock abundance, and exposure of bedrock. Rocks possess much higher thermal inertia than sand or dust—that is, it takes longer time to heat rocks up during the day and to cool them off at night. For example, sandy areas are much hotter at mid-day than adjacent rocks, and the sand cools off quickly after sunset whereas the rocks remain warm well into the evening.

The Earth's materials such as rocks, sand, dust, and bedrock may all contribute to the observed temperatures and the thermal inertias derived from them. Surface temperature variations are sensitive to the properties of the upper few centimeters of the surface. Therefore, thermal inertia is a measure of the geological properties of the surface on the same depth scale. Lowest thermal inertia (purple to dark blue colors in the maps) probably represents loose, fine surface dust and very few rocks. Medium thermal inertia (green to yellow) could represent a combination of coarser loose particles, crusted fines, a fair number of scattered rocks, and/or perhaps a few scattered bedrock outcrops. Higher thermal inertia (red to white) could be a combination of: coarse sand, dune sand, strongly-crusted fines, abundant rocks, and/or scattered bedrock exposures.

3.7 INTERPRETATION OF THERMAL INFRARED IMAGES

The capability of thermal infrared (TIR) remote sensing draws less attention though the present day remote sensing technology has got a powerful tool in the collection, analysis, and modeling of environmental data. They sense thermal emissions from the land, water, ice and the atmosphere. Many of the meteorological satellites include at least one thermal channel. Thermal Remote Sensing has got now recognition on utility of thermal IR bands in understanding key aspects of land surface processes and land atmosphere relationships. Interpretation of Thermal infrared image is quite different from that of an aerial photograph or a normal image produced by a multispectral scanner or charge coupled device (CCD). The main difference lies in the fact that unlike visible or near IR region, the tone of a thermal image expresses surface temperature and not reflectance. The thermal image analyst must think about emissivity, thermal inertia, source of heat, evaporative cooling, etc. while interpreting. He should consider minutely how the solar or terrestrial energy interacts with the various terrain features and how the optical scanners function as they record the terrain's emitted thermal infrared electromagnetic radiation. Finally, the analyst must understand how the background noise due to both the sensor system and the atmosphere that creates haze making the data less useful or lead to incorrect image interpretation.

Thermal Infrared (IR) images show the temperature of the land, the sea or the tops of the clouds above them. Warm temperatures (0-30°C) generally mean land or sea without cloud cover. As the temperature decreases it implies that clouds are getting higher and denser. Very cold temperatures mean that cloud tops are very high, which can imply strong convective storm activity. IR imagery is derived from emission from the Earth and its atmosphere at thermal-infrared wavelengths (10-12 μm) and provides information on the temperature of the underlying surface or cloud. However, since the emitted radiation must traverse the Earth's atmosphere before reaching the satellite, it is modified during passage by atmospheric absorption and re-emission. IR images are available 24 hours a day because temperatures can always be measured, regardless of day or night.

Thermal IR images depict radiant temperature contrasts of a given ground area as tonal variations. Warmer features appear in brighter tones, and cooler features represent in darker tones. Despite, one should know what time of day the image has been acquired. General guidelines for interpreting the terrain features in a thermal imagery are outlined below.

3.7.1 Water versus Soil and Rock

The day time acquired thermal imagery depicts water bodies in darker tone as they are cooler than soil and rock during the day; but at night, water being the warmest shows in lighter tones as compared to soil and rock. This is due to the fact that the heat energy is not transferred by the process of convection in soil and rock like water. Water bodies store heat more efficiently than others, i.e., they have a higher thermal capacity. Water has a much higher thermal capacity than soil and rock. The tonal signatures of water bodies in a thermal imagery give clues for the time of image acquisition. If water bodies in the imagery depict lighter tone than the surrounding terrain, then the time of data acquisition is the night; the image acquisition is the day time if water bodies appear to be darker than the adjacent terrain.

3.7.2 Vegetation

The trees appear in darker tones due to cooler during day hours and appear bright due to warmer at night. Transpiration is at maximum during the day and this process lowers leaf temperature. This type of vegetation appears warm in night time images because of the high water content of the leaves and because of convective air warming associated with night time inversions (i.e. air temperature at crown height is warmer than at ground level). Because trees are essentially decoupled from the ground, they are isolated from the influence of the ground's thermal properties.

Grass and other low-lying vegetations are warmest during the day but rapidly approach local air temperature after sunset. On a calm night the air next to the ground is apt to be cooler than a few meters aloft; consequently, lowlying vegetations will image in darker tones than deciduous or coniferous tree. Lowlying vegetation is well coupled to the ground's thermal properties.

3.7.3 Damp Ground

Areas of damp ground appear cooler than dry ground during both the day and night depicting dark tones in the thermal imagery. This is due to evaporative cooling of contained moisture. The amount of cooling increases with air temperature and decreases as humidity rises. The greatest contrast between damp and dry soils is, therefore, during the day time, particularly when the weather is hot and dry.

3.7.4 Consolidated versus Unconsolidated Materials

Moisture contained in soils at or near the surface has profound effect on the thermal behaviour of rock. Exposed rock surfaces and loosely consolidated materials generally appear in contrasting tones because of significant differences in thermal inertia or thermal diffusivity. In day time images, rock surfaces appear in darker tones than unconsolidated materials, whereas a tonal reversal occurs in night time images.

3.7.5 Pavement Materials

Materials such as packed dirt, asphalt, cement concrete, appear relatively warm (lighter tones) both day and night. They are generally good absorbers of solar radiation during the day, and because of their relatively high thermal capacities are able to radiate strongly for many hours after sunset. Packed Earth appears in the darkest tones of the three materials on night time images because it loses its internal heat at the highest rate.

3.7.6 Metal Surfaces

Bare metal surfaces appear in dark tones both day and night images because of their cold radiant temperatures. Their shiny surfaces have much lower emissivities than other substances found in aerial reconnaissance; consequently, they emit much less thermal IR radiation. Painted metal surfaces, however assume the emissivity properties of a particular paint (e.g. white and black paints).

3.7.7 High Temperature Sources

Thermal IR emissions from target such as forest fire, geothermal sources or active volcanoes are relatively unaffected by time of day. The emitted radiation from these targets remains fairly constant appearing hot (light tones) at all times. Even though the thermal radiation has no penetration capability, in certain instances, they are able to detect a cover feature. Buried heating lines can often be detected on night time winter images by virtue of conductive heat transfer to the overlying materials along the alignment of the subsurface lines. Night time images may also record small streams beneath a vegetation canopy. This occurs when heat energy from the water rises and warms the overhanging vegetation, effectively superimposing the drainage pattern on the vegetation.

3.8 DIURNAL TEMPERATURE CYCLE OF TYPICAL MATERIALS

The tonal variation in thermal images depends considerably on the time of acquisition of the image, whether it is during the warm part of the day or in the night or after the night in absence of the sun. For the best and effective use of thermal imageries, one should have a sound knowledge on diurnal cycle that is the daily cycle of the temperature of various surface materials.

The word "diurnal" comes from the Latin "diurnus", which means daily. The diurnal temperature cycle (**Figure 3.2**) consists of the daily rise and fall of temperatures caused by the progression of daytime and night-time. The same Figure depicts the corresponding rise and fall of temperature during 24:00 hours with respect to gain and loss of total incoming short wave and outgoing long wave energy. Most people have fundamental misconceptions of the diurnal temperature cycle. A regular pattern of temperature change can be seen over the course of a day. Temperature changes are driven by the difference in insolation vs. outgoing energy losses. The sun rises around 6:00 a.m., but the most rapid rise in temperatures doesn't occur until later in the morning. Ground warms the atmosphere. The most rapid rise in temperatures occurs at or before midmorning (9:00 a.m.). Temperature rises as long as incoming solar energy is greater than outgoing long wave energy. The insolation value may be highest around noon, but temperatures (kinetic) still continue to rise. Afternoon, insolation offsets outgoing emissions, so temperature continues to increase. The warmest temperatures occur during mid-afternoon. Temperature begins to fall when incoming solar energy is less than outgoing long wave energy. About 4:00 p.m. energy losses begin to offset insolation, highest temperature of the day. While the ground has received its full complement of radiation by 6:00 p.m., temperatures have already started to fall. Temperatures fall most rapidly (loss of insolation) as the sun is setting (6:00 p.m.). However, temperatures fall less rapidly as night wears on. Minimum temperature typically occurs near or just after dawn on clear nights. Energy losses exceed gains all through the night, so temperatures continue to fall until sunrise. Coldest temperatures are at around dawn (6:00 a.m.).

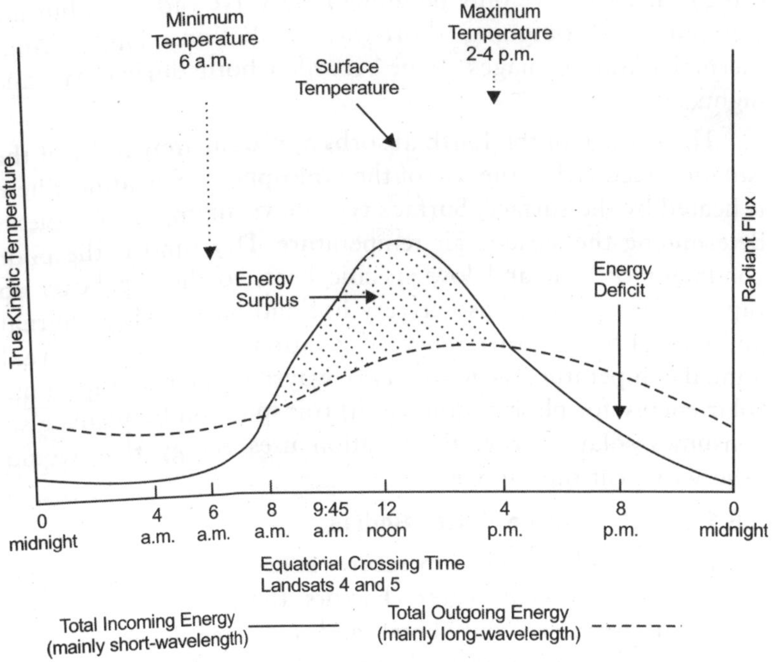

Figure 3.2

The diurnal cycle for the temperature record of the surface materials considers 24 hours. Starting at sunrise, the Earth receives short-wave solar radiation (0.4 – 0.7 μm) which continues throughout the day until sunset. From about 6:00 a.m. to 8:00 p.m., the incoming short-wave solar radiation interacts with the terrain materials. Much of it is being reflected back through the atmosphere to reach the optical space borne sensors. However, the terrain surface absorbs a fraction of it and later reradiates back to the sensor as long-wave thermal IR energy in the wavelength range between 3–14 μm. The outgoing long-wave radiation (OLR) reaches its peak during the day when the surface temperature attains a maximum. In general, OLR is lagging behind two to four hours the irradiance because of the time taken to heat the terrain. During the day hours, we experience surplus radiant energy due to combined reflected short-wave energy and emitted long-wave energy. During night time, rather we have only outgoing long-wave radiation but not incoming and outgoing short-wave radiation. This permits thermal infrared images to be recorded both during day and night.

The surface of the Earth absorbs approximately 50% of the isolation received at the top of the atmosphere. The atmosphere is heated by the surface. Surface type plays an important role in determining the surface air temperature. Dry sand is the poor conductor of heat and low specific heat, so the top heats up rapidly. Desert locations get very hot and have a large diurnal and annual temperature range. Vegetation modifies the diurnal annual temperature range in two ways: (i) plants trap some solar energy for photosynthesis; (ii) transpiration by plants also consumes solar energy. Evaporation uses energy that would otherwise heat the surface.

3.8.1 Effect of Large Water Bodies

Diurnal and annual temperature ranges are less for locations near large bodies of water. Factors that contribute to the difference between continental and maritime locations are: (i) the specific heat of water is about 3 times greater than land (heats up and cools down more slowly), (ii) evaporation of water consumes energy, (iii) mixing and transparency of water allows the solar radiation to be distributed throughout a large depth, and (iv) proximity of warm and cold ocean currents can also affect the temperature.

Figure 3.3 illustrates the typical diurnal radiant temperature curves for soil and rock, vegetation, water, moist soil (or damp terrain) and the metal objects found at the surface within a normal twenty-four-hour period, beginning and ending at local midnight. From these curves one can estimate the relative gray levels that a thermal sensor could record, as a function of the material and the time of day. Given two thermal images of the same locale, taken 12 hours apart, about noon and about midnight, we might determine the identities of co-registered pixels, based on their temperatures and thermal inertias. Some surfaces such as water effectively store up incoming energy and release it slowly throughout the day as emitted energy. Such surfaces demonstrate little temperature fluctuation. Other surfaces such as bare soil/rock and concrete do not have high thermal storage capacities, and thus, have wide temperature variations. A thermal crossover occurs when two or more surfaces at a given time has exactly the same radiant temperature in which case the surfaces cannot be separated through thermal infrared imagery.

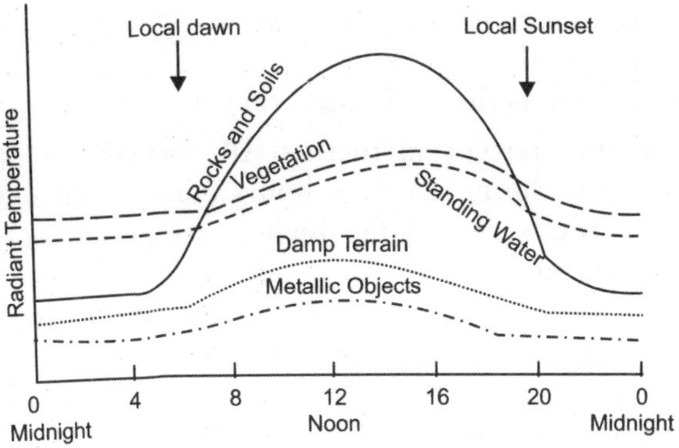

Figure 3.3

Let us observe the thermal imagery of a terrain surface containing soil, rock and water acquired at about 2:00 p.m. and at 4:00 a.m. in the morning. During daytime hours, the soil and rock would appear brighter than water due to their high radiant temperature. The soil and rock appears to be warmer than much of its surrounding during the early hours of the evening. Rock and soil continues to radiate energy that they absorbed during the day until midnight showing a continuous decline in brightness. However, on the contrary, water maintains its relatively stable temperature profile due to its high thermal capacity. Thus, at about 4:00 a.m., temperature of water is well above the surrounding soil and rock and metal objects like concrete. This is what the night thermal imageries show water body in brighter tone than the surrounding soil and rock and even the vegetation. If we compare the radiant temperature profile of vegetation with that of soil and rock, then we find that vegetation appears to be cooler (darker) than soil and rock on 2:00 p.m. daytime imagery and warmer (brighter) than soil and rock on predawn imagery (4:00 a.m.). Vegetation tends to be slightly warmer than water during daytime till mid-night. During predawn hours, it is reverse. If we closely examine the radiant temperature profile of the moist soil, it depicts a stable diurnal temperature cycle, like metal objects. This is due to the premise that the more water in the soil, the higher its thermal capacity. The metal objects in the thermal imageries appear to be the darkest due to their low emissivity. Therefore, for proper interpretation of thermal IR imagery, the time of acquisition should be known as the signatures vary during day and night.

High spatial resolution thermal IR imagery of an urban area shows that during daytime hours commercial areas such as business centres record the highest temperatures. Service, transportation, and industrial areas have the next highest temperatures. The lack of vegetation coverage is apparent in these areas. The coolest daytime temperatures are associated with water surfaces, vegetation, and agricultural land use. The agricultural land is covered by some type of vegetation and is not bare soil. Residential areas consisting of a mixture of buildings, grass, and tree cover record intermediate temperatures.

During the night, commercial, service, industrial, and transportation areas cool off rapidly; however, the predawn temperatures for these areas are still higher than those recorded for vegetation and agriculture. As would be expected, water with its high thermal storage capacity is the warmest surface during the predawn hours.

The environmental factors that influence the thermal imagery are as follows:

- Surface winds: produces confusing thermal patterns. High velocity winds lower thermal contrast. Wind shadows, wind streaks, wind smears.
- Rain, dew, frost: objects in contact with rain—cooler more uniform temperatures.
- Clouds and fog: masks thermal IR emissions from surface features. Total cloud cover causes overall reduction of thermal contrast.
- Dark and smoke: penetrated by thermal IR.
- High humidity: surface objects appear to have cooler and more uniform temperatures.
- Temperature inversions: hill and valley terrain.
- Best thermal images: clear skies with no surface wind and low humidity.
- Time of day: Night-time imagery is necessary for most geologic applications because the thermal effects of differential solar heating and shadowing area greatly reduced. On daytime images topography is typically the dominant expression because of these differential solar effects. As the radiant temperatures are relatively constant in the predawn hours, thermal imagery obtained during such hours is preferably for interpretation purposes.
- Surface moisture and dense vegetation: will mask other features on thermal images. Thermal IR images are much useful in arid and semi-arid terrain.

3.9 ADVANTAGES OF THERMAL IMAGERIES

i. Geology

Denser rocks such as basalt and sandstone have higher thermal inertias and on night thermal IR images they show warm signature compared to less dense rocks such as siltstones, cinders etc. Hence, differentiation of rock types is possible. Faults may be marked by cooler linear anomalies caused by evaporative cooling of moistures trapped along the fault zone. Folds may be indicated by thermal patterns caused by outcrops of different rock types. Surface temperatures of volcanic terrain may be mapped.

ii. Military

Military applications of thermal mapping are varied and often classified. Unusual concentration of troops, weaponry, military vehicles, jungle trails, etc. can be identified by the tonal changes in the thermal imagery.

iii. Hydrology

Cool underground springs that discharge into warmer courses of water may be discovered by thermal imagery. Hot industrial effluents that pollute water may be detected and the diffusion pattern may be mapped. Hot springs are clearly identifiable in IR imagery.

iv. Agriculture

Thermal images are utilized for the identification of crop species and soil types, for detecting crop diseases, for making animal censuses and for determining relative moisture content of various soils.

v. Botany

Leaf temperatures can be remotely measured using 3 to 5 μm band. Such knowledge is useful for assessing plant health, age, relative water supply or degree of irrigation. Thermal damage or frost damage to fruit groves can also be assessed.

vi. Forestry

Forest fires beneath a forest cover can be identified with the thermal imagery. During extensive fires damage caused can be assessed even under excessive smoke conditions. Besides, quantitative information about forest canopy structure, biomass, age, and physiological condition have been extracted from thermal infrared data.

SUMMARY

Thermal IR images record the radiant temperature of materials. Emissivity is the measure of its ability to radiate and to absorb thermal energy. Diurnal temperature range is a function of thermal inertia, which is directly related to density of materials. Clouds and surface winds produce confusing thermal patterns. Daytime images record the differential solar heating and shadowing of topographic features. Geologic and other interpretations require night-time images. Surface moisture and dense vegetation will mask other features on thermal images. Thermal IR images are mostly useful in arid and semi-arid terrain.

4

Microwave Remote Sensing

4.1 INTRODUCTION

The portion of the electromagnetic spectrum extending from wavelength of about 1 mm to 1 m is termed as the Microwave region. The upper limit and the lower limit are rather arbitrary. In this region the natural radiations emanating from the objects are a result of the atomic or molecular vibrations and are thus characteristic of the molecular structure and temperature of the object. Because of their long wavelengths, compared to the visible and infrared, microwaves have special properties that are important for remote sensing. Longer wavelength microwave radiation can penetrate solids and unconsolidated earth material. This makes it possible to detect radiations emanating from below the soil, snow and ice covers, etc. and thus open up possibilities for many new applications. This region provides an all weather remote sensing capability as the radiations are not appreciably affected by haze, cloud, rain, smoke, hail or falling snow. This property allows detection of microwave energy under almost all weather and environmental conditions so that data can be collected at any time. The microwave intensities being low, passive microwave remote sensing results in comparatively poorer temperature resolution as well as spatial resolution. There is another very important feature associated with microwaves. The surface which appears rough to us in visible light may be quite smooth in the microwave region, as the microwaves have much longer wavelengths.

Microwave remote sensing has got the capability of extending the human visibility to beyond what our eyes can see. That is, in contrary to the conventional aerial photography, the present day remote sensing technique gives scope to utilize the microwave region beyond the visible and infrared regions. Microwave region covers a wide range of frequencies from 1 GHz to 300 GHz . It is able to provide information on soil moisture, snow cover, surface conditions below vegetation cover, etc. Microwave remote sensing provides unique information on sea wind and wave direction, which are derived from frequency characteristics, Doppler Effect, polarization, back scattering, etc. that cannot be observed by visible and infrared sensors. Radiometry is a measurement of electromagnetic radiation. A microwave radiometer is a highly sensitive receiver capable of measuring low levels of microwave radiation. When a scene such as terrain is observed by a microwave radiometer (through its antenna beam); the radiation received by the antenna is partly due to self-emission by the scene and partly due to reflected radiation originating from the surrounding such as the atmosphere. Through proper choice of the radiometer's parameter (wavelength, polarization and viewing angle), it is sometimes possible to establish useful relations between the magnitude of the energy received by the radiometer and specific terrestrial or atmospheric parameters of interest.

Over the past three decades, microwave remote sensing has evolved into an important tool for monitoring the atmosphere and surface of the planetary objects with special emphasis on observations of the planet Earth. The term microwave remote sensing encompasses the physics of radio wave propagation and interaction with material media, including surface and volume scattering and emission; the techniques used for designing microwave sensors and processing the data they acquire; and translation of the measured data into information about the temporal or spatial variation of atmospheric or surface and medium parameters or properties. Just like active remote sensing and passive remote sensing in the optical region, microwave sensing also falls into two groups (Reeves, 1979) according to their modes of operation: active microwave remote sensing and passive microwave remote sensing. Active microwave sensors are those that provide their own source of illumination and therefore contain a transmitter and a receiver, while passive sensors are simply receivers that record the radiation emanating from the scene under observation. Active microwave sensors include radar imagers, scatterometers, and altimeters, while passive microwave sensors are often referred to as microwave radiometers to detect and record the naturally emitted microwave energy within its field of view. This emitted energy is related to the temperature and moisture properties of the emitting object or surface.

Active microwave sensors use their built-in (own) energy source—microwave energy. A radar system transmits pulses of energy in the direction of interest to illuminate the target and records the strength and origin of "echos" or reflection backscattered from objects on the ground within the system's field of view. Active microwave sensing systems are of two categories: imaging sensors and non-imaging sensors. The most common form of imaging active microwave sensors is RADAR (Radio Detection and Ranging). The term RADAR was coined in 1930s. These imaging radars are divided into two categories. The first category is real aperture and the second one is synthetic aperture systems. Some of them are Side Looking Airborne Radar (SLAR) and Synthetic Aperture Radar (SAR). In the real aperture system, resolution is determined by the actual beam width and antenna size. The synthetic aperture system utilizes signal processing techniques to achieve narrow beam width in the long track direction which provides better resolution. The function behind it is that the sensor transmits a pulse of microwave energy towards the target of interest to illuminate it and detects the backscattered energy emanating from the target. The intensity of the backscattered energy is measured to discriminate between different targets and the time delay between the transmitted and reflected signals determines the distance (or range) to the target. Non-imaging microwave sensors include altimeters and scatterometers. They record data in one linear dimension, unlike the two-dimensional representation of imaging sensors. The radar altimeter is an instrument on a satellite that measures the distance from the satellite to the earth's terrain using a radar signal. A radar altimeter sends out pulses of microwave signals to the earth's surface and records the microwave pulse scattered back from the Earth's surface. The height of the surface can be measured from the time delay of the return signals. Generally altimeters look straight down at nadir below the platform and thus measure height or elevation (if the altitude of the platform is accurately known). Radar altimeters are frequently used by commercial aircraft for approach and landing, especially in low-visibility conditions as well as by military aircraft flying extremely low over terrain to avoid radar detection. A radar altimeter, which is carried aboard satellites, collects elevation signals from land and ocean surfaces. These data are used to determine topographical features; and sea surface height estimation and retrieval of wind speed.

Microwave scatterometers, the non-imaging sensors are used to make precise quantitative measurements of the amount of energy backscattered from targets. According to a narrow definition, a microwave scatterometer may be a space borne sensor to measure the two dimensional velocity vectors of the sea wind, while according to the wider definition, it also involves airborne sensors, as well as ground based sensors to measure the surface backscattering as well as volume scattering, such as rain radar. A wind scatterometer can be used to measure wind speed and direction over the ocean surface. It sends out pulses of microwaves along several directions and records the magnitude of the signals backscattered from the ocean surface. The magnitude of the backscattered signal is related to the ocean surface roughness, which in turn, is dependent on the sea surface wind condition, and hence the wind speed and direction can be derived. Microwave scatterometers are classified as two types, pulse type and continuous wave type (CW). The pulse type uses wide band which has restrictions in obtaining a licence to operate and to avoid obstructions. CW type has the advantage that the band width can be reduced to 1/100 times that of the pulse type and the price becomes cheaper.

The amount of energy backscattered is dependent on the surface properties (roughness) and the angle at which the microwave energy strikes the target. Scatterometry measurements over ocean surfaces can be used to estimate wind speeds based on the sea surface roughness. Ground-based scatterometers are used extensively to accurately measure the backscatter from various targets in order to characterize different materials and surface types. This is analogous to the concept of spectral reflectance curves in the optical spectrum.

Passive sensors record microwave energy emitted by the atmosphere, reflected from the surface, emitted from the surface, or transmitted from the subsurface. There are two typical microwave scanning radiometers; the conical scanning type and the cross track scanning type. The former is used for the microwave channel which is influenced by the ground surface, while the latter is used for the channel which can be neglected by the influence of the ground surface. Microwave radiometers or passive type microwave sensors record the natural microwave emission from the earth. It can be used to measure the thermal radiation of the ground surface and/or atmospheric condition and estimates the total water content of the atmosphere within its field of view. Because the wavelengths are so long, the energy available is quite small compared to optical wavelengths. Thus, the field of view must be large to detect enough energy to record a signal. Most passive microwave sensors are, therefore, characterized by low spatial resolution.

The major field of applications of passive microwave remote sensing includes meteorology, hydrology, and oceanography. Meteorologists can use passive microwaves to measure atmospheric profiles to determine water and ozone content in the atmosphere. Hydrologists use passive microwaves to measure soil moisture since microwave emission is influenced by moisture content. Oceanographers measure sea surface roughness, surface wind vector, currents, sea ice, and as well as detection of pollutants, such as oil slicks.

4.1.1 Frequency Range of Radar

Radar operates in part of the microwave region of the electromagnetic spectrum, specifically in the frequency interval from 40,000 to 300 mega hertz (MHz). **Table 4.1** lists the various radar wavelengths and corresponding frequencies, specified by a band nomenclature.

TABLE 4.1: Radar wavelengths and frequencies

Band	Wavelength (cm)	Frequency (MHz)
Ka	0.8 – 1.1	40,000 – 26,500
K	1.1 – 1.7	26,500 – 18,000
Ku	1.7 – 2.4	18,000 – 12,500
X	2.4 – 3.8	12,500 – 8,000
S	7.5 – 15	4,000 – 2,000
C	3.8 – 7.5	8,000 – 4,000
L	15 – 30	2,000 – 1,000
P	30 – 100	300 – 1,000

4.2 THE RADAR PRINCIPLE

Radars consist of a transmitter, a receiver, one or two antennae and lots of signal-processing circuitry. This is a basic overview of how the principle of radar works:

(i) The transmitter generates a signal which is radiated through the antenna.

(ii) The signal is an electromagnetic radiation of wavelength in the centimeter range as a source of illumination to detect remote target, which is capable of propagating through space.

(iii) If an object happens to be in the way then part of the signal is reflected back to the radar.

(iv) The receiver picks up this echo, using the transmit antenna or a dedicated receiving antenna.

(v) The signal processor detects the echoes, sorts them out from among noise and other types of interference, takes measurements with regard to the object's location in space, its speed and perhaps some more characteristics, and prepares them for output to a display.

4.2.1 The Radar Equation

The fundamental relation between the characteristics of the radar, the target, and the received signal is called the radar equation. The radar equation relates P_r = received power to:

(i) parameters of the radar system

(ii) targets

(iii) scattering properties

(iv) target location

A radar image is a two dimensional representation of the power (or voltage) returned to the radar from a specific area on the ground presented as a picture element (pixel). This returned power is usually quantized to a radiometric scale of 256 values (8 bits) for presentation on image processing systems (Leberl, 1990). RADARSAT is quantized to 11 bits. To understand how to interpret radar images, we should understand the nature of the power scattered back toward the radar antenna. In most simple case, this can be stated verbally as suggested by Moore (1983):

$$P_r = (P_t) \times (E_t) \times (S_l) \times (A_a) \quad \text{... (i)}$$

where,

P_r = Power received

P_t = Power per unit area at target

E_t = Effective scattering area of target

S_l = Spreading loss of reradiated signal

A_a = Effective antenna receiving area

where, power received per unit area at the receiver is the energy scattered back from the terrain-backscatter. The spreading loss occurs because the signal starts from the backscattering point source on the ground and spreads out in all directions, so that the power per unit area is less at a greater distance than it would be near the scatterer. Therefore, the strength of the backscatter towards the receiver is a product of the power per unit area illuminating the target, times the effective scattering area of the target, and then the retransmission of this wave back toward the receiving antenna. The actual size of the receiving antenna also makes a difference.

The fundamental radar equation is derived by combining these word quantities to create the mathematical expression (Moore, 1983; Kasischke et al., 1984):

$$P_r = \frac{(P_t . G_t . \sigma . A_r)}{(4\pi)^2 . R^4} \qquad \text{... (ii)}$$

where P_r is power received, P_t is the power transmitted toward the target, G_t is the gain of the antenna in the direction of the target, R is the range distance from the transmitter to the target, σ is the effective backscatter area of the target (or radar cross section), and A_r is the area of the receiving antenna. The above equation can be broken down into seven parts for even more clarity:

$$P_r = P_t . G_t . \left(\frac{1}{4\pi R^2}\right) . \sigma . \left(\frac{1}{4\pi R^2}\right) . A_r \qquad \text{... (iii)}$$

The above equation relates that the power received by the radar system is a function of a pulse of electromagnetic energy (P_t), that has been focused down to an angular beam width by the antenna so that the flux becomes higher by a factor of G_t over a spherically expanding wave. The focused energy illuminates an area on the ground that has a cross section of σ. The radar cross section is defined as the equivalent of a perfectly reflecting area that reflects isotropically (spherically). The energy backscattered from the ground once again spherically expands from the source. Finally, the receiving antenna area (A_r) intercepts a portion of the reflected wave and records it.

Most radars use the same antennas for transmitting (A_t) and receiving (A_r). Consequently, the gain factors of the antennas may be combined using the relationship between gain and receiving aperture as (Moore, 1983):

$$G = G_t = G_r = \frac{(4\pi A_r)}{\lambda^2} \qquad \text{... (iv)}$$

where λ is the wavelength of the radar system. Substituting this value in (i) or (ii), modified radar equation becomes (Carver et al., 1985):

$$P_r = \frac{(P_t . G^2 . \sigma . \lambda^2)}{(4\pi)^3 . R^4} \qquad \text{... (v)}$$

Thus it can be noted here that the radar equation can be viewed as a product of system parameters and other environmental terrain parameters that produce the backscatter cross section (σ). Because the system parameters are well known, their effects are typically removed from the radar images i.e., the system parameters may be set to unity (1).

The amount of radar cross section, σ, reflected back to the receiver, per unit area (A) on the ground is called the radar backscatter coefficient (σ°) is computed as:

$$\sigma^\circ = \left(\frac{\sigma}{A}\right) \qquad \text{... (vi)}$$

where σ is the radar cross section. The radar backscatter coefficient determines the percentage of EM energy reflected back to the radar from within a resolution cell e.g. 10 × 10 m. The actual σ° for a surface depends on a number of terrain parameters, like geometry, surface roughness, moisture content, and the radar system parameters (wavelength, depression angle, polarization), see Leberl (1990). The σ° is a dimensionless quantity characteristic of the scattering behaviour of all the elements contained in a given ground cell. The back scattering coefficient σ°, according to Elachi (1987) is defined as the ratio of the energy received by the sensor over the energy that the sensor would have received if the surface had scattered the energy incident on it in isotropic fashion. This is expressed in decibels (dB). Backscattering coefficient describes the terrain contributing factor to the radar image tone, and the radar cross sections per unit area (resolution cell).

The total radar cross section of an area (A) on the ground therefore becomes (σ°A), and the final form of the radar equation for an area extensive target becomes (Henderson and Lewis, 1998):

$$P_r = P_t\left(\sigma^\circ A\right)\left(\frac{G^2 . \lambda^2}{(4\pi)^3 . R^4}\right) \qquad \text{... (vii)}$$

A digital SAR image is created that consists of a two dimensional array of pixels with the intensity of each pixel proportional to the power of the microwave pulse reflected back from the corresponding ground cell (Waring et al., 1995). The reflected radar signal is proportional to the backscattering coefficient (σ°) of a given ground cell.

4.3 RADAR SYSTEM COMPONENTS

The instruments and principles used in a radar system for data acquisition is quite different from those used in passive remote sensing systems working in the optical wavelength region of the electromagnetic spectrum. A typical radar system consists of the following components:

(i) a pulse generating device

(ii) a transmitter

(iii) a duplexer (coordinates sending and receiving of active microwave energy)

(iv) a directional antenna (shapes and focuses each pulse into a stream)

(v) a receiver (converts and amplifies the returned pulses into video signals)

(vi) CRT Display or Digital recorder (a recording device which stores them digitally for later processing).

The pulse generating device sends a pulse of electromagnetic energy at a specific wavelength (frequency) to the transmitter. When in sending mode, the duplexer then sends the polarized pulse of energy through the antenna toward the earth's surface.

Each pulse lasts only microseconds (typically there are about 1,500 pulses per second). Pulse length is an important factor along with bandwidth in setting the system resolution, the distance traveled during the pulse generation. The duplexer separates the outgoing and returned pulses (i.e., eliminates their mutual interferences) by blocking reception during transmission and *vice versa*. The antenna on a ground system is generally a parabolic dish.

4.4 WAVELENGTH, FREQUENCY AND PULSE LENGTH

Microwave sensors construct images of the earth in the microwave region of the electromagnetic spectrum. They carry onboard an instrument that sends a microwave pulse of specific wavelength and duration (pulse length measured in microseconds, μsec) through the antenna to the surface of the Earth and records the back scattered energy from the surface of the Earth. Synthetic aperture radar and microwave radiometer are flown on several satellites together or apart. Seasat has got two sensors on board a SAR and a SMMR (scanning multichannel microwave radiometer). Indian satellite IRS-P4 has got Multifrequency Scanning Microwave Radiometer (MSMR) which operates in four microwave frequencies both in vertical and horizontal polarisation. MSMR is designed to collect data on sea surface temperature, wind speed, cloud water content and water vapor content in the atmosphere above the ocean. Different sensors collect data utilising different bands, or 'windows', in the microwave region. Frequency and respective wavelength of bands in the microwave region most commonly used in the imaging radars are summarised in Table 4.1. The wavelengths are much longer than visible, NIR, MIR or TIR used in other remote sensing systems. Therefore, microwave energy is usually measured in centimeters rather than micrometers or microns (Carver, 1988).

Electromagnetic radiation travels in vacuum with the speed of light c = 3×10^{10} cm/sec. It has dual nature: waves and particles. Considering the radiation as waves, it is continuous sinusoidal wave with frequency denoted by υ i.e. it is the number of waves passing per second at a given point. The length of successive maxima is called the wavelength and is denoted by λ (cm).

$$c = \lambda \times \upsilon$$

$$\text{where, } \lambda = \frac{(3 \times 10^{10}\ \text{cm sec}^{-1})}{\upsilon} \text{ or } \upsilon = \frac{(3 \times 10^{10}\ \text{cm sec}^{-1})}{\lambda}$$

The following simple equation can be used to rapidly convert frequencies into units of radar wavelengths:

λ in centimeters = $30/\upsilon$ (in GHz)

From above, υ = c/λ

Since frequency is the number of wavelength in a distance of 3×10^{10} cm, it is usually a large number. For infrared, especially it is much more convenient to speak the number of waves in one centimeter and call the new unit wave number designated by (cm^{-1})

i.e. $(cm^{-1}) = 1/\lambda$

However, wave number becomes small number in microwave region, we use GHz for frequency. Similarly λ in cm are small, we represent μm for most part of the spectrum and for very short wavelength.

$$1 \text{ Å} = 10^{-8} \text{ cm} = 10^{-10} \text{ m} = 10^{-4} \text{ μm}$$

$$\text{and, } 1 \text{ μm} = 10^{-6} \text{ m} = 10^{-4} \text{ cm} = 10^{4} \text{ Å}$$

Conversion of units is given below:

$$1 \text{ cm}^{-1} = 3 \times 10^{10} \text{ Hz} = 30 \times 10^{9} \text{ Hz} = 30 \text{ GHz}$$

$$\text{Or, } 1 \text{ GHz} = 1 \times 10^{9} \text{ Hz} = 1/30 \text{ cm}^{-1}$$

Frequency: Frequency can be defined as the number of oscillations per unit time or number of wavelengths that pass a fixed point per unit time. In remote sensing, the term 'frequency' is most often used with radar. The frequency bands used by radar (radar frequency bands) were first designated by letters for military secrecy. In the microwave region, frequencies are on the order of 1 GHz to 100 GHz. For electromagnetic waves, in the microwave region, frequencies are on the order of 0.3 GHz -300 GHz, having wavelengths of 1 mm - 1 m respectively. The penetrating ability of microwave to clouds, precipitation, or land surface cover depends on its frequency. Generally, the penetration power increases for longer wavelength (lower frequency). The SAR backscattered intensity generally increases with the surface roughness. Any surface fluctuation of the order of 1 cm or less will be considered smooth. In SAR imaging, the reference length scale for surface roughness is the wavelength of the microwave. If the surface fluctuation is less than the microwave wavelength, then the surface is considered smooth. For example, little radiation is backscattered from a surface with a fluctuation of the order of 5 cm if a L-band (15 to 30 cm wavelength) SAR is used and the surface will appear dark. However, the same surface will appear bright due to increased backscattering in an X-band (2.4 to 3.8 cm wavelength) SAR image. Both the ERS and RADARSAT SARs use the C band microwave while the JERS SAR uses the L band. The C band is useful for imaging ocean and ice features. However, it also finds numerous land applications. The L band has a longer wavelength and is more penetrating than the C band. Hence, it is more useful in forest and vegetation study as it is able to penetrate deeper into the vegetation canopy. The short wavelength radar interacts mainly with the top layer of the forest canopy while the longer wavelength radar is able to penetrate deeper into the canopy to undergo multiple scattering between the canopy, trunks and soil.

Pulse length: It can be defined as duration of a burst of energy transmitted by a radar antenna, measured in microseconds. Pulse can be defined as a short burst of electromagnetic radiation transmitted by the radar. Also it is described as a group of waves with a distribution confined to a short interval of time. Such a distribution is described in the time domain, or in spatial dimensions, by its width and its amplitude or magnitude, from which its energy may be found. In radar, use is made of modulated or coded pulses which must be processed to decode or compress the original pulse to achieve the impulse response observed in the image. Coded pulses have a time-bandwidth product that is much larger than unity. The resolution that may be achieved after processing is determined by the bandwidth of the original pulse.

Wavelength: In a periodic wave, wavelength is the distance between two points of corresponding phase in consecutive cycles.

4.5 RADAR TERMINOLOGY

The various terms that we come across in radar geometry are explained below (**Figure 4.1**):

Depression angle: The angle at which the radar images the target as measured from the horizontal at the antenna is called the depression angle. The look angle is the complement of the depression angle. Rough surfaces produce diffuse scattering of relatively uniform intensity irrespective of the depression angle. Smooth surfaces give a stronger return at depression angles near vertical but little or no return occurs at lower depression angles.

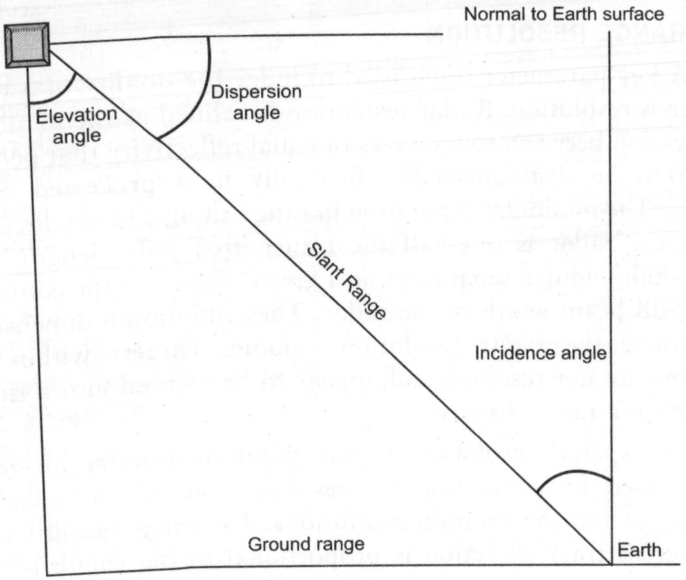

Figure 4.1

Slant range represents the distance measured along a line between the radar *antenna* and the target.

Nadir is a single point, or locus of points on the surface of the Earth directly below a sensor as it progresses along its line of flight. Nadir can be both a point and a line. In other words, when a straight-line is drawn between the sensor and the centre of the Earth, the nadir is the point where that line intersects the surface of the Earth. When the contiguous nadir points are joined along the ground, they form the nadir line. For radar, the nadir line corresponds to the beginning of the range.

Elevation angle is the angle between the slant range and the nadir. It is located between the slant range and the nadir.

Incidence angle is the angle between the radar beam and ground surface, which increases, moving across the swath from near to far range.

Ground range distance is the true horizontal distance along the ground corresponding to each point measured in slant range.

4.6 RANGE RESOLUTION

A key parameter often used to judge the quality of a radar image is resolution. Radar resolution is defined as the minimum separation between two targets of equal reflectivity that permits them to be distinguished individually in a processed radar image. The minimum separation in range along a given direction from the radar is one-half the transmitted pulse length. The minimum angular separation at a given range is approximately the 3-dB beam width of the radar. These minimum dimensions determine the radar resolution volume. Targets within the volume are not resolved and appear to be merged into a single target on a radar display.

The spatial resolution at any point in a radar image is determined by computing the resolution in two dimensions: the range and the azimuth resolutions. The range resolution in the across track direction is proportional to the length of the microwave pulse. The shorter the pulse length, the finer is the range resolution. Range or across-track resolution in slant range is determined by the pulse length (τ), the physical length of the radar pulse that is emitted from the antenna. If not given, the pulse length can be determined by multiplying the pulse duration (d) or the length of time in microseconds ($1\ \mu sec = 1 \times 10^{-6}$ sec) that the pulse was emitted from the antenna by the speed of the light ($c = 3 \times 10^8$ m/sec):

$$\tau = \delta c$$

The radar range resolution is the ability of radar to resolve between two targets on the same bearing, but at slightly different ranges in the range direction. More clearly, Range resolution is the ability of the radar to differentiate or resolve two targets that are close together in range. The range resolution is limited primarily by the width of the radar's transmitted pulse, the types and sizes of the targets, and the efficiency of the receiver and the indicator. For the radar to be able to distinguish two closely spaced targets in the across-track dimension, all parts of their reflected signals must be received at the antenna at different times. For this the targets must be separated by more than half the pulse length. In the diagram (**Figure 4.2**) targets A and B in the near range would most likely be resolved and so would appear as two targets. Conversely, targets C and D in the far-range would probably not be resolved and so would be seen as one target. The slant range resolution is determined by the duration of the pulse of the transmitted energy and is theoretically determined in the across-track dimension is equal to one half the transmitted pulse length. The pulse length traverses twice from the target to the sensor and back. Thus, while computing the slant range resolution, it is necessary to divide (τ.c) by 2. To scale it to ground range, 2 is multiplied by the cosine of the depression angle (g). Thus, the range resolution R_r is given by:

$$R_r = \frac{(\tau . c)}{2\cos\gamma} \qquad \text{... (viii)}$$

For a depression angle of 50° and pulse length of 0.1 μsec the range resolution R_r is calculated as 23 m. Hence, objects separated by less than 23 m in the range direction would not be resolved. At a depression angle 35°, R_r would be 18 m. The range resolution is poor for objects nearer the flight direction (**Figure 4.2a**).

4.7 AZIMUTH RESOLUTION

It is the ability of an imaging radar to resolve between two targets in the azimuth direction (direction parallel to the line of flight) or in the flight direction. In a radar image, azimuth is referred to as along-track direction, since it is the relative along-track position of an object within the antenna's field of view following the radar's line of flight. The azimuth direction is perpendicular to the range direction. It is determined by the angular width of the terrain strip illuminated by the radar beam. To be resolved, targets must bè separated, in the azimuth direction by a distance greater than the beam width on the ground (see **Figure 4.2b**, targets A and B). Azimuth resolution is controlled by the angular width of the radiated microwave

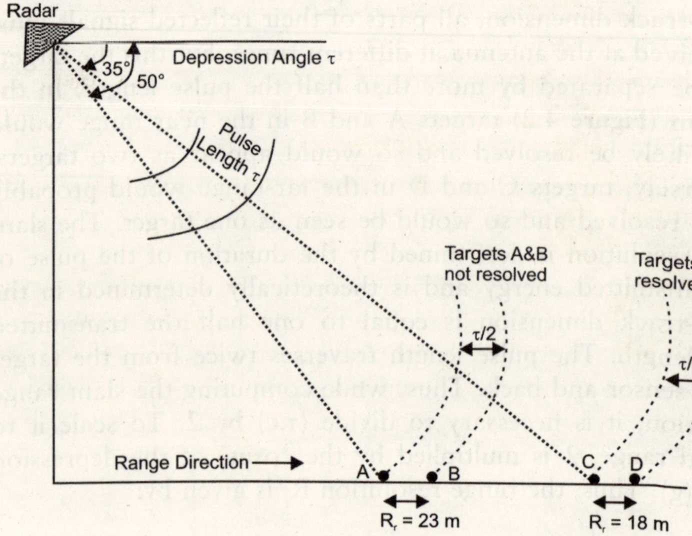

Figure 4.2a

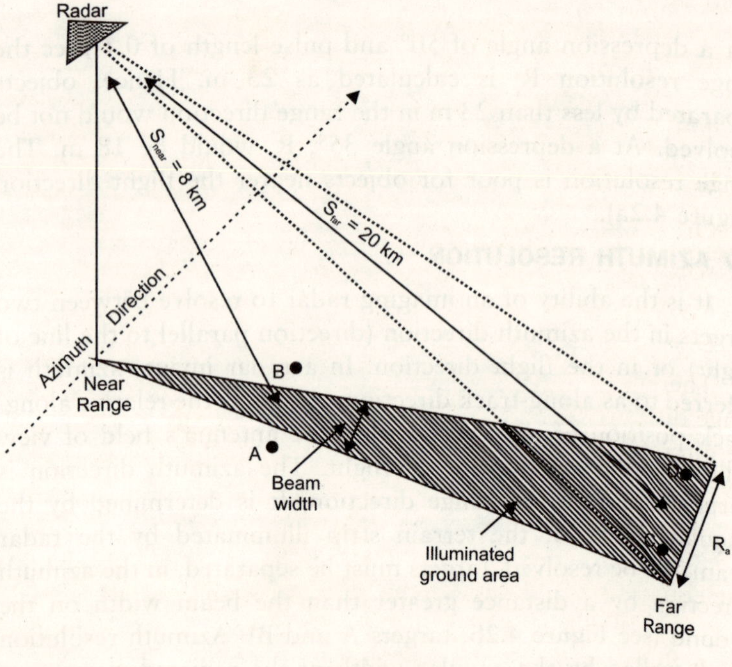

Figure 4.2b

beam and the slant range distance as only targets separated by more than one beam width can be displayed separately. The azimuth resolution increases with increasing distance from the sensor and therefore, targets near the antenna require less separation than those further out. As the beam width is narrower in the near range, azimuth resolution is greater in the near range. At greater distances, more separation is required. For a Real Aperture Radar, two targets in the azimuth or along-track resolution can be separated only if the distance between them is larger than the radar beam width. Therefore, the beam width is taken as the measure of azimuth resolution. For example, in the Figure 4.2 targets A and B can be resolved, whereas targets C and D cannot be resolved and will appear as a single object. Azimuth resolution R_a is determined by the angular width of the terrain strip illuminated by the radar beam.

The equation for determining azimuth resolution (R_a) is given by

$$R_a = \frac{(0.7 \times S \times \lambda)}{D} \qquad \text{... (ix)}$$

and,

Beam width (β) is given by, $\beta = \frac{\lambda}{d}$

Or, $R_a = 0.7 \times S \times \beta$

where S = slant range distance

λ = wavelength

D = antenna length

There is a trigonometrically relationship between the slant range distance (S) and the depression angle (γ) and the height of the aircraft or spacecraft above the datum (H):

$$S = \frac{H}{\sin\gamma} \qquad \text{... (x)}$$

The equation for computing the azimuth resolution becomes:

$$R_a = 0.7\left(\frac{H}{\sin\gamma}\right)\left(\frac{\lambda}{D}\right) \qquad \text{... (xi)}$$

The relationships expressed in the above equations show that (1) azimuth resolution decreases in proportion to increasing range (i.e. resolution is best in near range, where the width of the beam is narrowest) and (2) a long antenna or a short operating wavelength will improve azimuth resolution. One could build an extremely long antenna to place it on the aircraft to achieve very high azimuth resolution but there is a limit to the length of the antenna (practically about 2 m) which can be carried outside an aircraft and shorter wavelengths will not have enough power to penetrate a moist atmosphere or cloud as much of the energy would be scattered. The equation can be used to compute the azimuth resolution at any location between the near and far range. For a slant range of 8 km, wavelength of 0.86 cm and antenna length of 490 cm azimuth resolution (R_a) would be 9.8 m in the near range. For a slant range of 20 km R_a would be 24.6 m.

4.8 SYNTHETIC APERTURE RADAR

The real aperture radar uses an antenna of the maximum practical length to produce a narrow beam width in the azimuth direction as the beam width is inversely proportional to the size of the antenna, i.e. the longer the antenna, the narrower the beam. The synthetic aperture radar employs a relatively small antenna that transmits a moderately wide beam but the effective length, which is achieved by electronic data processing, is many times the accrual length. The beam width is uniform after processing in the case of synthetic aperture radar, while it is fan shaped in the real aperture radar. Synthetic aperture radar systems display ground range while real aperture systems display slant range. In synthetic aperture radar (SAR) imaging, antenna sends pulses of microwave towards the earth ground. The microwave energy scattered back to the spacecraft is measured. The SAR makes use of the radar principle to form an image by utilising the time delay of the backscattered signals.

The equation for the azimuth or along-track resolution for a point target in a SAR_a as given by Henderson and Lewis (1998):

$$SAR_a = \frac{D}{2} \qquad \text{... (xii)}$$

Here the factor 2 in the denominator is due to the two-way path of the radiation, since radar is an active instrument. This is an unexplained result. That is, the azimuth resolution is independent of range or radar wavelength and improves with smaller size of the antenna.

4.9 RADAR RETURN AND IMAGE SIGNATURE

When microwave energy strikes a terrain surface it gets reflected from the terrain to the radar antenna. The backscattered energy reaching the radar antenna is called radar return. The following parameters strongly affect the radar return:

(a) System properties:
  (i) Polarization
  (ii) Depression angle and incident angle
  (iii) Wavelength or frequency

(b) Terrain properties:
  (i) The dielectric constant of surface material which depends strongly on the moisture content.
  (ii) Surface geometry such as: surface roughness, slope, feature orientation relative to the radar beam direction, etc.
  (iii) Types of land cover such as: soil, vegetation or man-made objects.

4.10 POLARISATION

Polarisation, a fundamental property of the electromagnetic radiation refers to the orientation of the electric field component with respect to the plane of incidence. It depends on the phase difference between the electric and magnetic fields. Imaging radar systems typically transmit a plane polarized radar pulse. The electric field associated with such a pulse oscillates in a single plane perpendicular to the direction the wave is traveling. Most multipolarisation radars are designed to have two antennas to transmit/receive vertical as well as horizontal polarized microwave radiations, and some radar can receive either vertical or horizontal polarized radiation.

Thus one can have four kinds of both transmitting and receiving polarization combinations, namely VV for vertical transmit and vertical receive, VH for vertical transmit and horizontal receive, HV for horizontal transmit and vertical receive and HH for horizontal transmit and horizontal receive. Imaging radars can have different polarization combinations. However, more commonly used linear polarization combinations are HH, VV, HV and VH. The first letter corresponds to the transmitted polarized radiation, whereas the second letter corresponds to the received radiation. Using these measurements, it is possible to compute the response of the objects to any other polarization using Stock's matrix. Such a technique of determining the response to the complete polarization ellipse is known as radar polarimetry (Zebker and VanZyl, 1991).

Both the polarization state and power of a synthetic aperture radar backscatter vary as a function of the transmit polarization of the radar and environmental properties of the illuminated target. With polarimetric radar, this variation attributed to the target's bio- and geo-physical parameters. In spite of the fact, that radar is practically an all-weather instrument, radar imagery may suffer from precipitation and other environmental parameters. Moreover, the radar imagery is often noisy and not of visual quality comparable to high-resolution optical imagery. Radar polarimetry allows to monitor (i) cloud covered areas day and night, (ii) enhance the SAR data in order to improve target to background contrast or target to target separation, (iii) diminish unwanted signals like speckle, topographic and scatterer orientation effects, or, to derive more detailed digital elevation models (Tares and Hallikainen, 2003).

A comparison of like polarized (HH or VV) and cross polarized returns might reveal differences leading to terrain identification. When a polarized radar pulse is scattered by the rough surface of a material such as soil or rock (surface scattering) most of the scattered energy that returns to the antenna has the same polarization as the transmitted pulse. But, when the pulse strikes vegetation, it penetrates to varying degrees depending on wavelength and has numerous encounters with leaves, twigs, and branches that produce many scattering events. This volume scattering causes partial depolarization of the radar signal, so that some of the scattered waves vibrate in various directions. Depolarization somewhat reduces the strength of the like polarized signal returned to the radar antenna, but vegetation still produces strong like polarized signals at appropriate wavelengths. Short wavelength radar signals (X and C bands) interact primarily with the uppermost leaf canopy, and do not penetrate beyond it. Longer wavelength radars (L and P bands) penetrate deeper to interact with stems of smaller plants and twigs and branches of trees. Water and tree appear brighter in like polarized and darker in cross polarized imagery. Grasslands appear darker in like polarized image and brighter in cross polarized image. Regarding changes in the complex dielectric constant due to moisture variation, HH polarization is the least sensitive, whereas, HV and VV are the most sensitive. Thus, HV and VV polarized returns may contain information concerning soil and snow pack moisture levels.

4.11 RADAR WAVELENGTH AND SURFACE ROUGHNESS

Strength of the radar return is determined by the relationship of surface roughness to radar wavelength and to the depression angle of the antenna. Surface roughness is an important physical parameter quantifying the scattering of electromagnetic radiation. Materials differ from one another in their surface roughness. Roughness, in this sense, refers to minute irregularities that relate either to textures of the surfaces or of objects on them (such as, closely-spaced vegetation that may have a variety of shapes). For radar images surface roughness is a measurement of the average vertical relief of small-scale irregularities in the surface cover of the terrain from a reference plane surface. The surface roughness is measured relative to the radar wavelength and is on the order of centimeters. A surface appears "rough" to microwave illumination when the height variations become larger than a fraction of the radar wavelength and appears smooth if height variations become less than a fraction of the radar wavelength, following the Rayleigh criterion of surface roughness. For example, an ocean surface with capillary waves with 3 cm amplitude would be considered rough at C-band but smooth at L-band.

The roughness of a surface has a very important effect on the microwave backscatter from the surface. The surfaces which are very smooth, exhibit specular reflection and virtually they do not have backscattered energy. They image slightly rough and a small fraction of the incident energy is scattered into a wide range of angles. These smooth surfaces reflect most of the microwave energy away from the radar; the backscattered portion is small, so the surfaces look dark in radar images. For rough surfaces, the scattering pattern spreads out more and more, so the amount of energy scattered back to the radar increases. As a rule, the amount of backscatter is proportional to the roughness. Therefore, in a radar image, rough surfaces are brighter than smooth surfaces.

On the basis of the foregoing discussions, surfaces may be grouped into three roughness classes:

(i) Smooth when all the energy is specularly reflected.
(ii) Rough when the energy is diffusely scattered at all angles.
(iii) Intermediate rough when a portion of the energy is specularly reflected and the rest diffusely scattered.

The radar return is a function of the relationship between the incident wavelength (λ), the depression angle (γ) and the surface irregularity (h) above the reference plane surface. The modified Rayleigh criterion (Peake and Oliver, 1971) for the surface to be smooth is given by:

$$h < \frac{\lambda}{25 \sin\gamma}, \text{ and}$$

The surface is considered to be rough if:

$$h > \frac{\lambda}{4.4 \sin\gamma}$$

Inland water bodies tend to be relatively smooth, with most of the energy being reflected away from the radar and only a slight backscatter towards the radar. On the contrary, land surfaces tend to have a higher roughness. Water bodies appear in darker tone on radar imagery except in the case of wind stress or current that increases the water surface roughness, which results in high backscatter. In important applications such as flood extent estimation, coastal erosion/accretion, etc. the difference between respective properties of land and water can be extremely useful in the microwave region.

4.12 DIELECTRIC CONSTANT

The dielectric constant of a material is a measure of the ability of the material such as vegetation, soil, water, rock, etc. to conduct electric energy. It is a measure of both the conductivity and reflectivity in terms of the electrical response of materials. This electrical property describes a material's capability (capacity) to hold a charge, which also measures the material's ability to polarize when subjected to an electric field. Radar waves penetrate deeper into materials with low dielectric constants and reflect more efficiently from those with high constants. Moist soils have values typically between 30 and 60. Thus, variation in reflected-pulse intensities may indicate differences in soil moisture, other factors being constant. Variations among rocks are generally too small to distinguish most types by this property alone. Dielectrics are things that do not conduct electricity well, if at all. Dry air is a great example of a dielectric. Materials have different dielectric constants at room temperature. For example, air is about 1, paper is 3, and rubber is 7. In the microwave region, most natural materials such a soil, rock, and even vegetation have dielectric constants from 3 to 8 in the microwave portion of the spectrum. Because of this small range, radar generally is considered to be largely insensitive to the electrical properties of a surface material in the absence of water. However, the effective dielectric constant of water approaches 80, at least 10 times higher than for dry soil. Thus, ground targets produce significantly stronger backscattered radar return when they are damp. The dielectric constant is the ratio of the electrical conductivity of a dielectric material to free space. The dielectric constant of a material affects how electromagnetic signals move through the material. A high value of dielectric constant makes the distance inside the material look bigger. This means that light travels more slowly. It also 'scrunches up' the waves to behave as if the signal had a shorter wavelength. It consists of two parts: permittivity and conductivity. As discussed, they are highly dependent on the moisture content of the material considered. The amount of moisture content in the soil, rock and vegetation has a significant bearing on the amount of backscattered radar energy. As a result, a change in moisture content generally brings about a significant change in the dielectric properties of natural materials; increasing moisture is associated with an increased radar reflectivity. In case of damp soil, reflectivity of radar energy is more than that of dry soil. If the surface of the soil has high moisture content, then the incident energy will only penetrate a few centimeters into the soil column and be scattered more at the surface producing a stronger, brighter return.

The depth of penetration of radar waves below the surface varies inversely with the dielectric constant and of the material and directly with the radar wavelength. An increase in the dielectric constant increases the reflectivity of a surface. In case of vegetation, radar penetration depth depends on moisture, density and geometric structure of the plants.

4.13 RADAR PENETRATION AND FREQUENCY

Radar wavelengths influence penetrability below target tops to ground surfaces. Depth of penetration increases with wavelength. For longer wavelengths, penetration depth tends to be longer. The L (15-30 cm) and P (30-100 cm) band radar penetrate deeper than K (1.1-1.7 cm) or X (2.4-3.8 cm) bands. In case of forests, for example, shorter wavelength energy (X-band, $\lambda = 3$ cm) will only penetrate the first leaves on top of the trees thus from the canopy tops. The information content of the image is related to the top layer and the crown of the trees. Little energy reaches the ground. On the other hand, at longer wavelengths (L-band, $\lambda = 23$ cm), the radiation penetrates leaves and small branches; the information content of the image is then related to branches and eventually to tree trunks. Where surface scattering from the soil-vegetation boundary layer takes place, numerous pulses may be transmitted to the ground. Longer P-band radar would afford the greatest penetration through the vegetation and mainly reflect off large stems and the soil surface (Waring et al., 1995).

4.14 RADAR IMAGE CHARACTERISTICS

i. Image Compression in Near Range

Normally layover (the displacement of the top of an elevated feature with respect to its base on the radar image) may occur in the closer range from SAR, called near range, while in the far range more shadow may be seen. This means that care should be taken with the flight direction and range direction in interpretation of terrain features.

As illustrated in **Figure 4.3** the ground range distances AB and BC in near range will be compressed to A′B′ and B′C′ in the slant range representation. It is seen from the figure that the compression effect goes on reducing in the far range. This type of distortion occurs only in slant range display systems (Ka band).

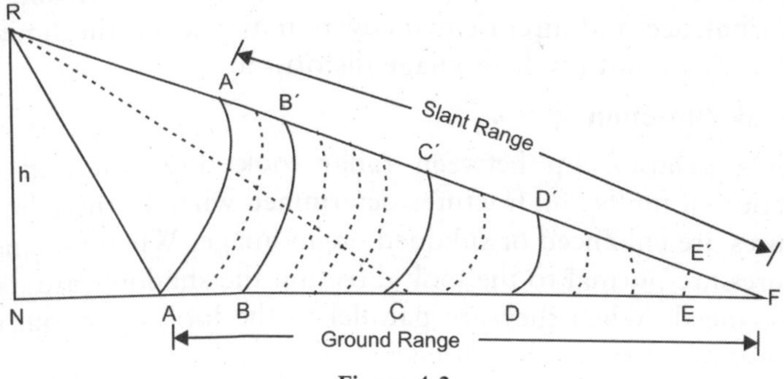

Figure 4.3

ii. Relief Displacement

Variations in the elevation of the terrain affect the relative travel times of the radar energy that returns to the sensor, producing unique geometric distortions that remain in radar images even after conversion to ground range. Geometric distortions such as foreshortening, layover, and shadowing exist in almost all radar imageries (Ford et al., 1980). Relief displacement in the radar image occurs when trees, tall buildings, or mountains are present in the scene. In radar relief displacement, the horizontal displacement of an object in the image is caused by the object's elevation is in a direction toward the radar antenna. The curvature of the transmitted radar pulse causes the top of a tall object having steep slope to reflect energy in advance of its base, resulting on displacement of the top, toward the near range direction. This is called "layover". In the near range, the depression angles being maximum, the resulting effect is called "foreshortening" as the facing slope is shortened. Thus, the elevation induced distortions in radar imagery are referred to as foreshortening and layover. Terrain slope inclined towards antenna will show strong reflections while terrain slopping away from the flight line shows no reflection and consequently appear as shadows. Shadows will increase in length as the angle of incidence increases from the near to far range. The radar shadows are areas of no information.

iii. Effects of Aircraft Motion

SLAR (Side Looking Airborne Radar) antennas have stabilization systems to compensate for normal aircraft roll, but air turbulence and aircraft maneuvers may exceed the limit of these devices and result in image distortion.

iv. Look Direction

The relationship between radar look direction and the direction of faults and fractures determined whether these linear features are enhanced or subdued on an image. When the linear features are normal to the look direction the shadows are most pronounced. When they are parallel to the look direction, the features are very much subdued.

v. Stereo Imagery

Radar flight lines may be spaced to give overlapping images which when viewed under the stereoscope, would give three dimensional view of the terrain. This technique is particularly useful in areas of no relief.

4.15 INTERPRETATION OF RADAR IMAGERY

From the foregoing discussions we can recap that the radar has got unique capability to collect data over virtually any region at any time, throughout day and night irrespective of weather conditions. Some radar waves can penetrate clouds, and under certain conditions, can also see through vegetation, ice and extremely dry sand. In many cases, inaccessible areas on the Earth's surface can be accessed through radar waves only. Interpretation of radar imagery is somewhat different from the interpretation of imageries and aerial photos obtained from optical sensors. These specific characteristics are the consequence of the imaging radar technique, and are related to radiometry (speckle, texture or geometry). The radar interpreter while examining an analog product he must keep in mind that the shades of gray level in the scene are related to the relative strength of the microwave energy backscattered by the landscape units. The imagery should be oriented in such a way that the radar shadows fall towards the observer.

The intensity of the backscattered signal varies according to roughness, dielectric properties and slope of the terrain. Thus the radar signal refers mainly to geometrical properties of the target and the imagery is interpreted using tone, texture, pattern, size and shape.

Tone in the radar imagery is the average intensity of the backscattered signal. Strong or bright signal returns appear as light tones are usually indicative of prominent cultural features or man-made features. Intermediate energy returns appear as medium gray tones on the imagery. It indicates areas of open country or areas of no return, as in the case of flat and smooth terrain of an air field. While, weak or low signal returns appear as dark (black) tones in the imagery commonly indicative of the presence of water and hydrologic features.

The shape of objects and features can provide distinctive clues for identification. It is the general form, structure, or outline of individual objects. It should be noted that the shape is as seen by the oblique illumination: slant range distance of the radar. Regular geometric shapes are usually indicators of human presence and use. Some objects can be identified almost solely on the basis of their shapes. Man-made structures like urban or agricultural land is identified by straight edge shapes. While natural features, such as forest edges, are generally more irregular in shape, except where man has created a road or clear cuts. Texture is the frequency of change and arrangement of tones. This is a micro image characteristic. The visual impression of smoothness or roughness of an area can often be a valuable clue in image interpretation. Smooth flat surfaces usually reflect very little radiation back to the antenna, so roads and still water will appear dark. Objects much smaller than the wavelength of the radar are invisible to the detector, so for a low frequency (L-band) radar, fields may also appear smooth. At shorter wavelengths, fields or woodland often give the image a textured appearance. The size of an object may be used as a qualitative recognition element on radar imagery. The size of known features on the imagery provides a relative evaluation of scale and dimensions of other terrain features.

4.16 ADVANTAGES OF RADAR IMAGERY

(a) Regional Investigations

Image strips covering a long belt may join together to form mosaics. These mosaics can be studied for the regional analysis of the terrain and geologic structures.

(b) Oblique Illumination

The oblique illumination produces a highlight and shadow effect thus enhancing faults and fractures.

(c) Suppression of Minor Details

Because of small scale and low resolution, the vegetation cover is suppressed in radar image. Minor cultural details are also suppressed so that the interpreter is not distracted in the search for larger terrain features.

(d) Detection of Lineaments

Oblique illumination, uninterrupted image coverage and the suppression of minor details help the interpreter to recognize subtle changes in tone, texture and topography that may be aligned on the image to form lineaments with structural significance.

(e) All Weather and Night Time Capability

Prevailing poor weather conditions in a place may inhibit satisfactory photographic coverage of the same. SLAR images can be acquired under almost all weather conditions and during night time also. Radar is also valuable for monitoring sea ice conditions at high latitudes during period of winter darkness.

5

Satellites and Sensors

5.1 INTRODUCTION

The evolution of remote sensing from using photo camera by balloonists and pigeons to the present state of the art has a great and remarkable history. It is not something entirely new. It is a spin off from the military reconnaissance activities and the space technology. As a technology, it began in the 1840s with newly invented photo-camera used to take pictures of the ground by the balloonists. First aerial photograph in the history was taken by the balloonist Felix Tournachon, nicknamed Nadar, who photographed the Paris suburb on a sunny autumn day using a tethered or captive balloon in 1858 A.D. Over one and half century, the photographic camera was the only means of performing the phenomenon of remote sensing. World War I and II have brought about a revolutionary change in aerial remote sensing, where aerial photography was a valuable reconnaissance tool. During World War II photo interpretation has got much recognition, while discriminating camouflaged military artilleries from the vegetation cover using Color IR films. All systems which are currently used on satellite platforms were originally developed for use with aircraft. The first free flight photo mission was carried out by Monsieur Triboulet in 1879. During the mid of 19th century kites were used as the platforms for aerial cameras to photograph the earth's surface. Aerial remote sensing has been realized since 1908 when aircraft was used as a platform by Wilbur Wright, who took first picture in France (LeMans) in 1908 and Italy (Centocelli) in 1909. Dr. Robert H. Goddard in 1926 constructed and tested successfully the first rocket using liquid fuel. In 1929 he launched the second rocket that carried scientific payload. It took the first picture of the Earth from a rocket. Among the first users of color film for aerial photography was Goddard in 1937. Color film was not used much for aerial photography until after World War II. During the 1950's the importance of color-infrared film has been considerably increased due to its capability in identifying different vegetation types and detecting diseased and damaged vegetation. Multispectral imagery has become familiar and steadily growing in popularity as a digital means for mission planning, thermal signature detection and terrain analysis. For the 20th century, radar has certainly proven to be one of the most amazingly useful developments and is vital to aviation. Radar technology can be said of a military technology that had important civilian uses. From then until the early 1960s, the aerial photograph remained the single standard tool for depicting the surface from a vertical or oblique perspective. Airborne camera systems are still a very important source of remotely sensed data (Lillesand and Kiefer, 1994). Remotely imaging the Earth from space at the top of atmosphere originated at the dawn of the Space Age (both Russian and American programs). The development of artificial satellites in the latter half of the 20th century allowed remote sensing to progress to a global scale. Since the early 1960s, numerous satellite sensors have been launched into orbit to observe and monitor the Earth and its environment. Most early satellite sensors acquired data for meteorological purposes. The first U.S. meteorological satellite, TIROS-1 (Television Infrared Observation Satellite), launched into orbit on April 1, 1960. The advent of Earth resources satellite sensors occurred when NASA launched Earth Resources Technology Satellite (ERTS-1), the first in a series of Landsat satellites in July 23, 1972 to provide repetitive global coverage of the earth's land masses. Subsequently renamed as Landsat, which was the first of the highly successful Landsat series of remote sensing platforms, the most recent being Landsat 7 (launched 15 April 1999). Earth observing and weather satellites such as Landsat, the Nimbus, NOAA, METEOSAT, SPOT, INSAT, IRS, and more recent missions such as RADARSAT, IKONOS and UARS provided global measurements of various data for civil, scientific, and military purposes.

5.2 PLATFORMS

Platforms we call the vehicles or carriers for remote sensing devices. Typical platforms range from balloons and kites for low altitude remote sensing to aircrafts and satellites for aerial and space remote sensing. As we go higher, larger area can be viewed by the sensor. Thus the altitude determines the ground resolution, which is a key factor for the selection of a platform. Sensor at higher altitude is useful for measuring features such as river course, mountain ranges, etc. providing with what is known as synoptic view. Since the dawn of civilization, the strategic importance of making observations from places at higher elevation than the surrounding areas has been well known. This is the reason why the forts were built on hills and elevated places so that in the event of an aggression the kings and generals could survey the battlefield below them and direct their troops while still being secure from the enemy (NRSA, 1995). However, the size of the objects appears smaller as we go higher and in fact an object may not be seen at all beyond a certain height.

The size of the objects that can be discriminated by a sensor depends on the resolving power of the sensor and height of the platform. Remote sensing platforms as discussed above may be balloons, kites, aircraft or satellites. Aircraft and satellites constitute two important types of present day platforms.

Aerial remote sensing started a few decades ago with photographic cameras and the technology is now well established with the development of sensors such as multispectral scanners, magnetometers and synthetic aperture radars. Aircraft remote sensing offers certain advantages such as flexibility of operation over any desired area at any convenient time and high resolution data for applications which need data in large scales.

5.2.1 Geostationary Satellites

A satellite generally orbits around the Earth in an elliptical path. Orbital period of a satellite can be defined as the time it takes to make one full orbit. When satellite moves round the globe, it traces out a path on the earth surface that is called its ground track. Remote sensing satellites are often launched into special orbits such that the satellite repeats its path after a fixed time interval. This time interval is called the repeat cycle of the satellite. Satellites used for remote sensing are generally of two types depending upon the altitude: (i) Geostationary and, (ii) Sun synchronous.

A geostationary satellite is any satellite which is placed in a geostationary orbit (**Figure 5.1a**). Geostationary orbit is a circular orbit orientated in the plane of the Earth's equator. As the name suggests, geostationary satellites are stationary with respect to the Earth, placed at an altitude of approximately 35,800 kilometers (22,300 miles) directly above the Earth's equator (0° latitude), that revolves in the same direction the

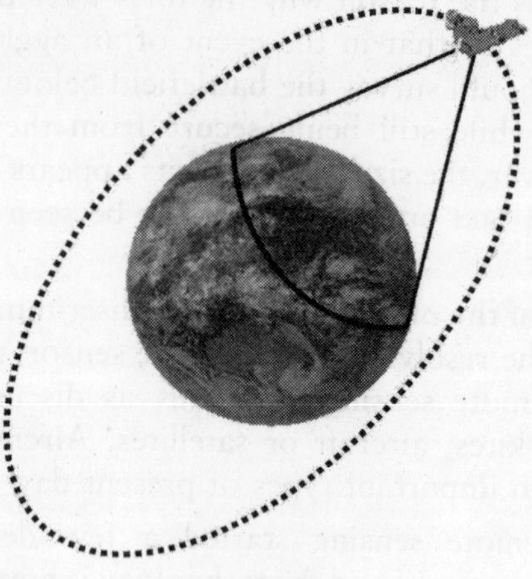

Figure 5.1a

earth rotates (west to east). This altitude is significant because it produces an orbital period equal to the Earth's period of rotation. These satellites do not see the poles at all, and to get global coverage of just the equatorial regions, a network of 5-6 satellites are needed. Satellite locations may differ by longitude only, since the latitude is always zero. They appear to be stationary in the sky to an observer at the ground. At this altitude, one orbit takes 24 hours, the same length of time as the earth requires rotating once on its axis. Single geostationary satellite can provide one-third coverage of the entire planet. Satellites in geostationary orbit maintain a constant position relative to the surface of the Earth. The orbit path traced by the satellite is called the Clarke Belt, in honor of Arthur C. Clarke. If a satellite in a geostationary orbit is in a certain place above the Earth, it will stay in that same spot above the Earth. Due to their large distance from the Earth, high resolution imaging is difficult. In 1945, Arthur C. Clarke has popularised the use of geostationary orbit for communications satellites. Since then, these have been extensively used for communications and meteorological applications. India's INSAT series of satellites are Geosynchronous providing for simultaneous domestic communications and Earth observation functions. The life period of such satellites are fixed at ten to fifteen years. Geostationary orbits are often referred to as geosynchronous.

5.2.2 Sun Synchronous Satellites

In order for acquisition of high quality remote-sensing data, satellites are better placed in lower Earth orbits unlike geostationary. The sun-synchronous orbits are normally just referred to as polar orbits (**Figure 5.1b**). The advantage of putting the resource satellites in the polar orbits lies in the fact

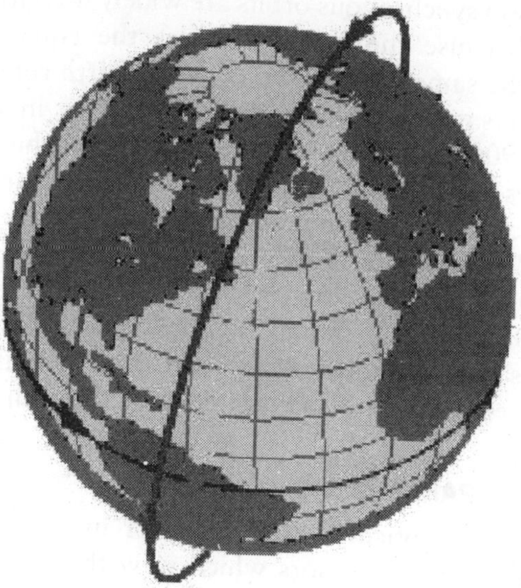

Figure 5.1b

that the Earth rotates beneath, thus, a single rotating satellite can have a global coverage. Its circular orbit implies a constant satellite velocity, which is important for having a regular scanning resolution along the satellite ground track. Therefore, sun-synchronous orbit can now be defined as an orbit that always passes over the same part of the Earth at approximately the same local sun time each day. This consistent lighting is a useful characteristic for satellites that image the earth's surface in visible or infrared wavelengths. These satellites are launched into orbits at high inclinations to the earth's rotation, such that they pass across high latitudes near the poles. The satellite's orbital plane appears to be fixed with respect to the sun. However, the times are different for ascending and descending passes. The point where the satellite, traveling northwards, passes directly over the equator is called the ascending node. The descending node describes the southward crossing (Stewart, 1985; Rees, 1990). The low altitude of a sun-synchronous orbit permits good ground resolution. The ground trace of the satellite can be made to repeatedly cover an area at intervals of a fixed number of days by suitably selecting the orbit parameters. Such polar sun synchronous orbits are widely used for monitoring the Earth because the satellite passes the equator and each latitude at the same time each day as the Earth rotates below it. Typically, a satellite in such an orbit moves in a near-circle about 700-900 km above ground and each orbit takes about 100 minutes permitting (1) large ground swath, and (2) a good ground resolution. At different altitudes they travel at different speeds. Most of the Earth observing missions use sun-synchronous satellites in low near polar orbits (NOAA, Landsat, SPOT, IRS, etc.). IRS (Indian Remote Sensing Satellites) series of satellites are low Earth, near polar orbit satellites dedicated for the observation and management of the country's natural resources.

5.3 SENSOR PARAMETERS

Sensors are optical devices used for making observations. Each satellite carries sensors which view the Earth in different parts of the electromagnetic spectrum and measure the amount of visible light, thermal (*infrared*) and microwave radiation reflected or emitted from the Earth surface. Examples are LISS, AVHRR, CZCS, SeaWiFS, MODIS, etc. The scanner underneath the satellite scans the Earth directly from left to right. The data will then be transmitted to Earth as one scan-line. By the time the scanner comes back to the start, the satellite will have progressed, so a different part of the Earth is seen. Each sensor is only sensitive to a small part of the spectrum, so multiple sensors are fitted to each scanner. For example, the AVHRR scanner has two sensors in the visible region and two in the infrared region (plus two in between). Sensors are usually sophisticated lenses with filter coatings to focus the area observed on a plane in which the detectors are placed. These detectors are sensitive to a particular region in which the sensor is designed to operate and produce outputs which are representative of the observed area.

Sensors which use lenses in the visible and reflective infrared region are called optical sensors. The performance of Optical Sensors is principally governed by important instrumental parameters such as spatial, spectral and radiometric resolution. These resolutions vary from one sensor to the other and are prefixed before the launch of the satellites. Each sensor has its own capability to sense the electromagnetic radiation emanating from the terrain features. High resolution sensors are now a days of greatest use for mapping and monitoring the natural resources and human dimensions of global environmental change. Each sensor is designed for the specific purpose associated with monitoring the effects of natural events and human actions on land cover. The prime requisites for such purposes are high frequency of coverage, specific combinations of spectral information, or high spatial resolution. In many cases, the data of two or more sensors offer complementary characteristics, making joint use of data advantageous. To cite an example, in IRS-1C, the PAN+LISS-III merged products are used in order to exploit the dual advantage of the spectral resolution of LISS-III and the high spatial resolution of PAN (5.8 m).

If a sensor is designed to record radiation from the natural sources such as the sun, as reflected by the Earth's surface, is called Passive sensor. Examples of such sensors are cameras (photographic and TV), multispectral scanners and radiometers. Electro-optical sensors are passive imaging instruments that record the solar radiation reflected off the earth surface. They are called passive instruments because they do not transmit their own source of energy rather they are using natural source of radiation. This means they can only operate in daylight. However, in thermal infrared band, an electro-optical sensor senses not the sun's energy rather the radiant temperature emitted from generating sources such as power plants. Sensors which illuminate the targets with their own source of energy and measure the reflected/scattered radiations from the target are called active sensors. Examples are active radiometers and radars. Synthetic Aperture Radars send out their own microwave radar signal, and receive the return signal after it reflects off the earth's surface. The information details in SAR imagery differs in some ways from information due to the spatial and spectral details of electro-optical imagery. SAR operation in longer wavelengths makes it distinction from the electro-optical systems, as it acquires images through clouds, fog, haze and darkness.

Each is divided further into non-scanning and scanning systems. A sensor belonging to a combination of passive, non-scanning and non-imaging category is a type of profile recorder, for example a microwave radiometer, magnetic sensor, gravimeter, Fourier spectrometer etc. A sensor belonging to a combination of passive, non-scanning and imaging category is a camera, which can be monochrome, natural color, infrared etc. Sensors as a result of a combination of parameters viz. passive, scanning and imaging are classified further into (i) image plane scanning, for example TV cameras and solid state scanners, and (ii) object plane scanning, for example optical-mechanical scanner and scanning microwave radiometers. An example of an active, non-scanning and non-imaging sensor is a profile recorder such as Microwave Radiometer, Microwave Altimeter, Laser Water Depth Meter, and Laser Distance Meter. An active, non-scanning and imaging scanners are classified into two categories: (i) object plane scanning—e.g. Real Aperture Radar and Synthetic Aperture Radar which can produce high resolution imagery, day or night, even under cloud cover, and (ii) image plane scanning, e.g. passive phased array radar.

Spaceborne sensors which are more familiar and widely used in remote sensing are the camera, solid state scanner, such as the CCD (charge coupled device), the multi-spectral scanner and the radars.

5.3.1 Spatial Resolution and IFOV

The detail of information extracted from an image as well as the total area on the ground imaged by the sensor largely depends upon the resolving power of the sensor and the distance between the target and the sensor. The concept resolving power of the sensor or resolution is a complex attribute that describes how much detail in an image is visible to the human eye for identification. This may be thought of as the ability of the eye or sensor to see fine detail. Resolution is thus our ability to resolve objects on the image. It measures the degree of discrimination or separation between objects. In general resolution can be defined as the ability of an entire remote-sensing system, including lens antennae, display, exposure, processing, and other factors, to render a sharply defined image. Resolving power or minimum resolvable distance of an optical device is the angular separation between two objects in an image. Higher resolution means more image detail.

Spatial resolution is defined as the size of the smallest object that can be discriminated by the sensor. Spatial resolution is a measurement of the minimum distance between two objects that will allow them to be differentiated from one another in an image (Sabins, 1978; Jensen, 1996). This is a function of sensor altitude, detector size, focal size and system configuration. For aerial photography the spatial resolution is usually measured in resolvable line pairs per millimeter on the image. For other sensors it is given as the dimensions, in meters, of the ground area which falls within the instantaneous field of view of a single detector within an array—or pixel size (Logicon, 1997). The ability to resolve or separate, small details is one way of describing what we call spatial resolution. For any remotely sensed image, there is a three-way trade-off between spatial resolution, aerial extent, and data volume. The greater the sensor's resolution, the greater the data volume and smaller the area covered. If we held data volume constant, we can have a small area coverage with high-resolution image and large area coverage with a low resolution image. Sabins (1987) defined spatial resolution as "the aoility to distinguish between two closely spaced objects on an image" and points out that spatial resolution are "not the size of the smallest object that can be seen". He also points out that other terms such as detectability; recognizability, signature, and texture have a place in the qualitative analysis of images. In fact, area coverage and resolution are independent and these factors determine the scale of imagery. Spatial resolution of the image is very important when choosing the best source for a given application. High resolution imagery will generally be more expensive than low resolution imagery, but it will also allow you to see much more detail in the landscape. High-resolution images are sharp, and many small details can be seen, but even the best pictures have a limit to their resolution. The spatial resolution of images acquired by satellite sensor systems is usually expressed in meters. For example, we often speak of one of the imaging systems on IRS 1C PAN as having "six meter" resolution. This means that two objects, six meters long or wide, sitting side by side, can be separated (resolved) on an IRS 1C image. Spatial resolution is directly related to image pixel size. In remote sensing, each pixel represents in the real world by the terms ground resolution element and ground resolution distance. As an example, the IRS-1C has resolution ranging from 5.8 to 188.3 m. In this case as smaller the resolution distance, the better one can resolve image spatial contents.

Instantaneous Field of View (IFOV)

The field-of-view (FOV) is the range of angles from which the incident radiation can be collected by the detector. The field of view may be resolved into its horizontal and vertical components, labeled as HFOV and VFOV respectively. A term specifically denoting the narrow field of view designed into detectors, particularly scanning radiometer systems, so that, while as much as 120° may be under scan, only radiation from a small area is being recorded at any one instant, that is IFOV. "IFOV or instantaneous field of view now can be defined as angular subtends at a given instant of the limiting detector aperture at the second principal point of the system. Commonly, the IFOV is expressed as the dimension(s) of the "foot-print" of the detector on the ground at a given instant. The instantaneous field of view (IFOV) alone does not adequately define the spatial response of a scanner system.

The instantaneous field of view (FOV) of a nadir scanning satellite instrument can be described by the intersection of a cone with a sphere. The FOV normally is circular at nadir and increases in size both along and across track as the scan angle increases. The across track distortion is due to the fact that the FOV edge closer to nadir is also farther away from the horizon, which causes a stretching of the FOV. The along track distortion is simply due to the fact that FOV angular size is fixed, but the distance from the satellite to the viewing location on the earth increases as the scan moves away from nadir. The diameter of the circular ground area viewed by the sensor, D, is a function of the instantaneous field of view, β, of the scanner measured in milliradians (mrad) and the altitude of the scanner above ground level, H, where,

$$D = H \times \beta$$

There is an inverse relationship between having high spatial resolution and high radiometric resolution when collecting thermal infrared data. The larger the radiometer instantaneous field of view, β, the longer the dwell time that an individual detector views the terrain within the IFOV during a single sweep of the mirror. A larger IFOV provides good radiometric resolution, which is the ability to discriminate between the very small differences in radiant energy exiting the terrain element. In fact, the radiant energy signal measured may well be the much stronger than any noise introduced from the sensor system components. When this takes place, we say that we have a good signal noise ratio. Of course, the larger the IFOV, the poorer is the ability to resolve fine spatial detail. Selecting a smaller IFOV will increase the spatial resolution, but the sensor will dwell a shorter time on each terrain element during a sweep of the mirror, resulting in poorer radiometric resolution and perhaps a poorer signal to noise ratio (Jensen, 2004).

5.3.2 Spectral Resolution

The spectral resolution is the ability of the sensor to resolve the energy received in a given spectral bandwidth to characterize different constituents of earth features. The number and width of the spectral bands in which an image can be taken determine the spectral resolution of a sensor. Improvement in spectral resolution can be made by reducing the bandwidth but this will also cause a reduction in signal strength (luminosity). Narrow bandwidths in certain regions of the electromagnetic spectrum allow us to discriminate various features more easily. The finer the spectral resolution, the narrower is the wavelength range for a particular channel or band. Consequently, we need to have more number of spectral bands; each having a narrow bandwidth and these bands should together cover the entire spectral range of interest.

Remote sensing systems record the reflected or emitted energy emanating from the objects on the earth surface in several discrete wavelength ranges (so-called spectral bands or channels) at various spectral resolutions. The finer the spectral resolution, the narrower the wavelength range for a particular channel or band, the better different objects can be detected and distinguished. Advanced multi-spectral sensors, called hyperspectral sensors, detect hundreds of very narrow spectral bands throughout the visible, near infrared and mid-infrared portions of the electromagnetic spectrum. Their very high spectral resolution facilitates fine discrimination between different targets based on their spectral response in each of the narrow bands (*Source*: Manual of Concepts on Land Cover and Land Use Information Systems, Doc. ESTAT/LAND/31rev1, p. 30). Therefore, a sensor even if is sensitive to a large portion of the electromagnetic spectrum, is said to have poor spectral resolution if its sensitivity contained in a small number of broad spectral bands. On the contrary, another sensor sensitizing in the same portion of the electromagnetic spectrum is said to have greater spectral resolution if it detects energy in many small spectral bands. Like spatial resolution, the goal of finer spectral sampling is to enable the analyst, human or computer, to distinguish between scene elements. Different remote sensing instruments record different segments, or bands, of the electromagnetic spectrum and thus the sensors are unique in this regard.

An image produced by a sensor system can consist of one very broad wavelength band, a few broad bands, or many narrow wavelength bands. The name usually used for these three image categories are panchromatic, multispectral and hyperspectral, respectively.

Aerial photograph taken using black and white film records an average response over the entire visible wavelength range (blue, green, and red). Because this film is sensitive to all visible colors, it is called panchromatic film. A panchromatic image reveals spatial variations in the gross visual properties of surface materials, but does not allow spectral discrimination. Imaging is performed in a single very broad spectral band, corresponding to the visible part of the electromagnetic spectrum to provide a synoptic overview of the scene, commonly at a higher spatial resolution than other sensors onboard. Despite varying wavelength ranges, such bands are commonly referred to as panchromatic bands. For most current panchromatic sensors, this single band usually spans the visible to near-infrared part of the spectrum, as 0.50-0.75 μm in case of IRS ID PAN. Panchromatic data is represented as black-and-white imagery. The panchromatic band in SPOT 1, 2 HRV covers 0.51 to 0.73 μm. For SPOT 4 HRVIR, the panchromatic band has a narrower bandwidth centered at the red band (0.61 to 0.68 μm). This band has a spatial resolution of 10 m in contrast to the 20 m resolution of the multispectral sensor bands. The panchromatic band of the ETM+ sensor of Landsat 7 satellite covers a wide spectral range of 0.52 – 0.90 micrometer (green, red, infrared), with a spatial resolution of 15 m (versus 30 m for the sensor's multispectral bands). Due to bandwidth limitations, many remote sensing systems have strengths in either spatial or spectral resolution, an inevitable trade off. It is, therefore, important to have capabilities to merge data and take advantage of enhancing the spatial resolution of multispectral images using a higher resolution panchromatic image.

A significant advance in sensor technology is allowing sensors in several bands to form multispectral images. In remote sensing multispectral imaging refers to imaging a given area in several narrow bands to obtain better identification and classification of objects. For instance, while one set of detectors measure reflected visible red energy, another set measures near infrared energy. These multiple reflectance values are integrated to create color images. The present day multispectral remote sensing satellites record reflectance values in three to seven different bands at once. The sensor is a multichannel detector with a few spectral bands. Each channel is sensitive to radiation within a narrow wavelength band. The resulting image is a multilayer image which contains both the brightness and spectral (colour) information of the targets being observed. Most satellite systems use from three to six spectral bands in the visible to middle infrared wavelength region. Some systems also employ one or more thermal bands. Bands in the infrared range are limited in width due to presence of many atmospheric absorption bands that significantly degrade the signal. These broadband multispectral systems allow discrimination of different types of vegetation, rocks and soils, clear and turbid water, and some man-made materials. A three band sensor with green, red, and near infrared bands is effective at discriminating vegetated and non-vegetated areas. The IKONOS satellite from Space Imaging (4 m resolution) and IRS LISS II sensor (36 m resolution) add a blue band to provide complete coverage of the visible light range, and allow natural color band composite images to be created. The Landsat TM (4 and 5) and Landsat 7 ETM+ sensors add two bands in the middle IR. Landsat TM band 5 (1.55 – 1.75 μm) and band 7 (2.08 – 2.35 μm) are sensitive to variations in the moisture content of vegetation and soils. Band 7 also covers a range that includes spectral absorption features found in several important types of minerals. An additional TM band (band 6) records part of the thermal IR wavelength range (10.4 – 12.5 μm).

In the recent days the hyperspectral imaging technology is more familiar and widely used for imaging the earth's natural resources to cope up with the recent advances in remote sensing and geographic information. It refers to a spectral sensor that measures reflectance in many individual bands, often hundreds, in narrow spectral bands of as little as 0.1 μm in width to provide even greater spectral resolution. The theory behind hyperspectral sensing is that measurement of reflectance in numerous narrow portions of the spectrum can detect very subtle characteristics and differences among surface features—especially in vegetation, soil and rocks. No hyperspectral sensors are now operated on a commercial satellite. These are passive sensors that acquire simultaneous images in many relatively narrow, contiguous and/or discrete spectral bands through the ultraviolet, visible and infrared portions of the electromagnetic spectrum. Reflectance spectra have been used for many years to obtain compositional information on the Earth's surface. It is now possible to acquire hyperspectral information via aircraft and satellite platforms. Hyperspectral sensors acquire imaging data at high spectral and moderate spatial resolutions, enabling reflectance, radiance and emittance spectra to be constructed that permit physical measurements of the Earth's surface. With this increased amount of spectral data, scientists have the ability to derive specific information about the earth's surface from an image pixel using diagnostic or characteristic absorption features in its spectral signature. The signature is generated by the information contained in the numerous spectral bands acquired by the sensor. The number of images collected by these sensors can be so great; the datasets are often referred to as "data cubes". The general problem is that the bands used for multispectral imaging are so broad that the surface material's characteristic spectral reflection and absorption peaks cannot be resolved; researchers have to rely on gross approximations of the spectra. Typical sensors produce from 24 to as many as 288 spectral bands for each image. Because of the scarcity and expense of the data, much is yet to be learned about the usefulness of hyperspectral imagery. It appears that its greatest utility is in mineral exploration in deserts. The various absorption windows in the atmosphere may prove to be useful hyperspectral features for monitoring atmospheric changes as well as surface features.

5.3.3 Radiometric Resolution

The actual information content in an image is determined by the radiometric characteristics of a sensor. It is commonly expressed as the number of bits (binary digits) needs to store the maximum level. The radiometric resolution of an imaging system describes its ability to discriminate very slight differences in energy. The finer the radiometric resolution of a sensor the more sensitive it is to detecting small differences in reflected or emitted energy. In a digital sensor radiometric resolution is determined by the number of discrete levels into which a signal may be divided. It can be defined as "dividing the total range (B to W) of the signal output of the sensor into a large number of just discriminable levels so as to be able to distinguish ground features differing only slightly in radiance or reflectance". Considering the effects of varying illumination the radiometric dynamic range of a sensor is determined by the maximum radiance value that the sensor can experience for a given band. Satellite data are represented either in analog or in digital format. Digital number or DN value is the reflectance value of a feature in the smallest slice of a band or portion of the electromagnetic spectrum. Digital image can be represented by a range of digital numbers or DN values from 0 corresponding to black to a selected power of 2 corresponding to white. This range corresponds to the number of bits used for coding numbers in binary format. Each bit records an exponent of power 2 (e.g. 1 bit = 2^1 = 2). The maximum number of shades of gray levels available depends on the number of bits used in representing the energy recorded. For example (see **Table 5.1**), a 10 bit radiometric resolution signifies that detector's converted digital output signal into a range from 0 to 1023 (or 1024 quantisation levels). However, if we take a smaller bit number, the number of shades of gray levels would be much less. Image data are generally displayed in a range of gray tones, with black representing a digital number of 0 and white representing the maximum value (for example, 255 in 8-bit data). If we compare

TABLE 5.1: Understanding the concept of quantisation levels

Number of Bits	Number of quantisation levels	Range of quantisation levels
1	2	0 - 1
2	4	0 - 3
3	8	0 - 7
4	16	0 - 15
5	32	0 - 31
6	64	0 - 63
7	128	0 - 127
8	256	0 - 255
9	512	0 - 511
10	1024	0 - 1023

a small radiometric data of 2-bit with that of 8-bit image, we can observe a significant difference in the detail discrete shades of gray level. The dynamic range of the most common satellite data is 7 bits or 128 different levels (Landsat MSS, IRS), or 256 levels (8 bits) for Landsat TM, ETM. High resolution IKONOS data is collected as 11 bits per pixel (2048 gray tones). This means that there is more definition in the gray scale values and as a viewer you can see more detail. There is a trade off between radiometric resolution and signal to noise ratio. Lower radiometric resolution implies a low-quality instrument with a high level of noise compared with a high-quality, high signal-to-noise-ratio instrument. Higher radiometric resolution may also conflict with data storage and transmission rates. Some applications (e.g. SST measurement) do not require a very high radiometric resolution while other observations (e.g. vegetation) require as many spectral channels as possible for good interpretation of the data.

Quantisation levels are frequently given in terms of the number of bits rather than the number or range of levels. These values are related by:

$$2^{\text{Number of bits}} = \text{Number of quantisation level}$$

In image processing, quantisation levels are usually referred to as Digital Numbers (DN).

Human eye can only perceive 20 to 30 different shades of gray levels. This implies present day satellite imageries that provide far from 30 shades of gray lcvels are not visually discernible. The human visibility for identification and interpretation can only be extended by using digital techniques to derive maximum discrimination from the available radiometric resolution. When the radiometric resolution of data acquired by one system are compared with data acquired by a higher radiometric resolution instrument (e.g., Landsat TM with 8-bit data) then the lower resolution data (e.g., 6-bit) should be decompressed to 8 bits for change detection purposes. However, the precision of decompressed brightness values can never be better than the original, uncompressed data in an image.

The atmospheric contribution due to absorption and scattering decrease the radiometric resolution of an image by reducing the discrimination between different radiation levels, especially at shorter wavelengths.

5.3.4 Temporal Resolution

In addition to spatial, spectral, and radiometric resolution, the concept of temporal resolution is also important to consider in a remote sensing system. Temporal resolution refers to how often an area can be imaged. It is simply the frequency of temporal coverage of a sensor/platform system for data acquisition over an area and is only relevant in time-series studies or those requiring an averaged or mosaic image. It is the revisit period of the satellite that refers to the length of time it takes for a satellite to complete one entire orbit cycle. The revisit period of a satellite sensor is usually in the range of several days. The temporal resolution equals this time period after which a remote sensing system records an image of the exact same area at the same viewing angle a second time. It depends on the (i) orbital period of the satellite, (ii) latitude of the target, (iii) swath width of the sensor, and (iv) pointing ability of the sensor.

During revisit of the satellite, some areas of the Earth are re-imaged frequently. This is due to (i) some degree of overlap in the imaging swaths of adjacent orbits and (ii) the increase in this overlap with increasing latitude. Thus, the actual temporal resolution of a sensor depends on a variety of factors, including the satellite/sensor capabilities, the swath overlap, and latitude. Geo-stationary satellites can provide continuous sensing while normal orbiting satellites can only provide data each time they pass over an area. A sensor such as Landsat Thematic Mapper will provide 30 meter pixels, but can only image a given area once every 16 days. On the other hand, the Advanced Very High-Resolution Radiometer can image the entire Earth every day, but has 1.1 kilometer pixels. Even higher temporal resolution is obtained by geostationary satellites such as the GOES weather satellites. The GOES satellites provide the same view of the earth every 30 minutes at 1 kilometer spatial resolution. As with spatial resolution, the required temporal resolution is dependent upon the application.

The ability to collect imagery of the same area of the Earth's surface at different periods of time is one of the most important elements for applying remote sensing data. Spectral characteristics of features may change over time and these changes can be detected by collecting and comparing multi-temporal imagery. For example, during the growing season, most species of vegetation are in a continual state of change and our ability to monitor those subtle changes using remote sensing is dependent on when and how frequently we collect imagery. By imaging on a continuing basis at different times we are able to monitor the changes that take place on the Earth's surface, whether they are naturally occurring (such as changes in natural vegetation cover or flooding) or induced by humans (such as urban development or deforestation). Remotely sensed images can be used to identify nutrient deficiencies, diseases, water deficiency or surplus, weed infestations, insect damage, hail damage, wind damage, herbicide damage, and plant populations.

5.4 IMAGING SENSOR SYSTEMS

The two regions of the electromagnetic spectrum which are currently being exploited the most by the satellite sensors are optical wavelength region (visible, infrared, thermal) and microwave region. Sensor gathers information from the target in the form of electromagnetic energy converts it into signal and transmits it into the ground in a suitable form of interpretable information. All the imaging sensors are classified based on technical configuration of the system and the capability of detecting the signal from the targets under different illumination and environmental conditions. These are (i) Multispectral imaging sensor systems, (ii) Thermal sensing systems, and (iii) Microwave imaging systems.

5.4.1 Multispectral Imaging Sensor Systems

Human eye is sensitive to the visible part of the electromagnetic spectrum and is able to identify different objects by their colors by which they could be distinguished. In a similar way a sensor can identify the object by making measurements of its response to distinctive bands of the electromagnetic spectrum. Multispectral imaging thus, refers to viewing a given area or an object in a spectrum of narrow bands to obtain better identification and classification of objects. Thus the concept of multispectral sensing has been brought into remote sensing for mapping and monitoring of earth resources.

Sensors can be of non-imaging and imaging types. Non-imaging sensor gathers the reflected and/or emitted energy from the target then integrates and converts it into signal and reports the result as electrical signal strength or some other quantitative attribute, such as radiance. Solid-state image sensors with higher resolution are used in many commercial applications and also for other light imaging uses. These sensors typically consist of detectors arranged in arrays called charge-coupled devices (CCDs) or CMOS (complimentary metal oxide semiconductor) image sensors.

Sensors (charge coupled devices) are electronic devices that are capable of transforming a light pattern (image) into an electric charge pattern (an electronic image). An individual CCD is an extremely small silicon (micro) detector, which is light-sensitive. Many individual detectors are placed on a chip side by side either in a single row as a linear array or in stacked rows of linear arrays in two dimensional space. The CCD consists of several individual elements that have the capability of collecting, storing and transporting electrical charge from one element to another. This together with the photosensitive properties of silicon is used to design image sensors. From 3,000 to more than 10,000 detector elements (the CCDs) can occupy a linear space less than 15 cm in length. The number of elements per unit length, along with the optics, determines the spatial resolution of the instrument. Smaller sizes represent smaller areas on the target surface.

A charge-coupled device sensor detects the photons reaching the camera and transfers the charge of one pixel after another to the output amplifier, line after line. The output amplifier converts the charge into a corresponding voltage value. These values are stored in the CCD array. The electronic processing system will then transform this output signal of the CCD into a suitable form used to make analog or digital images. The electronic processing system employed certain functions at the processing stage such as: (i) pre-amplification and (ii) onboard digitizing of the signal and the reference radiation source used in calibration. The final output is then sent to a ground receiving station, either by direct readout (line of sight) or through a satellite relay system like TDRSS (Tracking and Data Relay Satellite System).

The size of the CCD array determines the image resolution of the camera. A camera of 5 mega pixel image resolution implies an image in camera that contains 5×10^6 pixels, e.g., 2500×2000 pixels. To form such an image of huge dimension, we need as many CCD devices stored in the form an array of 2500×2000. For a color image to generate, we need $2500 \times 2000 \times 3$ sensors (each for red, green and blue). This is really a huge number. In order to accommodate such a huge number of detectors, the size of the camera is to be increased relatively. This results in a great problem while registering the pixel values and is encountered by the usage of Mosaic Focal Plane Array Technology. According to MFPA, each pixel detector is covered with a wavelength-specific optical filter. Since only one spectral component is sensed at each pixel, the other spectral components must be estimated from neighboring pixels. The capturing process follows a pattern usually called the Color Filter Array (CFA). The CFA is used to sample a multi-band image into a mosaiced image.

The multispectral imaging sensor systems operate over several wavelength bands in the optical wavelength region, e.g. from ultraviolet to visible and thermal infrared (0.2-14 μm). Such sensors which image in a small number of broad spectral bands have been used since Landsat 1, launched in 1972.

The sensor operates in a scanning or a non-scanning mode. Scanning implies movement of both the entire sensor and scene viewing apparatus or the light or radiation detectors across the scene over a time interval. Non-scanning refers to holding the sensor fixed on the scene or target of interest as it is sensed in a very brief moment. Scanners are mostly of two broad categories: (i) optical-mechanical and (ii) optical-electronic.

An optical mechanical scanner is a multispectral radiometer. It records two dimensional image using a combination of the motion of the platform and a rotating or oscillating mirror scanning perpendicular to the flight direction. Optical-mechanical scanners are composed of an optical system, spectrographic system, scanning system, detector system and reference system. Optical-mechanical scanners can be carried on polar orbit satellites or aircraft. Multispectral scanner (MSS) and thematic mapper (TM) of LANDSAT, and Advanced Very High Resolution Radiometer (AVHRR) of NOAA are the examples of optical-mechanical scanners.

In optical-electronic scanners the sensed radiation moves directly through the optics onto the linear or array detectors. Optical-electronic scanners are composed of (i) optical system that includes lenses, mirrors, apertures, modulators and dispersion devices, (ii) detectors providing an electrical signal proportional to the irradiance on its active surface, generally some type of semiconductors, and (iii) signal processor performing specified functions on the electrical signal to provide the desired output data. The concept of a high resolution Modular Optoelectronic Multispectral/Stereo Scanner (MOMS) initiated with the German Space Agency since 1979. MOMS-01 introduced the CCD push broom line-scanning detector technology on a space borne platform. The MOMS-02, a combined stereo and multi-spectral payload instrument introduced stereoscopic along-track imaging with a three-line imaging system. The (MOS) is a spaceborne imaging spectrometer for the VIS/NIR-spectral range. The Modular Optoelectronic Scanner (MOS) on board the Indian Remote Sensing Satellite IRS-P3 consists of three instruments: the spectrometer MOS-A, the spectrometer MOS-B and the camera MOS-C.

The most commonly used scanning system in space platforms today is the multi-spectral scanner (MSS), which collects data over a variety of different spectral wave bands. Multispectral scanner images are acquired by means of two basic processes: (i) across track and (ii) along track scanning (**Figure 5.2**). Across track systems are also familiarly known as whisk broom scanner systems, whereas along-track systems are known as push broom scanners.

I. Across Track Scanner

Across track scanner is an optical tool with an oscillating mirror that moves back and forth across a satellite's direction of travel, creating scan line strips that are contiguous or that overlap slightly, thereby producing an image. Unlike viewing the entire scene at a glance, they scan the Earth through a rotating or oscillating mirror in a series of lines that subtend right angles to the direction of motion of the sensor platform. The forward motion of the sensor platform enables the mirror to cause new ground strips to be covered by successive scan lines. This type of scanning in obtaining a two dimensional image of the Earth's surface due to successive scans is referred to as whisk broom scanning. **Figure 5.2a** illustrates how the scanning operation takes place with the motion of the sensor platform. In this type of scanners, the data are recorded repeatedly line by line from one side of the sensor to the other with the help of rotating mirror. Low altitude sensor like airborne sweeps typically between 90° and 120° whereas the high altitude space sensors require sweeping angle of between 10° and 20° to cover a large area. A small IFOV is necessary in order to capture spatial details. The sensor that records the energy at any instant within its IFOV explains the spatial resolution of the imaging system.

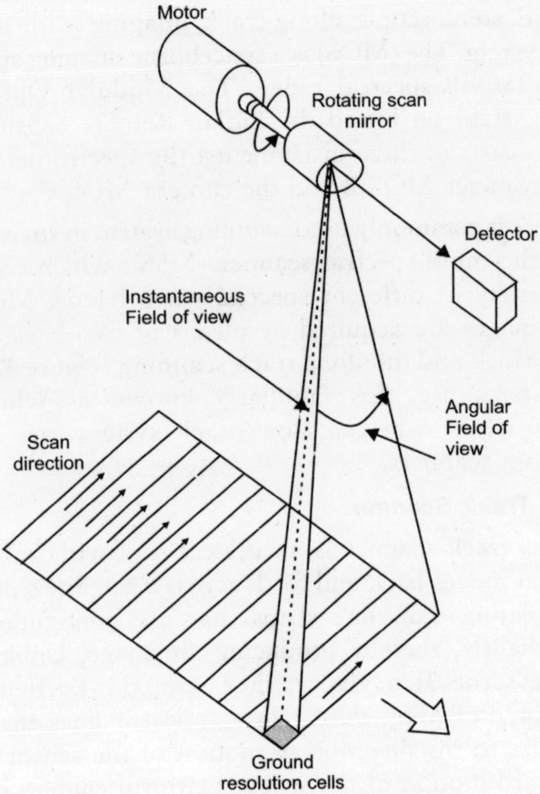

Figure 5.2a

ii. Along-track Scanner

The other type of multispectral scanning system is along-track scanning or push broom scanning. However, in this case, the system scans line by line in the direction of movement of the sensor. The operation of an along-track scanning system is shown in **Figure 5.2b**. No scanning mirror looking off at varying angles is found with this system. It employs an array of small sensitive detectors stacked side by side, each having some tiny dimension on its plate surface (CCDs) to acquire a complete line of the image at a single time. As the platform moves in the forward direction, the scanner scans the Earth to build up a two-dimensional image by recording successive scan

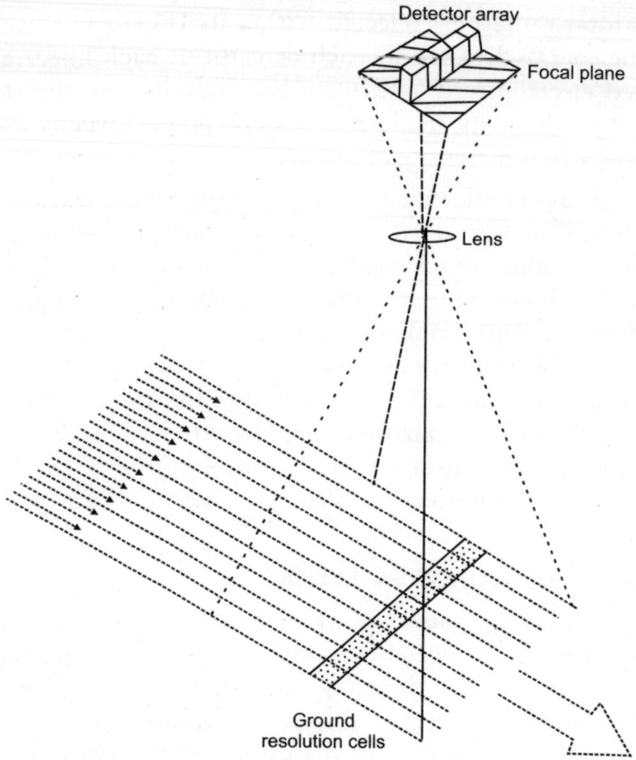

Figure 5.2b

lines perpendicular to the flight direction. Array of detectors in push broom scanner and its motion can be compared with bristles of a broom being pushed for cleaning (scanning) surface along a floor.

This type of imaging system is commonly used in optical remote sensing satellites such as SPOT. The push broom scanner contains an array of linear detectors as many as 6000 in SPOT HRV. Each individual detector projects an "instantaneous field of view (IFOV)" on the ground to measure the energy for a single ground resolution element (GRE). Thus the size and IFOV of the detectors determine the spatial resolution of the system. A separate linear array measures each spectral band or channel. The signal recorded by a detector element is proportional to the total radiation collected within its IFOV. For each scan line, the energy detected by each detector of each linear array is sampled electronically and recorded digitally. As the detector array flies along its track, the row of pixels sweeps along to generate a two-dimensional image.

There are significant advantages in using along-track scanners with linear arrays over across-track mirror scanners. Each detector in along-track scanner while measuring the energy from each pixel on the ground takes comparatively longer time (dwell time). This implies it gets more time to record more energy facilitating the system to improve its radiometric resolution. Despite, the increased dwell time also facilitates smaller IFOVs and narrower bandwidths for each detector. Thus, along with improved radiometric resolution, finer spatial and spectral resolution can also be achieved while preferring push broom type of scanners to whisk broom.

5.5 EARTH RESOURCE SATELLITES

Earth resource satellites are dedicated by various National Space Agencies to cover the global land area for national land survey, agriculture, forestry and fishery, environmental monitoring and protection, disaster protection and assessment, and coastal monitoring, etc. focusing on observation around the world and resource exploitation. The first remote sensing weather satellite (TIROS-1) and the Earth resource satellite (Landsat-1) were launched into the orbit in the years 1960 and 1972 respectively. Since then, numerous orbiting satellites of various advanced sensors have been launched to achieve very high resolution satellite imageries. They provide data to capture subtle biophysical and geological processes and their spatio-temporal changes, improving our knowledge on monitoring and modeling of the Earth's atmosphere, land and ocean. This section provides information about those satellite sensors likely to be of greatest use for studying the Earth's natural resources. Each of these sensors enables direct observation of the global land surface in a consistent fashion, with data having unique characteristics of spectral, spatial, and temporal resolution. To varying degrees, the data gathered by each sensor are suited for the variety of specific interests associated with monitoring the effects of natural events and human actions on land cover. Such interests may alternatively require high frequency of coverage, specific combinations of spectral information, or high spatial resolution. The present day achievements in the field of space science are the flown of commercial very high spatial resolution (VHSR) satellite sensors capable of acquiring multispectral (MS) images at spatial resolutions of 1 m (panchromatic) and 4 m (multispectral) to address the large dimension of such specific interests.

5.5.1 Indian Remote Sensing Satellite (IRS) Program

The Indian Government established the Department of Space in 1972 to promote development and application of Space Science and Technology for socio-economic benefits. Indian Space Research Organisation (ISRO), Bangalore is the primary agency under the Department of Space, for executing space programs. During the seventies, India undertook demonstration of space applications for communication, broadcasting and remote sensing; designing and building experimental satellites—Aryabhata, Bhaskara, APPLE and Rohini, and experimental Satellite Launch Vehicles—SLV-3 and ASLV. Today, India has established space systems that form an important element of the national infrastructure. The Indian Satellite based Remote Sensing Program commenced with the successful launch of demonstration flights of Bhaskara I on June 07, 1979 and subsequently an improved version Bhaskara II in November 20, 1981. These two experimental satellite programs in the time frame of 1976-82 have provided considerable experience in space technology and applications. Considering the multiple use of newly emerging technology of remote sensing in the areas of agriculture, water resources, forestry and ecology, geology, watersheds, urban planning and waste management, marine fisheries and coastal management, natural disaster management etc., Department of Space, Government of India began developing a series of satellites called IRS Series program. In July 1980, India, as the seventh nation in the world, proved its orbital capability launching indigenous communications, meteorological and Earth resources satellites. India's present achievements in the realization and launch of civilian satellites are tremendous. It has a lead in the civilian remote sensing field in the world having a spectrum of Indian Remote Sensing Satellites like IRS-1A, IRS-1B, IRS-1C, IRS-1D, IRS-P4, IRS-P6, IRS-P5. The IRS series of satellites have got the image capability of high, medium and coarse resolution ranging from 360 m to 2.5 m (CARTOSAT-1), 24 days to every 30 minutes repetitivity and 1 million scale to cadastral level map preparation. The following is the sequence of achievements in the Indian Space era commenced with the launch of IRS-1A on March 1988 and also with panchromatic and multispectral imaging capability, catering to the needs of the country in managing its natural resources.

IRS-1A and IRS-1B

Exploration of natural resources, particularly Earth resources, is vital for socio-economic development of any country. With ever-increasing pressure on resources, it has become essential to monitor the existing resources for optimum utilization. Management of resources has become a critical requirement with increased industrial development and growing population. The state-of-the-art technology of space based Earth observation systems offer timely and accurate information on various land and water resources. The capability of space technology for obtaining systematic, synoptic, rapid and repetitive coverage on a time and cost effect manner has virtually revolutionized natural resources management and has led towards its use of sustainable development of natural resources in consonance with maintaining the environment. This aspect is gaining momentum in recent times in view of the rapidly increasing population and impacts of human activities on climate and ecology. Several nationwide, Remote Sensing application projects have been completed during the year under the overall umbrella of National Natural Resources Management System (NNRMS). Rapid advances in computer processing and data handling capability combined with advances in sensor technology have resulted in the realization of progressively increasing spectral and spatial resolutions and in reducing the turn-around-time, crucial in disaster forecast and management (Chandrasekhar et al., 1992). The Indian Remote Sensing Satellites serve an important national goal in terms of providing continuous, operational services for the management of the natural resources of the country.

India's first indigenously developed operational Remote Sensing Satellite; IRS-1A was successfully injected into the sun-synchronous orbit on March 17, 1988. The satellite was launched from Baikonur launch site by "Vostok" Launch Vehicle, manufactured in "TsSKB-Progress". This launch marked the beginning of Indian Earth remote sensing system and India entered into the space world joining with the technologically sound developed countries. It was the major milestone in the history of remote sensing in the country. IRS-1A was placed in its orbit at an altitude of 904 km with equatorial crossing time of 10:25 a.m., which enable the study of natural resources at various regions under the same illumination condition. IRS-1A and 1B, the first two satellites in the series, together generated resources information in a variety of application areas, such as, crop acreage and production estimation of major crops, drought monitoring and assessment, disaster mapping and assessment, hydro-geo-morphology, irrigation command area status monitoring, snowmelt run-off estimation, land use and land cover mapping, urban planning, biodiversity characterization, forestry, wetland mapping, environmental impact analysis, coastal studies, etc. To get back continuous data, an identical satellite IRS-1B, the follow up of IRS-1A was launched into the same orbit in August 1991. IRS-1B was the second satellite in the Indian remote sensing series. It was launched from Baikonur cosmodrome, Kazakhstan. The repetitivity cycle of IRS-1A was 22 days. The satellites, IRS-1A & 1B were placed in a sun-synchronous orbit of 904 km in such a way as to provide a combined repetitivity cycle of 11 days thus enhancing the scope for monitoring dynamic processes associated with natural resources and environment. IRS-1A failed in 1992, while IRS-1B continued to operate through 1999. In order to sustain India's own space program, at an early date India could realize the built up of indigenous technological capabilities. As a consequence, India has two very capable launch systems at the start of the 21st century, namely PSLV (Polar Satellite Launch

Vehicle) and GSLV (Geosynchronous Satellite Launch Vehicle) (Sangatramani and Sangatramani, 2003). The successful launch and operationalization of IRS-1B and the continued satisfactory functioning of IRS-1A beyond its design life of three years are noteworthy millstones in the Indian Remote Sensing program. The payload and orbital parameters of IRS-1B are the same as that of IRS-1A. IRS-1B was declared operational on September 16, 1991 and has provided excellent data during the period 1991-2001, outliving its designed life. Many nation level mapping projects were carried out using the data.

LISS-I (Linear Imaging Self Scanner) was a payload for the IRS-1A satellite. This camera operated in four spectral bands. It operated in a push-broom scanning mode using a CCD array. Each band used 2 CCD's which were staggered in the focal plane. It was again used in IRS-1B. It used 7 bit quantization, and had a swath of 148 km. Images of LISS-I were extensively used in forestry, crop acreage, yield estimation, drought monitoring, flood monitoring, etc.

LISS-II was similar to LISS-I, but with higher spatial resolution and smaller swath. It was on payload in three satellites: IRS-1A, IRS-1B, and IRS-P2. The IRS-1A and 1B have payloads consisting of two types of advanced imaging sensors, called as Linear Imaging and Self Scanning (LISS) sensors. LISS-I has a resolution of 72.5 m while LISS-II A & B have a resolution of 36.25 m. The sensors scan the ground trace in the push broom mode. The composite swath of the two LISS-II cameras is 145 km. These cameras operate in four spectral bands in the range of 0.48 to 0.86 μm. The cameras make use of linear arrays coupled devices (CCD) as detectors with one CCD array for each spectral band. A few improvements have been carried out on IRS-1B based on the operational experiences of IRS-1A and that included time tagged command capability to enable multiple payload operations anywhere in the world. It had following payload and orbital parameters (**Table 5.2**).

Orbital parameters and sensor specifications are given in **Table 5.3**.

Spectral range and Applications of IRS-1A/1B/P2 are given in **Table 5.4**.

TABLE 5.2: Orbital Parameters and Sensors

Orbit	Polar Sun synchronous
Altitude	904 km
Inclination	99 degrees
Local Time	9:40 a.m.
Repetitivity	22 days
Orbits/day	14
Period	103 minutes
Sensors	LISS-I, LISS-II

TABLE 5.3: Orbital Parameters and Sensors

Parameters	LISS-I	LISS-II
Focal length	1620	324
FOV (microrad)	9.4	4.7 + 4.7
IFOV(microrad)	80	40
No. of elements of CCD	2048	2048 × 2
Ground Resolution (m)	72.5	36.25
Spectral Range (μm)	0.45 – 0.86	0.45 – 0.86
No. of Spectral Bands	4	4
Swath (km)	148	74 × 2
Radiometric Resolution	128	128
Data Rate	5.2	2 × 10.4

TABLE 5.4: Spectral Range and Applications of IRS-1A/1B/P2

1.	0.45 - 0.52	Coastal environment studies; chlorophyll absorption
2.	0.52 - 0.59	Green vegetation, Rock/soil Discrimination
3.	0.62 - 0.68	Strong chlorophyll absorption leading to discrimination of plant species
4.	0.77 - 0.86	Delineation of land and water geomorphic studies

IRS-P2

IRS-P2 of 870 kg mass launched on 15 October 1994 from Sriharikota aboard the PSLV-D2, the second development flight of the Polar Satellite Launch Vehicle. It carried the LISS-II system with a ground resolution of 32 m across-track and 37 m along-track having the same spectral characteristics as IRS-1A and 1B. The electro-optic module (EOM) consists of four telescope cameras in these four bands. While the optical characteristics of these band assembles are same as that of

IRS-1A/1B, the focal plane has two CCD detectors (each of 2048 element length). Two CCD arrays are separated by 25 mm on the focal plane which when translated to the imaging geometry works out to be 64 km on the ground. The total swath width is 131 km, the repetitivity was fixed at 24 days and the CCD array is tuned to four spectral bands between 0.45 and 0.86 μm. There was an improvement over IRS-1A/1B in the sense that a single Electro-Optic Module (EOM) practically carries out the function of two such EOMs in IRS-1A/1B. The IRS-P2 has completed one year of successful operation in orbit. Specifications are the same as the IRS-1C.

IRS-1C/1D

IRS system commissioned in 1988, has the world's largest constellation of seven potential remote sensing satellites—IRS-1B, IRS-1C, IRS-1D, IRS-P3, IRS-P4 (OCEANSAT-1), IRS-P5 (CARTOSAT-1) and IRS-P6 (RESOURCESAT-1) offering space based data in a range of spectral bands, spatial resolutions and swaths. IRS-1C/1D has the following application areas (see **Table 5.5**).

TABLE 5.5: Applications of IRS-1C/1D

Application Areas	Themes
Agriculture	Crop acreage and product estimation, Large area crop inventory, Crop yield forecasting, Crop condition assessment, Vegetation dynamics.
Forestry	Forest cover mapping, Bio-mass estimation.
Urban	Urban sprawl studies, Infrastructure/facility management, Urban monitoring products.
Land and Soil	Land use/land cover mapping, Change detection, Erosion prone area mapping, Land capability assessment, Watershed mapping, Salt affected soil mapping, Land degradation analysis, Soil resource mapping, Soil suitability analysis.
Geology	Geological mapping, Geomorphologic mapping, DEM Generation, Hydrogeomorphological mapping.
Water Recourses	Surface water inventory, Ground water targeting, Command area management, Environmental impact Assessment, Environmental Monitoring, Site suitability studies.
Disasters	Land slide hazard zonation, Agricultural drought monitoring and assessment.

The Indian Remote Sensing Satellite IRS-1C was successfully launched into polar orbit at a height of 817 km on December 28, 1995 by a Russian launch vehicle from Baikonur, Kazakhstan. It started operation in the first week of January 1996. On September 29, 1997, IRS-1D was successfully injected into near polar sun synchronous orbit by a PSLV-C1 launch vehicle from Sriharikota (SHAR, India). It was placed at an altitude of 780 km and started transmitting data in the middle of October 1997. The indigenous launch vehicle PSLV-C1, made a significant contribution to the Indian Space research program for the first time when launched the IRS-1D from the Indian soil after the two demonstration flights viz. IRS-P2 and IRS-P3.

The orbital and sensor characteristics of IRS-1C/1D are almost the same. The IRS-1C and IRS-1D are three axes body stabilized spacecraft. The IRS-1C satellite consists of three sensors viz. panchromatic (PAN), Linear Imaging and Self Scanning Sensor (LISS-III), and Wide Field Sensor (WiFS).

The Panchromatic camera operates in a single spectral broad band of 0.50-0.75 microns imaging a ground segment of pixel 5.8 m and a ground swath between 63-70 km (at nadir). This camera can be steered up to ±26 deg (steerable up to ±398 km across the track from nadir) which in turn increases the revisit capability to 5 days.

The LISS-III sensor operates in the spectral range of 0.52-1.70 μm in four bands of the Visible (B2: 0.52-0.59 μm), (B3: 0.62-0.68 μm); Near IR (B4: 0.77-0.86 μm) and Short wave IR (B5: 1.55-1.70 μm). The spatial resolution and swath in case of visible (two bands) and NIR (one band) regions is fixed at 23.5 m and 141 km, respectively. However, in case of SWIR, 70.5 m and 148 km are the spatial resolution and swath respectively. The revisit cycle of the satellite is fixed at 24 days. Land and water resources management is the major application area of this payload.

The Wide Field Sensor (WiFS) was designed to collect data in 0.62-0.68 μm and 0.77-0.86 μm spectral bands. It has a ground swath of 728-812 km with a spatial resolution of 169-188.3 m. Vegetation Index mapping is the major application of this payload. The selected sensor characteristics of PAN, LISS-III and WiFS of IRS-1C and IRS-1D are given in the Table 5.6 and Table 5.7 respectively.

TABLE 5.6: Key Parameters of the IRS-1C Sensors

	PAN		LISS-III	WiFS
Spatial Resolution	5.8 m	Band 2 (green)	23 m	
		Band 3 (red)	23 m	188 m
		Band 4 (NIR)	23 m	188 m
		Band 5 (SWIR)	70 m	
Swath-width	70 km	All Bands	142 km	810 km
Radiometric Resolution, Quantisation	6 bit	All Bands	7 bit	7 bit
Spectral Coverage	500 - 750 nm	Band 2 (green)	520-590 nm	
		Band 3 (red)	620-680 nm	620-680 nm
		Band 4 (NIR)	770-860 nm	770-860 nm
		Band 5 (SWIR)	1550-1700 nm	
Focal Length	982 mm	Band 2 (green)	347.367 mm	
		Band 3 (red)	347.367 mm	56.47 mm
		Band 4 (NIR)	347.367 mm	56.47 mm
		Band 5 (SWIR)	301.043 mm	
CCD Arrays (no. of arrays * no. of elements)	3 * 4096	Band 2 (green)	1 * 6000	
		Band 3 (red)	1 * 6000	2 * 2048
		Band 4 (NIR)	1 * 6000	2 * 2048
		Band 5 (SWIR)	7 * 300	
CCD Size	7 μm x 7 μm	Band 2 (green)	10 μm x 7 μm	
		Band 3 (red)	10 μm x 7 μm	13 μm x 13 μm
		Band 4 (NIR)	10 μm x 7 μm	13 μm x 13 μm
		Band 5 (SWIR)	26 μm x 26 μm	
Integration Time	0.88364 ms	Band 2 (green)	3.6047 ms	
		Band 3 (red)	3.6047 ms	28.8378 ms
		Band 4 (NIR)	3.6047 ms	28.8378 ms
		Band 5 (SWIR)	10.8142 ms	
Cross Track Field Of View (FOV) for Single Pixel (radiant)	0.0000071	Band 2 (green)	0.0000288	
		Band 3 (red)	0.0000288	0.0002302
		Band 4 (NIR)	0.0000288	0.0002302
		Band 5 (SWIR)	-	

Source: http://www.euromap.de/docs/doc_003.html.

TABLE 5.7: Key Parameters of the IRS-1D Sensors

	PAN		LISS-III	WiFS
Spatial Resolution	5.8 m	Band 2 (green)	23 m	
		Band 3 (red)	23 m	188 m
		Band 4 (NIR)	23 m	188 m
		Band 5 (SWIR)	70 m	
Swath-width	63 - 70 km	All Bands	127 - 141 km	728 - 812 km
Radiometric Resolution, Quantisation	6 bit	All Bands	7 bit	7 bit
Spectral Coverage	500 - 750 nm	Band 2 (green)	520-590 nm	
		Band 3 (red)	620-680 nm	620-680 nm
		Band 4 (NIR)	770-860 nm	770-860 nm
		Band 5 (SWIR)	1550-1700 nm	
Focal Length	974.8 mm	Band 2 (green)	346.411 mm	
		Band 3 (red)	346.411 mm	56.57 mm
		Band 4 (NIR)	346.411 mm	56.57 mm
		Band 5 (SWIR)	300.177 mm	
CCD Arrays (no. of arrays *no. of elements)	3 * 4096	Band 2 (green)	1 * 6000	
		Band 3 (red)	1 * 6000	2 * 2048
		Band 4 (NIR)	1 * 6000	2 * 2048
		Band 5 (SWIR)	7 * 300	
CCD Size	7 µm x 7 µm	Band 2 (green)	10 µm x 7 µm	
		Band 3 (red)	10 µm x 7 µm	13 µm x 13 µm
		Band 4 (NIR)	10 µm x 7 µm	13 µm x 13 µm
		Band 5 (SWIR)	26 µm x 26 µm	
Integration Time	0.8836458 ms	Band 2 (green)	3.6047576 ms	
		Band 3 (red)	3.6047576 ms	28.83806 ms
		Band 4 (NIR)	3.6047576 ms	28.83806 ms
		Band 5 (SWIR)	10.8142728 ms	
Cross Track Field Of View (FOV) for Single Pixel (radiant)	0.0000072	Band 2 (green)	0.0000289	
		Band 3 (red)	0.0000289	0.0002298
		Band 4 (NIR)	0.0000289	0.0002298
		Band 5 (SWIR)	-	

source: http://www.euromap.de/docs/doc_003.html.

IRS-P3

Satellite IRS-P3 launched successfully on 21 March 1996 from Sriharikota in South India by the third flight of the indigenously developed Polar Satellite Launch Vehicle (PSLV-D3). It was an experimental mission primarily earth remote sensing oriented. It was a follow-up mission to IRS-P2. The primary objectives of the mission were (i) the technological evaluation of the PSLV, (ii) scientific methodology and algorithm development in remote sensing applications using MOS and WiFS data.

It carried two remote sensing payloads: (i) Modular Opto-electronic Scanner (MOS), which was primarily devoted for oceanographic application and (ii) Wide Field Sensor (WiFS), similar to the WiFS of IRS-IC having an additional band (B5) in the middle infrared (SWIR) region. In addition to these, IRS-P3 has also an X-ray astronomy payload (Interface, 1996). Data acquisition started at NRSA, Shadnagar, Hyderabad on March 23, 1996. MOS data receiving foreign stations were (i) Neustrelitz (DLR, Germany), (ii) Maspalomas (ESA, Spain—since 1998), and (iii) Wallops Island (NASA, USA—since April 1999). The IRS-P3 spacecraft was in operation till 2003. There were only sporadic MOS data receptions in 2004. ISRO continued operations of the IRS-P3 spacecraft until the fall of 2004 (as per information provided by Andreas Neumann of DLR, Berlin, Germany).

IRS-P3 MOS is an experimental multispectral imaging spectrometer built by DLR, Germany. Its mission was dedicated to Ocean Color Remote Sensing and Study of the Atmosphere. MOS sensors consisted of two separate imaging spectro-radiometers MOS-A (4 bands) and MOS-B (13 bands) in the VIS and NIR and a CCD-line camera MOS-C (1 band) in the SWIR. The details of MOS-A, B and C sensors are given in **Table 5.8** (Neumann et al., 1996).

A comparison of IRS-P3 MOS bands with other sensors such as LISS (IRS), TM (Landsat), CZCS (Nimbus) and SeaWiFS (Seastar) shows that MOS sensor covers almost all the spectral bands covered by these sensors, besides covering additional bands not covered by these sensors (Interface, 1996).

IRS-P4 (Oceansat-1)

IRS-P4 launched in May, 1999 by ISRO's Polar Satellite Launch Vehicle (PSLV) from SHAR Centre, Sriharikota is known worldwide as Oceansat-1. Because it is the first satellite primarily built for ocean applications. The 1050 kg satellite is placed in a Polar sun synchronous orbit of 720 km height. It contains two payloads (i) Ocean Color Monitor (OCM) and (ii) Multi-frequency Scanning Microwave Radiometer (MSMR) which brought sea change in the ocean related applications.

TABLE 5.8: Spectral configuration of Modular Optoelectronic Scanner

Parameter	MOS-A	MOS-B	MOS-C
Spectral range (nm)	755-768	408-1010	1610
No. of Channels	4	13	1
Wavelength (nm)	756.6	408.0	1600
	760.6	443.6	
	763.5	484.6	
	766.3	520.8	
		570.5	
		615.3	
		650.3	
		685.3	
		749.7	
		814.1	
		868.3	
		942.5	
		1011.1	
Spectral half-width (nm)	1.4	10	100
FOV along-track (deg)	0.344	0.094	0.14
FOV across-track (deg)	13.6	14.0	13.4
Swath width	195	200	192
No. of pixels	140	384	299
Spatial resolution (m)	1570 × 1400	520 × 520	520 × 520
L_{min} (m Wcm^2 nm^{-1} sr^{-1})	0.1	0.2	0.5
L_{min} (m Wcm^2 nm^{-1} sr^{-1})	40	65	18
Quantisation	16 bit	16bit	16 bit
In-flight calibration	Internal lamp	Internal lamp and sun calibration	

A. Sensor Payloads

The two payloads of IRS-P4 are the OCM and MSMR. While OCM observes the surface in visible and near IR bands, MSMR collects radiation in microwave bands. The specifications chosen for OCM are such that it serves the applications in coastal zone area as well as land applications requiring narrow spectral bands and coarse resolution, without comparing ocean applications.

MSMR receives microwave radiation from Earth in 6.6, 10.65, 18 and 21 GHz band both in vertical and horizontal polarizations to measure sea surface temperature. The applications related to water vapor content, liquid water content and wind speed over oceans will also be served by MSMR.

A1. *Ocean Color Monitor (OCM)*

OCM is a solid state camera having eight narrow spectral channels operating in visible and near infrared band of electromagnetic spectrum. As these bands are narrow and also the total upwelling radiation from ocean surface is weak, the data handling system with high quantization resolution is required to take care of the total dynamic range of the signal from land and ocean. The instrument employs pushbroom scanning technology with linear CCD detector arrays (191 6K CCD) of 6000 elements (3700 active detectors, the rest are used to correct for dark current). A swath of 1440 km is provided with field of view of the optics is +43°. OCM optics is based on one lens per band (wide angle telecentric lens design, refractive system). It has a high spatial resolution of 360 m in the along-track and 236 m in the cross-track direction, with ground stations in India, USA, Germany and North Korea. Its equatorial crossing time is of around 12 noon and the repetitivity of 2 days. In order to avoid sun glint, the OCM sensor ids tilted in the along-track direction.

The OCM performance details as compared to OCTS, SeaWiFS and CZCS sensors show that the OCM performance is better than these sensors (Nayak and Navalgund, 1998). Projects under IRS-P4 OCM utilization program, the major achievements were the development of bio-optical algorithms and techniques for retrieval of various ocean parameters, validation programs through well coordinated sea truth campaigning. Standard OCM data products consist of Level 1 (geometrically and radiometrically corrected radiances), Level 2 (chlorophyll, sediments, yellow substances, diffuse attenuation coefficient, aerosol optical depth and normalized water leaving radiance) and Level 3 products (weekly and monthly averages generated on a trail basis). Level 1 and 2 data can be acquired by users directly from NRSA (at a nominal cost). Atmospheric correction procedures developed at SAC were compared with those of SeaDAS, and no major differences were found.

The various application areas (Kumar and Kumar, 1998; Desai et al., 2000) are the following:

(i) Identification of potential fishery zones.
(ii) Delineation of coastal currents and eddies.
(iii) Estimation of optical properties and phytoplankton abundance for marine resource and habitat assessment.
(iv) Observation of pollution and sediment inputs to the coastal zone and their impact on marine food.
(v) Sediment dynamics, dynamics of estuarine/tidal inlets, prediction of shoreline changes, coastal circulation and dispersion patterns.

IRS-P4 band application potentials as pointed out by Nayak and Navalgund (1998) are given in **Table 5.9.**

TABLE 5.9: Wavelengths and Applications

Band	Wavelength (nm)	Parameter	Rationale
1	412 (20)	Gelbstoff	Large absorption in smaller wavelengths
2	443 (20)	Chlorophyll	Chlorophyll absorption peak-435 nm
3	490 (20)	Chlorophyll ($<$ 1.5 mg/m^3)	Chlorophyll absorption suitable for small amount
4	520 (20)	Chlorophyll ($>$ 1.5 mg/m^3)	Hinge point and useful for large amounts
5	555 (20)	Suspended sediment	Away from chlorophyll and Gelbstoff, absorption maxima
6	670 (20)	Chlorophyll	Near the second chlorophyll absorption maxima (670 nm)
7	765 (40)	Aerosol path radiance	No water leaving radiance
8	865 (40)	Aerosol path radiance	No water leaving radiance

A2. Multi-frequency Scanning Microwave Radiometer (MSMR)

During the recent past, the passive microwave radiometry has got much recognition in its field of application. Seasat and Nimbus-7 experience demonstrates a large number of potential applications such as sea surface temperature, wind speed, atmospheric water vapor, and cloud liquid water on a global nearly all weather basis. MSMR is a payload devoted for retrieval of geophysical parameters in complementary with the OCM. Such parameters are: sea surface temperature (SST), sea surface wind speed (SSW), atmospheric water vapor (WV), and cloud liquid water content (CLW). Application areas are: sea state prediction, oceanic process studies, weather and climate studies, monsoon and atmospheric predictions, cyclone genesis and development, air-sea interaction, polar ice study, circulation and wave study, MLD study, rainfall estimation, latent heat flux estimation, soil moisture estimation etc. It is a dual-polarized passive microwave radiometer. It works on the principle of collecting radiation in microwave bands from earth surface which gives brightness temperature of the surface. Physical temperature when multiplied with the emissivity of the object (ocean water in this case) gives the brightness temperature of the scene. The radiation emitted by the ocean surface passes through the atmosphere, gets modified and sensed by MSMR.

At an orbit of 720 km and inclination of angle for 98.4°, swath of 1360 km and 2 days repetitivity are achieved. It features a conical scanning mechanism with a constant look angle of 43.13° and a scan rate of 11.16 rpm. It uses reflector antenna for forming the beam and conical scan to cover the required swath. The antenna is rotated continuously around the feed axis and the data is sampled over the scan angle during the forward scan. Rest of the scan is used for calibration using a cold sky horn and a hot reference load (Sarkar, 1998).

The MSMR instrument provides a two-day repetivity thus providing applications in fisheries, meteorology, sea-state, etc. Research applications, like rainfall, soil moisture, and snow cover are also being tested.

Major specifications of the MSMR are given in **Table 5.10**.

TABLE 5.10: Major specification of the MSMR

Frequency	6.6	10.6	18	21
Polarisation		V&H		
Orbit inclination		98.25		
Altitude (km)		720		
Antenna dia (cm)		86		
Incidence angle (in°)		42.13		
Spatial resolution (km)	120	75	45	40
Swath (km)		1360		
Temp. Resolution (°K)		1.0		
Dynamic Temp. Range (°K)		10.33		
Data rate (kbps)		5		
DC Power (W)		76		
Weight		65		

IRS-P6 (Resourcesat-1)

IRS-P6 is the tenth satellite of the IRS series launched by ISRO on October 17, 2003. It is the most advanced remote sensing satellite placed at an altitude of 817 km high polar sun-synchronous orbit by the 8th flight of India's PSLV-C5. The advantage of launching Resourcesat-1 lies in the fact that it gives enhanced quality data as compared to the previous ones and maintains the continuance of remote sensing data services provided by IRS-IC and IRS-ID.

The satellite is designed to provide both multispectral and panchromatic imagery of the earth surface. Similar to those of IRS-1C and IRS-1D, IRS-P6 carries three solid state cameras: LISS-IV, LISS-III and AWiFS.

LISS-IV is the advanced high resolution camera operating in three spectral bands, in the Visible: 0.52 to 0.59 microns (Green Band 2) and 0.62 to 0.68 microns (Red Band 3); and in the NIR: 0.76 to 0.86 microns (Band 4) with spatial resolution of 5.8 m and a swath of 70 km. The LISS-IV multispectral high-resolution camera is the prime instrument of this sensor complement. Its revisit capability is fixed at 5 days. LISS-IV can be operated in either of two support modes: (i) multispectral, and (ii) mono mode.

LISS-III however is a multi-spectral camera operating in

four spectral bands (B2, B3, B4 and B5), three in VNIR and one in SWIR with spatial resolution of 23.5 m. The instrument is identical to LISS-III on IRS-1C/1D and the new feature is the SWIR band (1.55 to 1.7 microns). The resolution of the SWIR band is 23.5 m on a swath of 140 km.

AWiFS has got enhanced capabilities compared to the WiFS camera on-board IRS-1C/1D. It operates in three spectral bands in VNIR and one band in SWIR with 56 m spatial resolution and 740 km of swath. The two VNIR bands are 0.62 - 0.68 μm, 0.77 - 0.86 μm and one SWIR band is 1.55 - 1.70 μm. The wide swath coverage enables AWiFS to provide a five-day repeat capability. The summary of the payloads of IRS-P6 are given in Table 5.11.

TABLE 5.11: Summary of the IRS-P6 instrument parameters

Parameter/Instrument	LISS-4	LISS-3	AWiFS
Spatial resolution or IFOV (Instantaneous Field of View)	5.8 m	23.5 m	56 m (nadir) (70 m a swath edge)
Spectral bands (μm)	B2: 0.52-0.59, (green) B3: 0.62-0.68, (red) B4: 0.77-0.86 (NIR)	B2: 0.52-0.59, (green) B3: 0.62-0.68, (red) B4: 0.77-0.86, (NIR) B5: 1.55-1.70 (SWIR)	B2: 0.52-0.59, (green) B3: 0.62-0.68, (red) B4: 0.77-0.86, (NIR) B5: 1.55-1.70 (SWIR)
Swath width	23.9 km in MS mode 70 km in PAN mode	141 km	740 km
Detector line arrays × No. of elements	1 × 12,288 PAN mode 3 × 12,288 MS mode	4 × 6,000	4 × 2 × 6,000
Data quantization	10 bit (selected 7 bit are provided to the data handling system)	7 bit (VNIR), 10 bit (SWIR)	10 bit
Square wave response at Nyquist	> 20	B2 > 0.40, B3 > 0.40 B4 > 0.35, B5 > 0.20	B2 > 0.40, B3 > 0.40 B4 > 0.35, B5 > 0.20
Power consumption	216 W	70 W	114 W
Instrument mass	169.5 kg	106.1 kg	103.6 kg
Date rate	105 Mbit/s	52.5 Mbit/s	52.5 Mbit/s

Applications

The launch of IRS-1C and IRS-1D satellites with the state of the art sensors provided a new dimension to the applications of the remote sensing data. The objectives of the IRS-P6 are:

(i) To provide continued remote sensing data services on an operational basis for integrated land and water resources management at micro level, with enhanced spectral and spatial coverage and stereo imaging.

(ii) To further carry out studies in advanced areas of user applications like improved crop discrimination, crop yield, crop stress, pest/disease surveillance, disaster management, etc.

With its enhanced multi-spectral/spatial coverage and stereo imaging capability, it gives scope to make study in the advanced areas of land and water resources management. To cite a few in the area of Agriculture: improved crop discrimination, crop yield estimation, damage estimation, surveillance of pests and diseases, precision farming, etc.; in the areas of Forestry: forest landscape analysis, inventory and updating, wildlife habitat analysis, biodiversity, fire damage, etc.; in the areas of Environment: hydrologic units, land unit map, oil spills, environmental impact assessment, etc.; and in the areas of Geology and Exploration: rock-type mapping, tectonic structural mapping, mining pollution analysis, landslide vulnerability/risk, geo-energy, etc.

IRS-P5 (Cartosat-1)

With the launch of above discussed IRS series of satellites, India claimed to be the leading country in the civilian remote sensing field in the world. IRS-P5 maintained this lead and also provides continuity of data to global users. It was launched on 5 May 2005 by PSLV-C6. IRS-P5 carries two state-of-the-art Panchromatic (PAN) cameras that take black and white stereoscopic pictures of the earth in the visible region of the electromagnetic spectrum. The swath covered by these high resolution PAN cameras is 30 km and their spatial resolution is 2.5 meters.

Also known as the Cartosat-1, the IRS-P5, has been devoted for cartographic applications. Two PAN Cameras operate in the 500 - 750 nm wavelength one fore-mounted with a tilt of +26 deg and the other Aft-mounted with a tilt of –5 deg from the yaw axis. It would be mainly useful for map making and terrain modeling. The details of salient features of the satellite are given in the **Table 5.12.**

TABLE 5.12: Salient features of Cartosat-1

Orbit height	~618 km
Orbit inclination	98.87°
Orbit period	97 min
Number of Orbits per day	15
Local Time of Equator crossing	10.30 a.m.
Orbital Repetitivity Cycle	126 days
Nominal Wait Time to Acquire Adj. Path	11 days
Max. Wait Time for Revisit	5 days
Data Rate	105 Mb/s
Solid state storage	120GB
Lift-off Mass	1,560 kg
Attitude and Orbit Control	3-axis body stabilized using Reaction Wheels, Magnetic Torquers and Hydrazine Thrusters
Power	5 sq. m. Solar Array generating 1100W (End Of Life) Two 24 Ah Ni-Cd batteries
Mission Life	5-6 years
PAN Fore and Aft camera: Spatial Resolution	2.5 m
PAN Fore/Aft camera: Swath-width	30 km/27 km
PAN Fore and Aft camera: Radiometric Resolution, Quantisation	10 bit
PAN Fore and Aft camera: Spectral Coverage	500-750 nm
PAN Fore and Aft camera: CCD Arrays (no. of arrays * no. of elements)	1 * 12288

Orbit : Circular Polar Sun Synchronous

5.5.2 Landsat Series

The satellite remote-sensing program of US (NASA) was the first civil Earth observing satellite program. For more than three decades, earth observing systems of NASA particularly Landsat series have been imaging the earth's surface. It started with the first Landsat satellite's launch in 1972 which has brought revolutionary change in the field of resource observatory satellite systems in the world. The first Landsat satellite was launched in 1972 and the most recent one Landsat 7, was launched on April 15, 1999. The sensors on the Landsat series have acquired millions of images. The images acquired throughout the world are used a unique resource for various applications in the field of global change, ocean studies agriculture, cartography, geology, forestry, regional planning, education and national security. The different capabilities, specifications and applications of the Landsat series are briefed below.

Landsat 1-2-3

The original name of the Landsat-1 was the Earth Resources Technology Satellite 1. It was a modified version of the Nimbus 4 meteorological satellite. It was launched on July 23, 1972, the first satellite of the United States' Landsat program. It carried the MSS (multispectral scanner) which imaged the Earth from a 900 km altitude with green, red and two IR spectral bands at 80 m resolution. Landsat 1 had a near-polar sun synchronous orbit and was also equipped with a three-camera return beam vidicon (RBV) system that could obtain visible and near IR photographic images of the Earth. Landsat 1 was finally removed from service in January 6, 1978 due to tape recorder faults, it had returned over 300,000 images.

The Landsat planning and program was the collaborative effort of NASA and Dept. of Interior (USGS). According to Dr. V.E. McKelvey the ERTS spacecraft represent the first step in merging space and remote-sensing technologies into a system for inventorying and managing the Earth's resources. Determination of the spectral wavelengths was primarily based on forestry and geologic applications that had traditionally used Color IR photography. The RBV was supposed to be the prime instrument, but the MSS data were found to be superior. Dr. Stan Freden played a major role in the Landsat program who prepared the model of the Landsat-1 as the project scientist for Landsats 1, 2 and 3. As project scientist, he was responsible for evaluating the performance of the Multispectral Scanner (MSS) instrument aboard each of the satellites, evaluating the quality of the MSS data. The MSS recorded data in four spectral bands: green, red, and two infrared bands.

The first 3 Landsat missions were also known as the Earth Resources Technology Satellite (ERTS) series. Images from Landsat 1 demonstrated the usefulness of remote sensing data for land surveys, land management, water resource planning, agricultural forecasting, forest management, sea ice movement, and cartography.

Landsat-2 was launched on January 22, 1975 and was removed from service on February 25, 1982. It was 3-axis stabilized using 4 wheels to +/-0.7° attitude control. Its quantization level was 6 bits. The characteristic of its orbit is that it is sun-synchronous, near polar orbit, placed at an altitude of 917 km. It circled the Earth every 103 minutes at an inclination of 99.2°. It completed 14 orbits a day with repeat cycle of 14 days and swath width of 185 km. Its equatorial crossing time was 9.30 a.m. It had two payloads: RBV and MSS.

The RBV system on Landsat 2 was operated primarily for engineering evaluation purposes. It operated in three visible spectral bands with 80 m spatial resolution. The spectral bands are: Visible blue-green (475-575 nm), visible orange red (580-680 nm) and visible red to NIR (690-830 nm).

The Multi-Spectral Scanner (MSS) operated in 4 spectral bands. Those were: B4 - visible green (0.5-0.6 μm), B5 - Visible red (0.6 to 0.7 μm), B6 - near Infrared (NIR) (0.7 to 0.8 μm) and B7 - NIR (0.8 to 1.1 μm). Each spectral band was provided with six detectors implies six scan lines for each spectral band. It scanned the ground segment with a pixel size of 57 × 79 m.

The US Landsat program launched its third satellite Landsat 3 on March 5th, 1978, with the primary goal of providing a global archive of satellite images. This mission was solely managed by NASA. It transmitted data till March 21, 1983 and was removed from service on 31 March, 1983 due to technical failure. It was placed into a polar orbit at an altitude of 920 kilometers. Its revisit period was fixed at 18 days. It had a multispectral scanner having 75 m spatial resolution and an additional thermal band.

The first three satellites Landsat 1-2-3, were since all alike, they are termed as the first Landsat series.

Landsat 4-5

Landsat 4 and 5 were called the second Landsat series. They contained two payloads: (i) MSS, and (ii) TM. They were placed in sun synchronous polar orbit at an altitude of 705 km. Revisit period was fixed at 16 days. Landsat-4 was launched on July 16, 1982 and operated till July 1987. However, Landsat-5 was launched on March 1, 1985 which is operational till today. The spectral configurations of the sensors as well as their application areas are as follows (see **Table 5.13** and **Table 5.14**):

TABLE 5.13: Spectral configuration and Applications of Landsat MSS. These high-resolution scanners have seven spectral bands and always cover a 185-by-185 km area

Band	Spectral band	Resolution	Use
1	0.5 - 0.6 μm	79 m x 82 m	Coastal zones, marine sediments
2	0.6 - 0.7 μm	79 m x 82 m	Roads and urban areas
3	0.7 - 0.8 μm	79 m x 82 m	Plant studies and mapping of earth/ water boundaries
4	0.8 - 1.1 μm	79 m x 82 m	Plant studies and mapping of earth/ water boundaries

TABLE 5.14: Spectral configuration and Applications of Landsat TM

Band	Spectral band	Resolution	Use
1	0.45 – 0.52 μm	30 m x 30 m	Ground/plants differentiation, coastal zones
2	0.52 – 0.60 μm	30 m x 30 m	Vegetation
3	0.63 – 0.69 μm	30 m x 30 m	Plant species differentiation
4	0.76 – 0.90	30 m x 30 m	Biomass
5	1.55 – 1.75 μm	30 m x 30 m	Snow/cloud differentiation
6	10.4 – 12.5 μm	120 m x 120 m	Thermal
7	2.08 – 2.35 μm	30 m x 30 m	Lithology

Landsat 6-7

They were categorized as the third series of Landsat satellites. The Landsat 6 mission was a failure one as the satellite was lost just after its launch on 3 October 1993. The other mission, Landsat 7 was launched in 1999 and is operational till date. It carried with a multispectral sensor known as ETM+ (Enhanced Thematic Mapper Plus). It is the enhanced version of previous TM sensor with an additional high resolution panchromatic wide band. They were placed at sun-synchronous polar orbit with an inclination of 98.2° at an altitude of 705 km, their period of revisit was fixed at 16 days.

Sensor characteristics of the Landsats are as follows (Table 5.15):

TABLE 5.15: Sensor characteristics of the Landsat 6-7

Band	Spectral band	Resolution	Use
1	0.45 – 0.515 μm	30 m x 30 m	Ground/plants differentiation, coastal zones
2	0.525 – 0.605 μm	30 m x 30 m	Vegetation
3	0.63 – 0.69 μm	30 m x 30 m	Plant species differentiation
4	0.75 – 0.90	30 m x 30 m	Biomass
5	1.55 – 1.75 μm	30 m x 30 m	Snow/cloud differentiation
6	10.4 – 12.5 μm	120 m x 120 m	Thermal
7	2.09 – 2.35 μm	30 m x 30 m	Lithology
PAN	0.50 – 0.90 μm	15 m x 15 m	

The Landsat series provided a wealth of information to the scientific community throughout world in various fields of earth resources application. The important achievement of the Landsat mission is the identification, distribution, growth, movement and break-up of ice sheets in Alaska and Arctic regions. Landsat 7 detects surface features as small as 15 meters, providing wealth of land resource information on earth resources for various uses.

5.5.3 NOAA Series

NOAA stands for National Oceanic and Atmospheric Administration. NOAA series are US weather satellites which successfully flown into their orbit after decades of strides made by the US scientists since the launch of TIROS-1 (Television Infrared Observation Satellite) on April 1, 1960. These satellites, in sun-synchronous, near-polar orbits (830-870 km above the Earth), are part of the Advanced TIROS series. In the year 1978, NOAA series of satellites were introduced and a family of NOAA satellites has been placed in polar orbit. Subsequently advanced series of such satellites have been generated introducing with NOAA-9. Most of the advanced series of NOAA satellites carry a version of the AVHRR sensor, usually with three bands in the VIS-NIR and two in the thermal region.

The NOAA satellites pass over the north and south poles. Observations indicate that in the sky of the UK and Western

Europe, NOAA satellites pass over three or four times a day, providing regularly updated images of the weather systems below them. The advanced optical sensors in the NOAA series observe our planet Earth day and night providing global data for NOAA's operational user requirements including short and long range weather forecasts. The operational system consists of two polar-orbiting satellites. The even numbered missions cross the equator from north to south at roughly 7:30 a.m. and 7:30 p.m. local solar time; and the odd numbered missions cross the equator north to south at roughly 2:00 a.m. and 2:00 p.m. local solar time. These satellites monitor the entire Earth and provide data on global weather and climate like temperature, humidity, ozone, cloud images, etc. Apart from these, the AVHRR data have been used extensively in studies of vegetation dynamics, flood monitoring, regional soil moisture analysis, dust and sand storm monitoring, forest wild fire mapping, sea surface temperature mapping and various geological applications including observation of volcanic eruptions and mapping of regional drainage and physiographic features. The major contribution of these satellites in the area of sea research is providing valuable input to forecasting models, especially for ocean areas, where conventional ground-based data are lacking.

The following **Table 5.16** lists the most known and used satellites and their sensors, with specifications about spectral, spatial, and temporal resolutions, what they can detect, and applications they can be used for.

NOAA-17 and NOAA-16 are classified as the NOAA Polar operational satellites. GOES-10 (West) and GOES-12 (East) are classified as the NOAA Geostationary operational satellites. Other, NOAA satellites serve as backups and continue to transmit available data.

The primary sensor onboard the NOAA satellites is the Advanced Very High Resolution Radiometer (AVHRR). The AVHRR sensor detects radiation in the visible, near and mid-infrared, and thermal infrared portions of the electromagnetic spectrum, over a swath width of 3000 km. The detailed specifications and their primary applications are given in Table 5.16.

TABLE 5.16: Satellites and Sensors

Satellite	Sensor	Spectral Resolution (Wavelength in μm)	Spatial Resolution	Temporal Resolution	What Can Be Detected? Spatial	What Can Be Detected? Temporal
LANDSAT 4, 5	MMS (Multispectral scanner system)	1: 0.5-0.6 (G) 2: 0.6-0.7 (R) 3: 0.7-0.8 (VNIR) 4: 0.8-1.1 (NIR)	80 m; 185 km swath width	16 days	Mapping coastal features in sediment-laden water Mapping roads and urban areas Vegetation studies and mapping land/water boundaries	Deforestation Urban and suburban development
	TM Thematic Mapper	1: 0.45-0.515 (B) 2: 0.52-0.60 (G) 3: 0.63-0.69 (R) 4: 0.75-0.90 (NIR) 5: 1.55-1.75 (Mid-IR) 6 (Thermal): 10.40-12.5 7: 2.09-2.35 (Mid-IR)	30 m (visible, near and mid-IR); 120 m (thermal IR); 185 km swath width	16 days	Soil/vegetation differentiation and coastal water mapping Vegetation mapping Plant species differentiation Biomass survey Snow/cloud differentiation Thermal mapping Geological mapping	Changes in heat islands Vegetation/land use patterns
LANDSAT 7 (1, 2, 3, 6 are inactive)	ETM+ (Enhanced Thematic Mapper)	1: 0.45-0.515 (B) 2: 0.52-0.60 (G) 3: 0.63-0.69 (R) 4: 0.75-0.90 (NIR) 5: 1.55-1.75 (Mid-IR) 6 (Thermal): 10.40-12.5 7: 2.09-2.35 (Mid-IR) 8 (pan): 0.52-0.90	30 m (visible, near and mid-IR), 15 m (panchromatic), 60 m (Thermal Infrared); 185 km swath width	16 days	Major Thoroughfares Large Buildings Forest Stands Agricultural Plots Coastline Advance/Retreat Rugged Topography Sea Ice Coverage	Changes in human infrastructure Development patterns Migration patterns Agricultural variations Urban/Rural interchange

TABLE 5.16 (Contd...)

Satellite	Sensor	Spectral Resolution (Wavelength in μm)	Spatial Resolution	Temporal Resolution	What Can Be Detected?	
					Spatial	Temporal
SPOT 1, 2, 4 and 5 (3 is inactive) Launched by France from 1986-1998. SPOT 5 was launched in May 2002	Two HRV-IR (High Resolution Visible, Infrared) push-broom sensors. Provides coverage between 87 degrees north and 87 degrees south	1: 0.50-0.59 (G) 2: 0.61-0.68 (R) 3: 0.79-0.89 (NIR) 4: 1.58-1.73 (SWIR) added on SPOT 4 Pan: 0.51-0.73	20 m (Visible, Near Infrared), 10 m (panchromatic); 60 km swath width	26 days	Agriculture (Resource mapping, production management, crop classification) Land use (Urban and suburban Land use, land mapping, energy, human infrastructure) Oceanography (water quality management) Water resources (Surface water, soil moisture and vapor-transpiration, lakes and rivers studies, wetlands and habitat mapping, resource assessment) Geological applications (mapping, economic geology, engineering geology, hazards and land morphology, oil and gas exploration) Engineering applications (terrain analysis, site investigation, water resources engineering, transport studies).	Deforestation Suburban/Urban land use changes Residential development Coastal pollution Water resource pollution monitoring Snow and Ice mapping Harvest forecasting Conservation monitoring Hazard prediction Landslide hazards Forest damage assessment

TABLE 5.16 (Contd...)

Satellite	Sensor	Spectral Resolution (Wavelength in μm)	Spatial Resolution	Temporal Resolution	What Can Be Detected? Spatial	Temporal
					Forest monitoring (inventory, forest management) and vegetation cover study (especially the VEGETATION sensor)	
	High Resolution Geometry (HRG), the high spatial resolution version of SPOT 4 HRV-IR	1: 0.50-0.59 (G) 2: 0.61-0.68 (R) 3: 0.79-0.89 (NIR) 4: 1.58-1.73 (SWIR) added on SPOT 4 Pan: 0.51-0.73	10 m (Visible), 20 m (Near Infrared), 5 m (panchromatic); 60 km swath width	26 days		
	VEGETATION instrument (on SPOT 4)	1: 0.43-0.47 (B) 2: 0.61-0.68 (R) 3: 0.78-0.89 (NIR) 4: 1.58-1.75 (SWIR)	1 km; 2200 km swath width	Daily		
IKONOS 1, 2 Launched in 1999 by the United States (IKONOS 2 failed)	MMS (Multispectral) and PAN (Panchromatic)	1: 0.45-0.53 (B) 2: 0.52-0.61 (G) 3: 0.64-0.72 (R) 4: 0.76-0.88 (VNIR) Pan: 0.45-0.90	4 m (visible), 1 m (panchromatic); 11 km swath width	26 days (680 km sun-synchronous orbit)	Roads, vehicles, buildings, infrastructure (panchromatic) Land use, agricultural uses, vegetation (color imager)	Changes in human infrastructure Development patterns Migration patterns Agricultural variations Urban/Rural interchange

TABLE 5.16 (Contd...)

Satellite	Sensor	Spectral Resolution (Wavelength in μm)	Spatial Resolution	Temporal Resolution	What Can Be Detected?	
					Spatial	Temporal
Quick bird Launched in October 2001	MS (Multispectral) and PAN (Panchromatic)	1: 0.45-0.52 (B) 2: 0.52-0.60 (G) 3: 0.63-0.69 (R) 4: 0.76-0.99 (NIR) Pan: 0.45-0.90	2.44 m (Multispectral); 61 cm (panchromatic); 16.5 km swath width	1 to 3.5 days depending on latitude at 70-centi-meter resolution	Roads, vehicles, buildings, infrastructure (panchromatic) Land use, agricultural uses, vegetation (color imager)	Changes in human infrastructure Development patterns Migration patterns Agricultural variations Urban/Rural interchange
NOAA-7 Launched in 1981 and deactivated 1986 due to a power failure	AVHRR (Advanced Very High Resolution Radiometer)	1: 0.58-0.68 (G and R) 2: 0.72-1.10 (NIR) 3: 3.53-3.93 (Mid-IR) 4: 10.3-11.3 (Thermal IR) 5: 11.5-12.5 (Thermal IR)	4.4 km (Global Area Coverage), 1.1 km (Local Area Coverage); 2800 km swath width	2 times per day; 8-day & monthly averaged data available	Day and night cloud top and sea surface temperatures Ice and snow conditions	Changes in climate and global land and sea temperatures Changes in snow and ice coverages
AVIRIS Airborne Visible Infrared Spectrometer (instrument on board of planes)	Hyperspectral airborne sensor Uses a scanning mirror in a ýwisk broomý manner	Contains 224 different detectors each with a wavelength sensitive range of 10 nm, allowing it to cover the entire range between 0.4 and 25 m.	20 m (high altitude), 4 m (low altitude); 11 km swath width	Only scheduled flights	Ecology (chlorophyll, leaf water, lignin, cellulose, pigments, structure, non-photosynthetic constituents) Geology (mineralogy, soil type) Cloud and Atmospheric studies (water vapor, clouds properties, aerosols, absorbing gases)	Snow and Ice Hydrology (melting, snow cover fraction) Commercial (agricultural correction) Ecology (changes in vegetation and

TABLE 5.16 (Contd...)

Satellite	Sensor	Spectral Resolution (Wavelength in μm)	Spatial Resolution	Temporal Resolution	What Can Be Detected? Spatial	Temporal
					Oceanography/Coastal and Inland	community maps)
					Waters (chlorophyll, dissolved organics, sediments, bottom composition, bathymetry)	Oceanography (changes in plankton coverage and chlorophyll)
					Snow and Ice Hydrology (grain size, impurities), Biomass burning (smoke, combustion products)	Forest Fires
					Environmental Hazards	
					Commercial	
ERS2 (Active)	AMI (Active Microwave Instrumentation) with SAR-Image Mode, SAR-Wave Mode, Scatterometer Mode and Radar Altimeter	5.3 GHz (C-Band) 13.5 GHz for the Radar Altimeter	30 m (SAR) 50 km (Scatterometer); 80-100 km swath width (SAR-Image mode); 5 km swath width (SAR-Wave mode), 500 km swath width (Scatterometer mode)	3-day, 35-day or 168-day cycles	All-weather instrument Ocean wave height/lengths, wind speed/direction, ice parameters, sea surface and cloud top temperatures, cloud cover and atmospheric water vapor.	Alterations and observations in ocean, land, ice, atmosphere, and climate Flood activity Changes in ocean activity, coastal regions and ice caps
ATSR-M (Along-Track Scanning Radiometer with Microwave	1.6, 3.7, 11, 12 (IR), 23.5 and 36.5 GHz (Microwave)	1 km (IR), 22 km (Microwave); 500 km swath width	3-day, 35-day or 168-day cycles	All-weather instrument Ocean wave	Alterations and observations in ocean, land, ice, atmosphere, and climate Flood activity	

TABLE 5.16 (Contd...)

Satellite	Sensor	Spectral Resolution (Wavelength .n μm)	Spatial Resolution	Temporal Resolution	What Can Be Detected? Spatial	Temporal
Sounder)				height/lengths, wind speed/ direction, ice parameters, sea surface & cloud top temperatures, cloud cover and atmospheric water vapor.	Changes in ocean activity, Coastal regions and ice caps	
GOME (Global Ozone Monitoring Experiment). Sensor is a double spectrometer	1: 0.24-0.295 2: 0.29-0.405 3: 0.40-0.605 4: 0.59-0.79	40 x 2 km 40 x 320 km; 960 km swath width				
AATSR (Advanced Along-Track Scanning Radiometer)	0.65, 0.85, 1.27, 1.6	0.5 km; 500 km swath width				

TABLE 5.16 (Contd...)

Satellite	Sensor	Spectral Resolution (Wavelength in μm)	Spatial Resolution	Temporal Resolution	What Can Be Detected?	
					Spatial	Temporal
SEASTAR	SeaWiFS (Sea-viewing Wide Field Sensor)	1: 0.402-0.422 2: 0.433-0.453 3: 0.480-0.5 4: 0.5-0.520 5: 0.545-0.565 6: 0.66-0.68 7: 0.745-0.785 8: 0.845-0.885	1.1 km (local area coverage) 4.5 km (global area coverage); 285 km swath width	1 day	Ocean color and chlorophyll Subsurface scattering Atmospheric correction Atmospheric correction Sea-surface temperature	Changes in phytoplankton Designed to provide global coverage of the oceans on a regular
TERRA Launched December 1999	ASTER (Advanced Space borne Thermal Emission and Reflection Radiometer)	14 bands, with wavelengths ranging from 0.52 to 11.65	15 m (VNIR), 30 m (SWIR), 90 m (TIR); 60 km swath width	4-16 days By request	Major Thoroughfares Large Buildings Forest Stands Agricultural Plots Coastline Advance/Retreat Rugged Topography Sea Ice Coverage	Infrastructure Changes Residential Developments Deforestation/ Reforestation Harvest Flood Area Landslides & Mass Movements
	MODIS (Moderate Resolution Imaging *Spectro-Radiometer)*	36 bands, with wavelengths ranging *from 0.405 to 14.38*	250 m (bands 1-2), 500 m (bands 3-7), 1000 m (bands 8-36); 2330 x 10 km swath width	1 to 2 days	Ideal for large scale changes in the biosphere, measures photosynthetic activity of land and marine plants Surface temperature measure-, ments Deforestation Forests,	Forest Fires Regional Harvest/ Cycles *Plankton Blooms* Sediment Plumes Maps extent of snow

TABLE 5.16 (Contd...)

Satellite	Sensor	Spectral Resolution (Wavelength in μm)	Spatial Resolution	Temporal Resolution	What Can Be Detected?	
					Spatial	Temporal
					Open Canopy Vegetation, Large Scale Agriculture Water Clarity, Atmospheric Aerosols, Smoke Plumes, Snow Cover, Ocean Temperature	and ice brought by winter storms and frigid conditions
	MISR (Multi-angle Imaging Spectro-Radiometer)	4 bands, with wavelengths ranging from 0.44 to 0.86	275 m; 360 km swath width	9 days	The amount of sunlight scattered in the atmosphere under natural conditions, Atmospheric aerosol particles (formed by both natural and human activities), Cloud Cover/Type, Vegetation Type	Smoke Plumes Regional Air Quality Climate Regional Forest Canopy Structure
	MOPITT (Measurement of Pollution in the Tropo-sphere)	2.3 (CH_4) 2.4 and 4.7 (CO)	22 km horizontally and 3 km vertically; 640 km swath width	3-4 days	Measurements of pollution in the troposphere Used to determine the amount of Carbon dioxide and methane in the atmosphere	Contributes to wider range weather forecasting
AQUA Launched May 2002	MODIS (Moderate Resolution Imaging Spectro-Radiometer)	36 bands, with wavelengths ranging from 0.405 to 14.38	250 m (bands 1-2), 500 m (bands 3-7), 1000 m (bands 8-36); 2330 x 10 km swath width	1 to 2 days	Ideal for large scale changes in the biosphere, measures photo-synthetic activity of land and marine plants Surface temperature measure-ments, Deforestation Forests, Open Canopy Vegetation, Large Scale Agriculture Water Clarity, Atmospheric Aerosols, Smoke Plumes, Snow Cover, Ocean Temperature	Forest Fires Regional Harvest/ Cycles Plankton Blooms Sediment Plumes Maps extent of snow and ice brought by winter storms and frigid conditions

TABLE 5.16 (Concld.)

Satellite	Sensor	Spectral Resolution (Wavelength in μm)	Spatial Resolution	Temporal Resolution	What Can Be Detected? Spatial	What Can Be Detected? Temporal
	CERES (Clouds and Earth's Radiant Energy System)	Short-wave: 0.3-5 Long wave: 8-12 Total: 0.3-> 200	20 km	Daily	Cloud/radiation flux measurements for models of oceanic and atmospheric energetic The cross-track mode continues measurements of Earth Radiation Budget Experiment and Tropical Rainfall Measuring Mission	Contributes to wider range weather forecasting
	AMSR/E (Advanced Microwave Scanning Radiometer)	12 channels and 6 frequencies ranging from 6.9 to 89.0 GHz (center frequency at 6.925, 10.65, 18.7, 23.8, 36.5 and 89.0 GHz)	Ranging from 56 km (at 6.925 GHz) to 5.4 km (at 89.0 GHz); 1445 km swath width	Daily	Cloud properties; radiative energy flux; precipitation; land surface wetness; sea ice; snow cover; sea surface temperature; sea surface wind fields	Contributes to weather forecasting and Climate Models
	AIRS (Atmospheric Infrared Sounder)	2,300 spectral channels in the range of 0.4 to 1.0 and 3.4 to 15.4	13.5 km (IR) and 2.3 km (VIS/NIR); 1650 km swath width	Daily	Measures atmospheric temperature and humidity; land and sea surface temperatures; cloud properties; radiative energy flux	
	AMSU (Advanced Microwave Sounding Unit) consists of two sensors: AMSU-A1 and AMSU-A2	15 discrete channels in the range of 50 to 89 GHz	40 km; 1650 km swath width	Daily	Measures atmospheric temperature and humidity	
	HSB (Humidity Sounder for Brazil)	4 channels: 1 at 150 GHz, 3 at 183 GHz	13.5 km; 1650 km swath width	Daily	Aimed at obtaining humidity profiles throughout the atmosphere	

Source: http://sedac.ciesin.columbia.edu/tg/

GOES Satellites

The GOES (Geostationary Operational Environmental Satellite) system is the joint efforts of NOAA and NASA. It is the follow up to the ATS series. The data are distributed in near real time for use in local weather monitoring and forecasting. These satellites together with the family of meteorological satellites spaced at approximately 70° longitude interval form a network providing near real time data on full coverage of the globe. They are placed at geosynchronous orbit about 35,800 km (22,300 miles) above the Earth. Such an elevated altitude allows the satellites to view a full-disc of the Earth.

The data acquired from GOES series of satellites are used for (i) rainfall estimation, (ii) estimation of snow fall accumulation and extent of snow cover, and (iii) mapping and monitoring of sea and lake ice. The important use of such imageries is issuing warning for winter storm and advisories for spring snow melting.

5.5.4 NIMBUS Series

NIMBUS series of satellites started with the launch of NIMBUS-1, first of its series on August 28, 1964. It imaged 27,000 cloud cover images. However, it operated successfully until September 22, 1964. The technology was initiated by NASA in early 1960s to meet the R and D needs of the global scientific community. They were the first three-axis stabilized meteorological satellites placed first in the sun synchronous orbit. Six more NIMBUS satellites were subsequently launched to provide continuous coverage of the Earth. The last in its series NIMBUS-7 was launched in 1978 that carried CZCS instrument. The Coastal Zone Color Scanner (CZCS) instrument collected ocean color data from November 1978 to June 1986. It collected information on sea surface temperature, chlorophyll, suspended matter and colored dissolved organic matter. Thus the CZCS is said to be one of the ocean color satellite sensors. The CZCS has led to a series of ocean color sensors such as MOS, OCTS, POLDER, and SeaWiFS launched in 1996-97, and MODIS, MISR, OCM, GLI, OCI, OSMI, MERIS, and POLDER-2.

CZCS was a conventional multi-channel scanning radiometer. Until 1996, all our practical experience on ocean color techniques has been achieved from the CZCS, the only color sensor. Although new color sensors were flown to space by various countries but the potential of the CZCS experiments are still being explored. The CZCS had very modest spectral resolution. By sensing the ocean color with very high signal-to-noise ratios, CZCS provided a modest goal of estimating a phytoplankton pigment index. The ocean color data collected by CZCS were thus used to map chlorophyll concentration in water, sediment distribution and yellow substance concentrations and its infrared channel was used to determine the temperature of coastal waters and ocean currents.

The spectral characteristics of CZCS are as follows:

Spatial resolution: 825 m

Temporal resolution: 6 days

Radiometric resolution: 8 bits

Swath width: 1566 km

Spectral bands: Band 1 = 433-453 nm, Band 2 = 510-530 nm, Band 3 = 540-560 nm, Band 4 = 660-680 nm, Band 5 = 700-800 nm, Band 6 = 10,500-12,500 nm.

5.5.5 European Remote Sensing Satellite-1

ERS-1 was launched on July 17, 1991 into a sun synchronous polar orbit at an altitude of 782–785 km. It carried number of earth-observation instruments viz. radar altimeter, infrared radiometer, microwave sounder, synthetic aperture radar and wind scatterometer. The objectives of this mission were to map the Earth's bathymetry and geoid over the seas, to study coastal zones, monitoring the movement of ice cover, global weather and ice cover.

5.5.6 Radarsat

Canadian Space Agency developed an advanced Earth observation satellite called RADARSAT-1 and launched on 4 Nov. 1995 to monitor environmental change and support natural resource sustainability. Its orbit was sun synchronous, placed at an altitude of 798 km with 98.6° inclination. The orbital period was 10.7 min and repeat cycle was 24 days. Its lifetime was for five years. It used an active microwave system, synthetic aperture radar to facilitate 24 hours data collection. The SAR system used a 5.6 cm wavelength which is known as C-band and had a HH polarization. The main application areas were land cover mapping, agricultural and forestry monitoring, disaster monitoring, landslide identification, flood monitoring and snow distribution mapping.

RADARSAT-2 the follow up of the RADARSAT-1 is Canada's next-generation commercial SAR satellite. It was launched in 1995 and the new satellite will be launched in March 2007. The specialty lies in RADARSAT-2 includes: (i) high resolution satellite imagery of 3 m, (ii) flexibility in selection of polarization, (iii) imaging options for left and right-looking, (iv) superior data storage, and (v) more precise measurements of spacecraft position and attitude.

RADAR instrument characteristics are: (i) frequency band—C band (5.405 GHz); (ii) Channel bandwidth—11.6, 17.3, 30, 50, 100 MHz; (iii) Channel polarization—HH, HV, VH, VV; (iv) SAR antenna dimensions-15 m × 1.5 m. Applications are: (i) agriculture, (ii) forestry, (iii) geology, (iv) hydrology, (v) disaster management, (vi) ice, (vii) marine surveillance, (viii) mapping.

5.5.7 Quickbird

The Quickbird satellite, at an altitude of 450 km was launched on October 18, 2001 from Vandenberg Air Force Base, California. Its orbital characteristics are sun synchronous, 93.5 orbit time, having inclination of 97.2°, and equatorial crossing time (descending) of 10:30 a.m.

The spectral configurations are given below:

(i) Resolution

Panchromatic:

Basic: 0.61 m at nadir, 0.72 m at 25° off nadir

Standard and orthorectified: resampled to 0.6/0.7 m GSD

Spectral bandwidth: 450 – 900 nm

Multispectral

Basic: 2.44 m at nadir, 2.88 m 25° off nadir

Standard and orthorectified: resampled to 2.4/2.8 m GSD

Spectral bandwidth: Blue 450-540 nm, Green 520-620 nm, Red 630-690 nm, NIR 760-900 nm.

It provides highly accurate commercial high resolution imagery of the Earth. It is designed to support applications ranging from mapping land and asset management to insurance risk assessment. This satellite of Digital Globe could able to provide sub-meter high resolution imagery which are geo-locationally accurate. Such high resolution data create maps in remote areas without the use of ground control points.

5.5.8 IKONOS

The IKONOS Satellite is a high-resolution satellite launched on 24 September 1999 by USA. This satellite has a polar, circular, sun-synchronous 681 km orbit and both sensors have a swath width of 11 km. Its weight is 1600 pounds (720 kg). It captures imageries of 3.2 m multispectral, Near-Infrared (NIR)/ 0.82 m panchromatic resolution at nadir.

The spectral configuration of IKONOS-2 is given below:

Band	1 m PAN	4 m MS & 1 -m PS
1 (Blue)	0.45-0.90 μm	0.445-0.516 μm
2 (Green)	*	0.506-0.595 μm
3 (Red)	*	0.632-0.698 μm
4 (Near IR)	*	0.757-0.853 μm

The above table indicates that the satellite collects data in a panchromatic band (0.45-0.90 μm) with spatial resolution of 1 m. It also covers the terrain in multispectral bands (blue, green, red, NIR) of 4 m resolution. The PAN and multispectral (MS) bands can be combined to produce pan-sharpened (PS) multispectral imagery with an effective resolution of 1 m data collection over 11 bit quantization levels. The temporal resolution is 3 to 5 days off nadir and 144 days for true nadir. Its single scene covers a swath of 11 km × 11 km. The IKONOS data are utilized in mapping of natural resources in rural and urban areas, natural disaster management, agricultural sector, geology and mining, forestry, change detection analysis etc. Its high resolution data also provides 3D-terrain modeling and analysis, and coastal monitoring.

5.5.9 SPOT Satellites

SPOT (System Probatoire de l'Observation dela terre) is an operational series of Earth-observing satellites which is competing with the Landsat Series. It was sponsored initially by the French government and managed by the French Space Agency. The first satellite of the series SPOT-1 was launched from Kourou Launch Range in French Guiana on February 21, 1986 aboard an Ariance Launch Vehicle (AIV). SPOT satellites orbit the Earth at an altitude near 825 km and an inclination of 98.7 degrees to produce a 26-day repeating ground track pattern with sun-synchronous conditions. SPOT 2 and SPOT 3 joined SPOT 1 on January 1990 and 26 September 1993 respectively. The SPOT 1, SPOT 2 and SPOT 4 constellation of satellites is a system which has been operational for over fifteen years. SPOT 4 has several advantages in that it offers 4 multispectral bands at 20 m resolution and one panchromatic band at 10 m resolution. It has the capacity to acquire stereo pairs (for DEM generation) and is able to revisit the same area every 5 days (effectively the revisit period) otherwise every 23 days. This is because Spot HRV sensor can tilt E-W (±28 deg) as it orbits. On 3 May 2002 Spot 5 was successfully launched and is now fully operational. Its revisit is 2 to 3 days depending on latitude. It has two HRG (High Resolution Geometric) instruments and one HRS (High Resolution Stereoscopic) instrument. HRG instrument covers a wide imaging swath of 60 km × 60 km at a resolution of 2.5 meters, while its HRS (High Resolution Stereoscopic) instrument, facilitates to produce high accuracy digital elevation model. Spot 5 is ultimately able to provide multispectral imagery at 10 m resolution and panchromatic imagery at 5 m resolution. The key applications of SPOT 5 are medium-scale mapping (at 1:25,000 and 1:10,000 locally), urban and rural planning, Oil and Gas Exploration and natural disaster management. The payload of SPOT is the HRV (High Resolution Visible).

SPOT 1-2-3 were having two identical HRV (very high resolution) sensors. Each HRV sensor can acquire the images in panchromatic mode (P mode: a single wide band in the visible part of the spectrum) or multispectral mode (XS mode: the green, red, and infrared bands of the electromagnetic spectrum) indifferently. The following is the spectral configuration of the HRV instrument:

Mode	Band	Spectral band	Resolution
XS-multispectral	XS1	0.50 - 0.59 µm	20 m x 20 m
	XS2	0.61 - 0.68 µm	20 m x 20 m
	XS3	0.79 - 0.89 µm	20 m x 20 m
P-panchromatic	PAN	0.51 - 0.73 µm	10 m x 10 m

Spectral configuration of HRIR sensors of SPOT 4 is as follows:

Mode	Band	Spectral band	Resolution
Multispectral	B1	0.50 – 0.59 µm	20 m x 20 m
	B2	0.61 – 0.68 µm	20 m x 20 m
	B3	0.79 – 0.89 µm	20 m x 20 m
	MIR	1.58 – 1.75 µm	20 m x 20 m
M-monospectral	PAN	0.61 – 0.68 µm	10 m x 10 m

Spectral configuration of VEGETATION sensor of SPOT 4 is as follows:

Band	Spectral band	Resolution
B1	0.43 – 0.47 µm	1165 m x 1165 m
B2	0.61 – 0.68 µm	1165 m x 1165 m
B3	0.79 – 0.89 µm	1165 m x 1165 m
MIR	1.58 – 1.75 µm	1165 m x 1165 m

HRG Sensors on SPOT 5

Two HRG (High Resolution Geometric) instruments are capable of generating data at four resolution levels with the same 60 km swath. The following is the spectral configuration:

Mode	Band	Spectral band	Resolution
Multispectral	B1	0.50 - 0.59 µm	10 m x 10 m
	B2	0.61 - 0.68 µm	10 m x 10 m
	B3	0.79 - 0.89 µm	10 m x 10 m
	SWIR	1.58 - 1.75 µm	20 m x 20 m
M-monospectral	PAN	0.51 - 0.73 µm	5 m x 5 m (or 2.5 m x 2.5 m)

HRS Sensor of SPOT 5

It has the stereo-capability of imaging and it provides quality digital elevation model products. The specifications are as follows:

Spectral band: panchromatic (0.51 – 0.73 μm); Resolution: 10 m, along the track sampling: 5 m; Imaging swath (centered on the satellite track): 120 km; maximum scene length: 600 km. However, Vegetation - 2 sensor remains unaltered in comparison to the one in SPOT 4.

5.5.10 MODIS Sensor

A new era in hyperspectral satellite remote sensing began when Moderate Resolution Imaging Spectroradiometer (MODIS) went into orbit with the launch of the TERRA (EOS AM-1) satellite on December 18, 1999. It helps monitoring continuously the environment providing data on aerosol density, trace gases, mapping the surface of clouds, land and sea on a variety of spectral bands. To make a study on the Earth's water cycle and our environment, a second MODIS sensor was put into orbit on May 4, 2002 by NASA onboard AQUA (EOS PM-1) from Vandenberg Air Force Base, CA. The space platforms TERRA and AQUA are at an altitude of 705 km having a sun-synchronous, near polar, circular orbit. The MODIS sensor is designed to monitor and provide integrated measurements of Earth's land, ocean and atmospheric processes. A particular phenomenon that MODIS is specifically designed to monitor is fire. MODIS has several characteristics that facilitate its ability for daily global satellite fire detection.

MODIS is operating in an array of 36 spectral bands. It provides a global data set every 1-2 days with a 16-day repeat cycle. The spatial resolution of MODIS (pixel size at nadir) is 250 m for channel 1 and 2 (0.6 μm - 0.9 μm), 500 m for channel 3 to 7 (0.4 μm - 2.1 μm) and 1000 m for channel 8 to 36 (0.4 μm - 14.4 μm), respectively.

The following **Table 5.17** provides the primary use of specific bands, specification of 36 channels, bandwidth and spatial resolution:

TABLE 5.17: MODIS performance in different spectral bands

Primary use	Band number	Central wavelength [nm]	Bandwidth [nm]	Spatial resolution [m]
Land/Cloud/	1	645	620 - 670	250
Aerosols/Boundaries	2	858.5	841 - 876	
Land/Cloud/	3	469	459 - 479	500
Aerosols Properties	4	555	545 - 565	
	5	1240	1230 - 1250	
	6	1640	1628 - 1652	
	7	2130	2105 - 2155	
Ocean Colour/	8	421.5	405 - 420	1000
Phytoplankton/	9	443	438 - 448	
Biogeochemistry	10	488	483 - 493	
	11	531	526 - 536	
	12	551	546 - 556	
	13	667	662 - 672	
	14	678	673 - 683	
	15	748	743 - 753	
	16	869.5	862 - 877	
Atmospheric	17	905	890 - 920	
Water Vapor	18	936	931 - 941	
	19	940	915 - 965	
Surface/Cloud	20	3750	3660 - 3840	
Temperature	21	3959	3929 - 3989	
	22	3959	3929 - 3989	
	23	4050	4020 - 4080	
Atmospheric	24	4465.5	4433 - 4498	
Temperature	25	4515.5	4482 - 4549	
Cirrus Clouds/	26	1375	1360 - 1390	
Water Vapor	27	6715	6535 - 6895	
	28	7325	7175 - 7475	
Cloud Properties	29	8550	8400 - 8700	
Ozone	30	9730	9580 - 9880	
Surface/Cloud	31	11030	10780 - 11280	
Temperature	32	12020	11770 - 12270	
Cloud Top Altitude	33	13335	13185 - 13485	
	34	13635	13485 - 13785	
	35	13935	13785 - 14085	
	36	14235	14085 - 14385	

6 Visual Image Interpretation

6.1 INTRODUCTION

Interpretation of aerial photo started since World War II. It is an important art or science of human skill for deriving meaningful information by recognizing different objects and their spatial relationships or physical phenomena as a whole in an aerial photo or an imagery. Such interpretation of aerial photo or satellite image is said to be a powerful scientific tool because by doing so we are extracting meaningful information beyond our visual perception, getting information on aerial and regional perspective, able to obtain a historical image record to document change, and three dimensional depth perception.

Pictures are the common and convenient means of conveying or transmitting information. About 75% of the information received by humans is visual. Satellites send us data which can be presented in this convenient format. An image can be thought of as a pictorial representation of an object. Variations in intensity of electromagnetic radiation from different Earth terrain are displayed as variations in brightness on black and white images. Sensors provide us several images of the same area by using different wavelength bands for image generation. Both photographic and non-photographic sensors are used to produce remotely sensed images. By studying the qualitative as well as quantitative aspects of images recorded by various sensor systems, like aerial photographs (black and white, black and white infrared, color, and color infrared, multiband photographs, satellite data both pictorial and digital) including thermal and radar imagery, an interpreter well experienced in his field can derive lot of information.

6.2 IMAGE INTERPRETATION

Interpretation is the critical process of making sense of the data. By looking at a portrait or photo or an image, a person by his imaginative power and careful observation tries to put meaning to the picture, here we can say it is a human skill that creates a visual interpretative analysis. A visual interpretative analysis is a tool that helps students begin to look closely and deeply at the way an illustration from a picture book, creates meaning that enhances and extends story. The logic behind interpretation of aerial photo or satellite imagery is exactly the same. Now, image interpretation can be defined as the act of examining images for the purpose of identifying objects and judging their significance (Philipson, 1997). Interpreters study remotely sensed data and attempt through logical processes to detect identify measure and evaluate the significance of environmental and cultural objects, pattern and spatial relationships (Reeves, Ed., 1975). Basically image interpretation is to photogrammetry as statistics is to mathematics. Visual interpretation was the backbone of remote sensing when aerial photographs were the only remotely sensed images available. Visual interpretation is basically the capability of a human interpreter in distinguishing the characteristics of a given feature from its background. A number of visual interpretation aids are available either to assist the human interpreter or to improve the interpretability of the data (Sahai, 1983 and Dubey, 1983).

A single photograph or an image of a terrain on the Earth surface can be a unique representation of the spatial arrangement of the visual attributes at a unique moment, not to be repeated again. These photographs or images serve as the valuable historical records of the spatial distribution of natural and man-made phenomena. These records now the scientists are using for change detection study of the terrain features and making models for future predictions. Visual interpretation is an established method of deriving information from imageries. A human interpreter uses not only tone but also the spatial pattern at different scales. Visual interpretation till today remains a predominant aid for interpretation in application areas involving mapping functions. It is, therefore, said to be an information extraction process.

When a human reccives and analyses visual information, it is called sight, perception or understanding. Anyone who looks a photograph or an imagery to examine closely to extract meaningful information, we can say he or she is a human image interpreter. When computer processes image data, it is called by various names like computer image processing, machine processing of remotely sensed data. Still it is the human who makes on screen interpretation of the visual image on computer after application of certain digital techniques to an image. A soil scientist, a geologist, or a hydrologist or a planner having training efficiency can recognize the vertical view presented by the ground objects on a aerial photograph or a satellite image, which enables him or her to detect many small or subtle features that an amateur would either overlook or mis-interpret. An interpreter is basically therefore a specialist trained in the study of photography or imagery, in addition to his or her own discipline.

6.2.1 Three Dimensional Depth Perception

We are acquainted with two dimensional perception of a landscape when viewing a single photograph. However, it is possible nowadays to have a three dimensional view of the terrain both in aerial photos and satellite images as if we are observing from a high altitude platform such as aircraft or spacecraft. The three dimensional view of a terrain can be created by taking a pair of photographs or images from two camera stations but covering some common area. The pair of photographs or images are called stereoscopic pair which when viewed in a certain manner gives an impression as if a three dimensional model of the common area is being seen. Human beings can distinguish depth instinctively. However, there are many aids to depth perception, for instance closer objects partly cover distant objects or distant objects appear smaller than similar objects nearby. These aids apply to monocular vision. For short distances binocular vision is more important and is of interest to photogrammetrists, for it is binocular vision which enables us to obtain a special impression of a model formed by two photographs of an object (objects) taken from different viewpoints. Normally, our eyes give us slightly different views which are fused physiologically by the brain, and result in a sensation of seeing a model having three dimensions. This three dimensional effect due to binocular vision, is very limited, however, decreasing rapidly beyond a viewing distance of one meter. According to Caylor and Lachowski (1988) our mind fuses this stereoscopic information into a three dimensional model of the landscape that we perceive in our minds as being real.

6.2.2 Basic Reasons for Application

Aerial photographs as well as satellite imageries obtained by remote sensing employing electromagnetic energy as the means of detecting and measuring target/objects characteristics, has applicability to various fields because of four basic reasons:

(i) It represents a larger area of the Earth from a perspective view and provides a format that facilitates the study of objects and their relationships.

(ii) Certain types of imagery and aerial photograph can provide a 3-D view.

(iii) Characteristics of objects not visible to the human eye can be transformed into images.

(iv) It provides the observer with a permanent record/ representation of objects at any moment of time. In addition, data is real-time, repetitive and when in digital form, is computer compatible for quick analysis.

6.2.3 Basic Principles (Images and their interpretability)

1. An image taken from the air or space is a pictorial presentation of the pattern of a landscape.
2. The pattern is composed of indicators of objects and events that relate to the physical, biological, and cultural components of the landscape.
3. Similar conditions, in similar circumstances and surroundings, reflect similar patterns and unlike conditions reflect unlike patterns.

4. The type and amount of information that can be extracted is proportional to the knowledge, skill and experience of the analyst's awareness of any limitations.

6.2.4 Factors Governing the Quality of an Image

In addition to the inherent characteristics of an object itself, the following factors influence image quality:

(i) Sensor characteristics (film types, digital systems)
(ii) Season of the year and time of the day
(iii) Atmospheric effects
(iv) Resolution of the imaging systems and scale
(v) Image motion
(vi) Stereoscopic parallax
(vii) Exposure and processing.

Factors governing interpretability:

(i) Visual and mental acquit of the interpreter
(ii) Equipments and techniques of interpretation
(iii) Interpretation keys, guides, manuals and other aids.

6.2.5 Visibility of Objects

The objects on aerial photographs or imagery are represented in the form of photo images in tones of gray in B/W photography and in color and false color photography in different colors/hues. This visibility of objects in the images varies due to:

(a) The inherent characteristics of the objects
(b) The quality of the aerial photography or imagery.

6.2.6 Inherent Characteristics of the Objects

In any photographic image forming process, the negative is composed of tiny silver deposits formed by the action of light on photo-sensitive film during exposure. The amount of light received by the various sections of the film depends on the reflection of electromagnetic radiation from various objects. This light after passing through the optical system, gives rise to different tones and textures.

In visual interpretation, an interpreter is primarily concerned with the recognizing changes in tonal values, thereby differentiating an object of a certain reflective characteristic from other. However, he must be aware that the same object under different moisture or illumination conditions, and depending on the wavelength of incident energy, may reflect a different amount of light. For this reason, a general key, based on tone characteristic of objects, cannot be prepared. In such cases, other characteristics of objects such as their shape, size and pattern, etc. help in their recognition.

6.2.7 Quality of Aerial Photography/Imagery

The quality of image interpretation depends on the quality of the basic material on which the images are formed. Normally, in visual interpretation, these images are formed on the photograph and represented in tones of gray or in colors of varios hues, chroma and values. A study of the factors affecting image quality and characteristics of images is essential from an interpreter's point of view.

The tonal or color contrast between an image and its background

Photographic tone contrast is the difference in brightness between an image and its background. Similarly in color photography color contrast is the result of all hue, values and chroma differences between the image and its background. The tonal contrast can be sufficiently increased with proper filters.

Image sharpness characteristics

Sharpness is the abruptness with which tone or color contrasts appear on the photograph or imagery. Both tone and sharpness enable an interpreter to distinguish one object from another. To a large extent, image sharpness is dependent on the focusing ability of the optical system. Image sharpness is closely related to the resolution of the optical system.

Stereoscopic parallax characteristics

Stereoscopic parallax is the displacement of the apparent position of an image with respect to a reference point of observation. Sufficient parallax is necessary in order to distinguish objects from their shadows. Parallax depends on the height of an object, flying height and the stereo base or its corollary, the forward overlap. Stereoscopic parallax can be improved by choosing the right base/height (B/H) ratio.

The abovc investigation appears to be over simplified as a number of other factors can be mentioned which obviously affect the image quality. However, for the purpose of simplification we may conclude that other factors influence image quality indirectly through their affect on tone, sharpness or parallax.

6.3 ELEMENTS OF IMAGE INTERPRETATION

Information on physical terrain or land resources at regular interval is needed by the legislatures, planners and other government agencies to determine its proper utilization and management. Remote sensing techniques having capability of providing regular, synoptic, multispectral and multi-temporal coverage are being used on operational basis in inventory and evaluation of terrain resources. Analysis of remotely sensed data has most frequently been carried out by human interpreters adopting the well tried methods of image interpretation.

In order for proper understanding of the satellite imageries/aerial photographs and subsequent application strategy, image interpretation methods, procedures for relating image pattern to ground conditions are essential. These methods can be analyzed in respect to the role of ancillary information in the image interpretation process. In general, those procedures that are comparatively independent of ancillary information can be applied in varied geographic settings. Because almost all interpretation procedures depend to some extent upon ancillary information, a detailed and integrated knowledge of the cultural and physical landscape is a prerequisite for both manual and automated interpretation procedures.

Imageries of large areas permit the observer to perceive the relation of objects and their surrounding which to an observer on the ground might not be apparent. Image interpretation differs from direct observations in aerial scope, perspective and temporal relationships.

Using visual cues, such as tone, texture, shape, pattern, and relationship to other objects, an observer can identify many features in an image. Methods to visually interpret satellite images are very similar to methods developed to interpret aerial photographs over 100 years ago.

Image is a pictorial representation of the energy reflected and/or emitted from the scene in different parts of the electromagnetic spectrum. This appears in many shapes, tones, sizes, etc. on an image depending upon the scene, source and sensor characteristics. The understanding of these image elements is essential to describe the perception and significance of the objects. Nine elements of image interpretation have been recognized out of which few are interconnected. The order of these elements which is given below is not strictly hierarchical.

Primary : Tone/Color

Secondary : Size, Shape and Texture

Tertiary : Pattern, Height and Shadow

Higher : Site and Association.

6.3.1 Tone/Color

Tone or hue refers to the relative brightness of color of objects on an image or each distinguishable variation from white to black. Color may be defined as each distinguishable variation on an image produced by a multitude of combinations of hue, value and chroma. Different objects on the Earth surface reflect and emit different amounts of energy in different wavelengths. These differences are recorded as variations of picture tone, color or density which enable discrimination of many spatial variables, for example, on land different crop types or at sea, water bodies of contrasting depth or temperatures. These differences are further enhanced optically or electronically to provide the interpreter with an expanded contrast. The terms light, medium or dark are used to describe variation in tones. Difference in tone/color between objects and its background is important for detection. Color may be an important clue to the object's identity—water, soil, vegetation, buildings, etc.

6.3.2 Shape

Shape refers to the general form, configuration or outline of the individual objects on the ground. The shapes of some objects are so distinctive that their images may be identified solely from this criterion. Natural shapes tend to be irregular, for example salt affected patches, boundary of undulating uplands or regular and uniform snow or glacier, whereas cultural (man-made) shapes are often geometric in nature with straight-lines, sharp angles, and regular forms, e.g. playground, settlements, road/rail, saltpans etc.

6.3.3 Size

Size refers to the spatial dimension of the object on the ground such as length, width, height, perimeter and area. It is one of the most important elements of image interpretation in discriminating objects and features (cars vs. trucks or buses, single family vs. multifamily residences, bush vs. trees, etc.). While judging the significance of objects and features, size is considered. The size of the crowns of trees can be related to board feet that may be cut for specific species in managed forests. Size should be considered in context of the image scale. Some object sizes are common experience, for example a car, tree, etc. The approximate size of many objects can be judged by comparison with familiar features (e.g. roads) in the same scene. The consideration of this element helps to distinguish the objects of similar shape but varying size.

6.3.4 Pattern

Pattern refers to the spatial arrangement of objects on the landscape. Patterns can be either man-made (city street grids, airport runways, agricultural fields road, rail, canal; clustered settlements; regular orchards; etc.) or natural (drainage, streams, creeks contiguous snow and sand, etc.). It is a very diagnostic characteristic of many features. Typical patterns familiarly used are: random, systematic, circular, centripetal, oval, curvilinear, linear, radiating, rectangular, hexagonal, pentagonal, octagonal, etc. (Jensen, 2004). Pattern is an important diagnostic element in geologic or geomorphologic analysis. Drainage anomalies and drainage pattern offer extremely valuable clues in deciphering structures and lithological units in a study area. Dendritic drainage patterns develop on flat bedded sediments, radial on domes, linear or trellis in areas with faults etc. Sudden change in the drainage pattern may have significant structural implications but differencc in stream pattern may suggest variation in lithological character of the underlying rocks. Patterns of vegetation are also commonly of geologic significance and may reflect structural conditions or lithologic character of rock type. It must be noted here that pattern is highly scale dependent.

6.3.5 Texture

Texture is the frequency of tonal change and its arrangement in an image. It is the visual impression of roughness or smoothness of an area created by tonal repetition in groups of objects which are often too small to discern as an individual object. Examples include tree crowns in a forest canopy, individual plants in an agricultural field, cars in a traffic jam; etc. The scale of the photo has an important bearing on the above definition of the texture. As it can be seen that in a large scale (1:500) photograph, we can easily discriminate individual leaves and branches of a tree's canopy and infer the texture of the area as coarse. On the other hand, when the same image seen on smaller (1:5000) and very smaller (1:50,000) scales, the arrangement of such leaves and branches will give an impression of intermediate and smooth texture respectively. Thus, the texture is scale dependent. As the scale of the image is reduced, the texture of any given object or area becomes progressively finer, and ultimately disappears. It is the ability of the interpreter how to perceive and describe it.

6.3.6 Shadow

Shadows are result of shape and size of objects, illumination condition and viewing geometry of the sensor. Shadows are especially useful in geomorphological studies where micro-relief features may be easier to detect under conditions of low angle solar illumination than when the sun is high in the sky. Unfortunately, deep shadows in areas of complex detail may obscure significant features, e.g. the volume and distribution of traffic on a city street. Distinctive shadows can give clues for identification of objects such as trees, buildings, animals, bridges and towers. Photographs without shadows often seem "blank" and difficult to interpret; however, too much shadowing may obscure features of interest.

6.3.7 Site

The location of objects with respect to terrain features or other objects is often useful in refining the identification and classification of certain picture context. For example, some tree species grow in swamp and some other in dry uplands. While in industrial areas the association of several clustered, identifiable structures may help us determine the precise nature of the local enterprise.

6.3.8 Association

Association refers to the fact that when we come across a certain phenomena or activity, we almost encounter other related or associated features of such phenomena or activity. In other words, some objects are so commonly associated with other objects that one tends to indicate or confirm the other. Examples are association of a playground to school, ravines to rivers, etc. Site, situation and association together used synergistically to arrive at a logical conclusion.

6.3.9 Aspect

The term aspect is defined as direction that a topographic slope faces usually measured in degrees from north. Aspect can be generated from continuous elevation surfaces. Aspects have a marked effect on the positioning of vegetation, settlements and cultivation.

6.3.10 Resolution

Clarity of an image is determined by the scale of a photograph and resolution of satellite imagery. Resolution depends upon aspects of the remote sensing system itself, including its nature, design and performance, as well as the ambient conditions during the sensing program and subsequent processing of the acquired data. An interpreter must have knowledge about the resolution of various remote sensing data products.

Image interpretation may be very general in its approach and objectives, such as in the case of terrain evaluation or land classification. On other occasions, it is highly specific, related to clear-cut goals in such fields as geology, forestry, transport studies and soil erosion mapping. In no instance should the interpreter fail to take into account features other than those for which he or she is specifically searching. Failure to give adequate consideration to all aspects of a terrain is, perhaps, the commonest source of interpretation error.

The interpretation of images is, therefore, an essentially deductive process, and the identification of certain key features leads to the recognition of others. Once a suitable starting point has been selected, the elements listed earlier are considered either consciously or subconsciously. The completeness and accuracy of the results depend on an interpreter's ability to integrate such elements in the most appropriate way to achieve the objectives that have been set for him or her.

6.4 TECHNIQUES OF IMAGE INTERPRETATION

The interpretation work should start methodically, should proceed from general considerations to specific details and should proceed from known to unknown. In carrying out this task, an interpreter may use other information (collateral) data in addition to imagery. Uses of interpretation keys are also commonly used in these tasks. The development of interpretation techniques has been mainly by the empirical method. The gap between the photo-image on the one hand and the reference level, i.e. the level of knowledge in a specific field, in the human mind on the other hand is bridged by the use of image interpretation. The techniques adopted for one discipline may differ from those adopted for another. The sequence of activity and the search method may have to be modified to suit the specific requirements.

Most appropriate band or combination of bands of satellite imagery should be selected for each interpretation task. For example red band is usually the best for detecting the cultural features like urban areas, roads, etc. because of contrast with the surrounding features and better atmospheric penetration of these wavelengths. Images in the NIR region are best suited for water body mapping. For the studies related to vegetation identification and classification the most useful procedure is to prepare color composites using various band combinations.

Image interpretation comprises at least three mental acts that may or may not be performed simultaneously:

(i) the measurement of images of objects

(ii) identification of objects imaged

(iii) appropriate use of this information in the solution of the problem.

In the visual interpretation, the methodology of interpretation for each separate discipline will depend on:

(a) kind of information to be interpreted

(b) accuracy of the results to be obtained

(c) the reference level of the person executing the interpretation

(d) kind and type of imagery or photo available

(e) instruments available

(f) a scale and other requirements of the final map

(g) external knowledge available and any other sensory surveys that have been or will be made in the near future in the same area.

For the scrutiny of the above list, it is found that no stereotyped approach can be prescribed for the techniques or the methodology of photo-interpretation. An interpreter must work out the plan of operations and the techniques depending on the project's special requirements.

In carrying out this task, an interpreter may use many more types of data than those recorded on the images he is to interpret. Many sources such as literatures, laboratory measurements, analysis, fieldwork, and ground and aerial photographs or imageries make up this collateral material.

6.5 INTERPRETATION KEYS

The criteria for identification of an object with interpretation elements are called an interpretation key. Image interpretation is a complex process consisting of physical investigation as well as human skill application. A professional expert sometimes depends solely on reference materials. The reference materials required for a particular theme forms an interpretation key (e.g. wetland, land use/land cover) which serves a useful aid for the interpreter while interpreting a satellite image or aerial photograph. An image interpretation key helps the interpreter to organize the information present in the image form and guides to correct identification of unknown objects. In the process, the distinctions and properties of the objects on the image are systematically listed and described. Keys may be developed to aid the interpretation of any type of imagery or photograph. The meaning of some of the basic interpretation elements especially tone/color may vary. Many types of interpretation keys are developed taking into account the purpose and ability of the interpreter. Generally, standardized keys must be established to eliminate the differences between different interpreters.

Image interpretation keys are categorized into four:

(i) *An item key*: concerned with the identification of an individual object or condition.

(ii) *A subject key*: collection of object keys concerned with the identification of principal objects or conditions within a given subject category.

(iii) *A Regional key*: compilation of items or subject keys dealing with the identification of objects or conditions characteristic of a particular region.

(iv) *An Analogous area key*: a subject or regional key which has been prepared for an accessible area and which by interpretation may be used in the interpretation of objects or conditions in inaccessible areas which exhibit similar characteristics.

6.5.1 Technical Level of Image Interpretation Keys

(i) *Technical key*: prepared for use by image interpreter who have had professional or technical training or experience in the subject concerned.

(ii) *Non-technical key*: prepared for use primarily by image interpreters who have had no professional or technical experience in the subject concerned.

Intrinsic character of image interpretation keys:

(i) *Direct key*: it is designed primarily for the identification of discrete objects or conditions directly from the image.

(ii) *Association key*: it is designed for the deduction of information not directly discernible on the image.

6.5.2 How to Organize the Image Interpretation Keys?

The image interpretation depends on the interpretation keys which an experienced interpreter has established from prior knowledge and the study of the current images. For the sake of organizing the interpretation keys, as stated above, two general types of keys (a) selective and (b) elimination, are recognized for identification of objects by selection or elimination.

Selective keys are arranged in such a way that the interpreter simply selects the example corresponding to the object he is trying to identify. A selective key illustrates and describes classes of phenomena which most closely fill his unknown. A typical interpretation key is the present land use/land cover classification system developed by NRSA, Hyderabad (**Figure 6.1**). Elimination keys are arranged so that the interpreter follows a prescribed step-wise process that leads to the elimination of all items except the one he is trying to identify. Most interpreters consider the latter type of key preferably.

6.5.3 Selective Keys

(a) *Essay key*: in this key, the objects or conditions are described in textural form using images for illustrations only.

(b) *File key*: an item key composed of one or more selected images, with notes concerning with their interpretation. This type of key is generally assembled for use by an individual interpreter.

(c) *Photo key*: an item key composed of one or more selected images, with notes concerning their interpretation, assembled for rapid reproduction and distribution to other interpreters.

LEVEL–I	LEVEL–II
1. Built-up land	1.1 Built-up land
2. Agricultural land	2.1 Crop land:
	(i) Kharif
	(ii) Rabi
	(iii) Kharif + Rabi
	2.2 Fallow land
	2.3 Plantation
3. Forest	3.1 Evergreen/Semi-evergreen forest
	3.2 Deciduous forest
	3.3 Degraded or scrub land
	3.4 Forest blank
	3.5 Forest plantation
	3.6 Mangrove
4. Wastelands	4.1 Salt affected land
	4.2 Waterlogged land
	4.3 Marshy/Swampy land
	4.4 Gullied/Ravinous land
	4.5 Land with or without scrub
	4.6 Sandy area (coastal and desertic)
	4.7 Barren rocky/Stony waste/Sheet rock areas
5. Water bodies	5.1 River/Stream
	5.2 Lake/Reservoir/Tank/Canal
6. Others	6.1 Shifting cultivation
	6.2 Grass land/Grazing land
	6.3 Snow covered/Glacial area

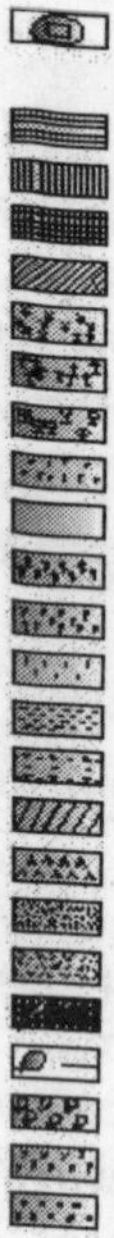

Figure 6.1

(d) *Integrated key*: key in which images and recognition features for any individual object or condition, within a subject or regional key, are so associated that by reference to the appropriate portion of the key the object or condition can be identified.

6.5.4 Elimination Keys

An elimination key provides a step by step method of identification, eliminating all incorrect choices. This involves the following steps:

(a) *Disk key*: key in which recognition features of the selected images are grouped or arranged on one or more disks so that when the recognition features are properly aligned, all but one object or condition of the group under consideration is eliminated from view.

(b) *Punch card key*: one in which the recognition features of the selected image are arranged in groups on separate punch cards. When the properly selected cards are superimposed upon a coded base, all but one object or condition of the subject group under consideration is eliminated from the view.

(c) *Dichotomous key*: one in which the graphic or word description assumes the form of a series of pairs of contrasting characteristics which permit progressive elimination of all but one object or condition of the subject group under consideration.

A modification of the elimination key is to allow probabilistic rather than absolute identification at any step or steps in a sequence of steps. A probabilistic key based on local *a priori* statistics, is necessary where identification cannot be completed.

6.6 SEQUENCE OF ACTIVITIES

While performing image interpretation, the following sequence of activities is followed:

1. *Deduction*: Deduction means selectively picking out an object or element of importance for the particular type of interpretation in hand. It is often coupled with recognition, in which case, the object is not only seen but also recognized.
2. *Recognition and identification*: Recognition and identification are sometimes together termed as photo-reading. However, they are fundamentally the same and refer to the process of classification of an object by means of specific or local knowledge within a known category upon an object's detection in a photo-image.
3. *Analysis*: Analysis is the process of separating or delineating a set of similar objects. In analysis, boundary lines are drawn separating the groups and the degree of reliability of these lines may be indicated.
4. *Deduction*: It may be directed to the separation of different groups of objects or elements and the deduction

of their significance based on converging evidence. The evidence is derived mainly from visible objects or from invisible elements, which give only partial information on the nature of certain correlative indications.

Classification

Classification establishes the identity of a surface or an object delineated by analysis. It includes modification of the surface into a pertinent system for use in the field investigation. Classification is made in order to group surfaces or objects according to those aspects that, for a certain point of view, bring out their most characteristic aspects.

6.7 CONVERGENCE OF EVIDENCE

Image interpretation is basically a deductive process. Features that can be recognized and identified directly lead the image interpreter to the identification and location of other features. Even though all aspects of an area are irreversibly intertwined, the interpreter must begin some place, he cannot consider drainage, landform, vegetation, and man-made features simultaneously. He should begin with one feature or group of features and then on to the others, integrating each of the facets of the terrain as he goes. For each terrain, the interpreter must find his own point of beginning and then consider each of the various aspects of the terrain in logical fashion. Deductive image interpretation requires conscious or unconscious consideration of the elements of image interpretation listed earlier. The completeness and accuracy of image interpretation are proportional to the interpreter's understanding of how and why images show shape, size, tone, shadow, pattern, and texture while an understanding of site, association and resolution strengthen the interpreter's ability to integrate the different features making up a terrain. For the beginners, systematic consideration of the elements of image interpretation should precede integrated terrain interpretation.

The principle of convergence of evidence requires the interpreter first to recognize basic features or types of features and then to consider their arrangement (pattern) in a real context. The accuracy of image interpretation is proportional to the interpreter's understanding of how and why images show shape, size, tone, shadow, pattern and texture. An understanding of site, association and resolution strengthens an interpreter's ability to integrate the different features making up a terrain. The greatest difficulty in interpreting images involves judging degrees of probability.

6.8 TYPE OF IMAGES

The present day sensors (airborne and spaceborne) have got the capability to record and display the object characteristics in an image beyond the visible region of the electromagnetic spectrum which are beyond the human visibility. This enables the interpreter to get images of the Earth through energy reflected and/or emitted in different wavelength bands by the objects on the Earth's surface. These images are produced in two formats. When the images are produced in the form of transparencies (film) or paper prints through photographic process, the image data is referred to as being in an analog format. Satellite retrieved images in a computer are represented as arrays of pixels (picture elements), with each pixel corresponding to a digital number, representing the brightness level of that pixel in the image. In this case, the data are said to be in a digital format. When remote sensing data are available in digital format, digital processing and analysis may be performed using a computer. Digital processing may be used to enhance data as a prelude to visual interpretation. Digital processing and analysis (supervised and unsupervised classification) may also be carried out to automatically identify targets and extract information completely without manual intervention by a human interpreter.

6.9 USE OF MULTI-IMAGE/HYPERSPECTRAL IMAGE

Due to the present advancement of sensor technology with increased spatial spectral and radiometric resolutions, the image interpreters get increased amount and type of information for interpretation. Sensor systems currently used are capable of providing the images with spectral signatures beyond the visible region of the electromagnetic spectrum i.e. in infrared and microwave regions. Available sensor systems can provide the interpreter with a variety of scales. It is now possible for the interpreter to use the multi-band, multi-date, multi-stage and multidisciplinary techniques to extract maximum possible information about the objects or conditions they react to various bands and environmental conditions. The application of satellite image interpretation has been demonstrated in many fields such as agriculture, forestry, geology, hydrology, land use etc. In some of the applications, such as agriculture, forestry, etc., the scene characteristics change with time. The use of multi-date data improves the interpretation.

6.10 PHYSIOGRAPHIC APPROACH TO HUMAN INTERPRETATION

The type of interpretation methods to be applied will depend upon the objectives of the mapping task and complexity of identification. When interpreting original or enhanced images, interpreters have as color variation along with image contextual information as basic data available. The interpretation of ground properties follows either directly or indirectly from there. The relationships between directly observable properties of images and primary interpreted properties is shown in **Figure 6.2** and hierarchical structure of human interpretation is shown in **Figure 6.3** (Townsend, 1981).

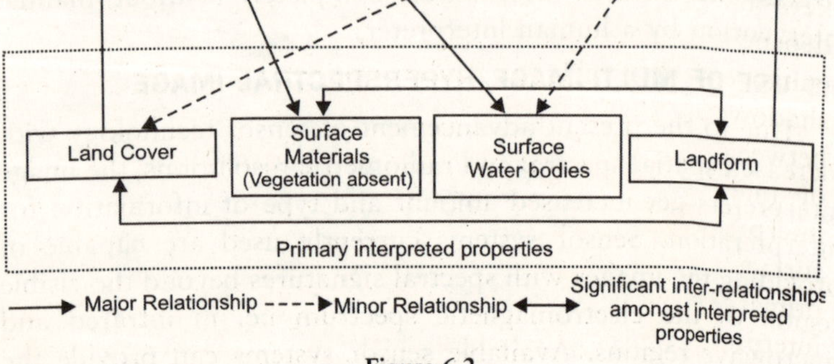

Figure 6.2

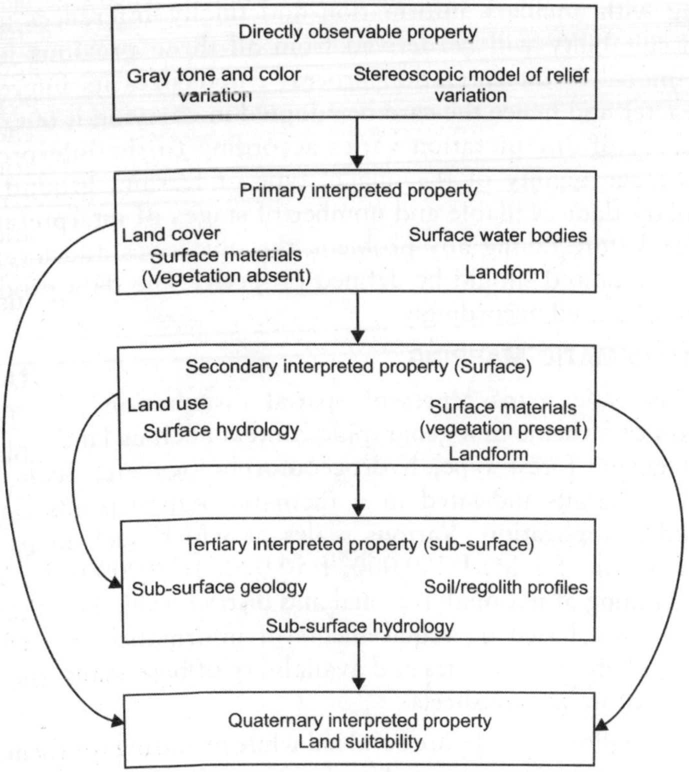

Figure 6.3

All four primary interpretated properties (Figure 6.2) depend on both of the directly observable properties. Thus land cover type will primarily be classified on the basis of interpretation keys and to lesser extent on relief. Similarly the landform will depend primarily on the stereoscopic model but this will be enhanced by gray tone variations, especially those due to shadows, the precision of our knowledge of relationships between them. On this basis we can derive two further levels of interpretation to create a hierarchical model of image interpretation (Figure 6.3). Secondary interpreted properties are surface properties, which are interpreted principally using the primary interpretations along with ancillary information. Tertiary interpreted properties are sub-surface characteristics derived both from the primary and secondary interpreted properties along with ancillary information and finally inferences about land suitability will be derived from all three previous levels. The model of interpretation process emphasizes its inferential character and hence the care be adopted in carrying it out. The accuracy of interpretation varies according to the interpreter's knowledge quality of the image, type of terrain, ground and ancillary data available and number of stages of interpretation. Hence before taking any problem, the objectives and levels of details required should be defined properly. The data products should be used accordingly.

6.11 THEMATIC MAPPING

Thematic maps represent spatial distribution of various classes of a theme in a geographic context such as land use and land cover, forest type, hydrogeomorphology etc. Scale and level of details indicated in a thematic map depends on its intended application. Various scales at which such maps are prepared are 1:1 M, 1:250,000, 1:50,000, 1:25,000 to be used for planning at national, regional and district level. These scales are selected based on requirements of information at various administrative hierarchies and availability of base maps, such as Survey of India toposheets.

The following steps are involved while preparing the thematic maps using remote sensing data:

(i) Selection and acquisition of remote sensing data
(ii) Base map preparation
(iii) Preliminary visual interpretation
(iv) Ground data collection and verification
(v) Final interpretation
(vi) Transferring interpreted details onto the base map
(vii) Final cartographic map preparation and reproduction, and
(viii) Class-wise area statistics calculation.

A flow chart showing methodology for mapping land use/land cover on 1:250,000 scale through visual interpretation techniques developed by NRSA, Hyderabad is shown in **Figure 6.4.**

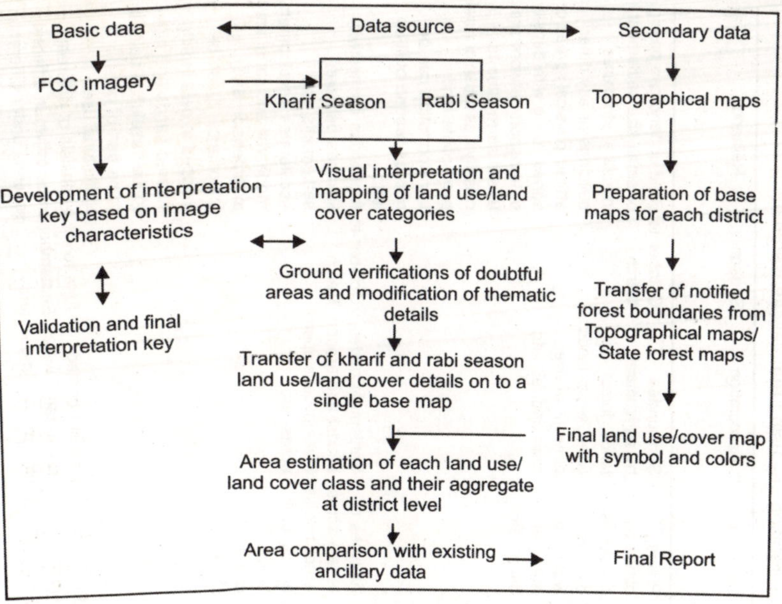

Figure 6.4

Acquisition period for selection of remote sensing data is dependent on the theme to be mapped. For example, for preparation of land use map of an area say district, both rabi and kharif season data would be required if both net and gross cropped area are to be mapped. For soil mapping ideal period would be when agriculture is devoid of standing crop cover.

Remote sensing data i.e. false color composite (FCC) images should be analysed visually to identify image characteristics related to various classes of a classification system and a preliminary interpretation key is prepared. Many image understanding systems rely heavily on *a priori* knowledge of their domain of application.

There are typical interpretation keys available with respect to land use/land cover mapping, coastal wetland mapping, and urban land use mapping (**Tables 6.1 to 6.3**). Using the interpretation key, the various land use/land cover categories, for example, in Gopalpur estuarine area of Orissa are identified and area statistics are calculated. Spatial distribution of the

TABLE 6.1: Land use/Land cover mapping image interpretation key using remotely sensed data

Sl. No.	Land use/ land cover category	Tone/ color	Size	Shape	Texture	Pattern	Location	Association	Season	Remarks
1.	Built-up land	Dark bluish green in the core and bluish on the periphery	Small to big	Irregular and disconti-nuous	Coarse and mottled	Clustered to scattered and non-contiguous	Plains, plateaus, on hill slopes, deserts, water-front, road, rail, canal, etc.	Surrounded by agricultural lands, forest cover, waste lands, network of river, road and rail, etc.	October to March	Build-up land can be big or small size settlement, industrial structures, buildings or any other artifact physical spread or sprawl along with density of transport network are useful surrogates to classify it as urban or rural perceptible land transformation can be noticed around built-up land.
2.	Transportation	Very dark to dark bluish green, light yellow for minor roads, red if vegetation along the road	Small in width for roads and narrow for rails	Regular with straight/ sharp and smooth curves	Smooth to fine	Linear to sinuous and contiguous	All types of terrain across water bodies, agricultural lands connecting settlements	Settlement nodes, amidst and around built-up developed areas, etc.	October to March	It provides connectivity linkages between settlements and accelerates development mode, rail and canal vary in dimension and importance. It can be mapped in detail using infrared bands and high-spatial resolution data. It forms part of non-agricultural use.
3.	Cropland	Light red to red	Vary in size	Regular to irregular	Medium to smooth	Contiguous to non-contiguous	Plains, hill slopes, valleys,	*Amidst irrigated* (canal, tank, well, etc.) and un-irriga-	*June to* September and	*It consists of different* crops grown *in different* seasons under different

TABLE 6.1 (Contd..)

Sl. No.	Land use/ land cover category	Tone/ color	Size	Shape	Texture	Pattern	Location	Association	Season	Remarks
							cultivable wastelands etc.	ted (rain fed/dry farming arable lands, proximity to rivers/streams, etc.	October to April	farming and land tenural systems. Mixed and multiple cropping patterns generate mix spectral response on the images.
4.	Fallow land	Yellow to greenish blue (depending on the soil type and moisture)	Small to large	Regular to irregular	Medium to smooth	Contiguous to non-contiguous	Plains, valleys, uplands, etc.	Amidst cropland as harvested agri-cultural fields, etc.	January to December	It consists of different arable lands left un-cultivated as seasonal/ temporary fallows for less than a year and as permanent fallows up to five years or more because of diverse season. Fallow lands devoid of vegetation, accelerates erosion.
5.	Plantation (Agriculture)	Dark red to red	Small to medium	Regular with sharp edges	Coarse to medium	Dispersed contiguous	Plains, foothills and uplands	Dry land or un-irrig-ated lands. Uplands occasionally amidst cropland, proximity to river and un-gentle hill slopes	January to December	Agricultural plantations consist of a variety of trees, orchards and groves. These occur throughout the year and are seen very prominent in the imagery those occurring in the forest areas (but outside the notified forest areas) are also treated as plantations like coffee, tea, arecanut etc.

TABLE 6.1 (Contd...)

Sl. No.	Land use/ land cover category	Tone/ color	Size	Shape	Texture	Pattern	Location	Association	Season	Remarks
6.	Evergreen/ semi-evergreen forest	Bright red to dark red	Varying in size	Irregular, discon-tinuous	Smooth to medium depending upon crown density	Contiguous to non-contiguous	High relief mountain/hill tops and slopes and within notified areas	High relief/hill slopes exposed to very heavy rainfall zones	January to December	These are closed (40% tree cover) or high density forest cover or conifers and other broad leaved forests. These coincide with the zone of high rainfall and relief. They provide shelter to wildlife and livestock. They influence the climate and water regime and protect the environment.
7.	Deciduous forest	Dark red ored	Varying in size	Irregular, discon-tinuous	Smooth to medium depending upon crown density	Contiguous to non-contiguous	Medium relief mountains/ hill slopes within notified area	Different forest types/ sub-types species which sheds leaves	January to April	These are broad leaved tropical forests which seasonally shed their leaves annually. Dry forest trees are subject to wild forest fires particularly during summer/autumn. These occur on the lower elevations and slopes than the evergreen/semi-evergreen forests.
8.	Degraded or scrub land	Light red to dark brown (subject to canopy cover)	Varying in size	Irregular discon-tinuous	Coarse to mottled	Contiguous to non-contiguous	Mountain slopes, isolated hills, and foot slopes	Hill slopes having skeletal soil/ different forest types/sub-types and where abiotic	January to April	It accounts for less than 20% of tree cover and are also called as open forests. The degradation is due to biotic and abiotic

TABLE 6.1 (Contd...)

Sl. No.	Land use/ land cover category	Tone/ color	Size	Shape	Texture	Pattern	Location	Association	Season	Remarks
							and within notified forest area	interference		disturbances caused to dense forest cover. It contributes to land degradation found on uplands and on foot slopes with this soil cover.
9.	Forest Blank	Light yellow to light brown	Small	Regular to irregular	Coarse to mottled	Non-contiguous dispersed/ scattered	Hill slopes and hill tops	Amidst forest cover and slopes inhabited and accessible	January to April	Forest blanks are openings without any trees cover in forest areas. These result due to tree felling for timber, grazing, cultivation, habitation, etc. it contributes to deforestation and soil erosion.
10.	Forest plantation	Light red to red	Varying in size	Regular to irregular	Smooth to medium	Contiguous to non-contiguous	Uplands, foot slopes, coastal plains and within the notified area	Forest plantations of different types and sub-types, area within soil cover and beach sands	January to December	Forest plantations consist of a variety of trees which occur both within and outside the notified forest areas. These provide timber, protect soil from erosion and provide greenery, to the environment. Occurring amidst forest, it gives a mixed spectral response.
11.	Mangrove	Bright red to red	Small	Irregular, discon-	Smooth to medium	Linear to contiguous	Coastal estuaries,	Coastal tidal waters and areas rich in	January to	Mangroves like evergreen forests are

TABLE 6.1 (Contd...)

Sl. No.	Land use/ land cover category	Tone/ color	Size	Shape	Texture	Pattern	Location	Association	Season	Remarks
				tinuous			tidal creeks and lagoons near delta coasts	coastal (saline) sediments	December	seen throughout the year. They provide good source of timber, protection of wild life and safeguard the coastal lands from massive erosion and tidal protection of waves. These are the areas amenable for aquaculture, pisciculture pearl culture. Their and spatial distribution is limited to very small patches.
12.	Salt affected land	White to light blue (subject to moisture content)	Small to medium	Irregular, discon-tinuous	Smooth to mottled	Dispersed, non-contiguous	River plains, valley coastal low lands and desert plains etc.	Irrigated agricultural lands with excess salt and poor drainage amidst crop and around tidal marshes, lagoons and inland coastal salt-lakes	January to March	Salinity results due to capillary action in dry climate or due to excess use of fertilizers, intensive irrigation and impeded drainage or due to brackish water near coastal areas. They hamper the growth of vegetation.
13.	Waterlogged land	Light to dark blue (subject to water spread and organic matter)	Varying in size	Irregular, discon-tinuous	Smooth to mottled	Dispersed, non-contiguous	*Flood plains/* low lands, coastal plains and along canals	Proximity to flood plains/low lands/ marshy/swampy grounds and near canal water	December *to March*	It is associated with low *land with high water table* and surface impeded drainage due to the flow of irrigation/flood/sea

TABLE 6.1 (Contd...)

Sl. No.	Land use/ land cover category	Tone/ color	Size	Shape	Texture	Pattern	Location	Association	Season	Remarks
										water. It often has weeds/ grasses occurring inland and as marshes/swamps near coasts.
14.	Gullied/ ravinous land	Light yellow to bluish green (subject to surface moisture and depth of erosion)	Varying in sizes depth	Irregular, broken	Very coarse to coarse	Dendritic to sub-dendritic	Along river and stream courses and drainage on sloping grounds	Severely eroded areas in plains, along streams and on sloping grounds made of loose sediments, areas having entrenched drainage, good rainfall and surface runoff	January to March	Gullies and ravines contribute to soil erosion and land degradation. Gullies with deep and steep sides develop into ravines. These result due to excessive surface runoff on loose sediments.
15.	Land with or without scrub	Light yellow to bluish green (subject to surface moisture and cover)	Varying in sizes	Irregular, discon- tinuous	Coarse to mottled (subject to vegetation cover)	Contiguous dispersed in patches	Terrains with varying lithology and landforms	Gentle relief with moderate slope in plains and foot hills and surrounded by agricultural lands	October to March	Thin veneer of soil cover on the top supports scrub and grasses or devoid of vegetation where surface erosion is dominant. Such lands occur more in dry lands, foothill areas, undulating uplands and also along fallows.
16.	Sandy area (coastal and desertic)	White to light yellow blue subject to moisture content and red for	Varying in sizes	Regular to Irregular	Smooth to mottled (subject to vegetation	Contiguous to linear	Deserts, river, beds and coastal onshore plains	River sand, shifting desert sands, sand dunes, coastal beach/dune sands and river beds/	October to March	Areas which have an accumulation of sands which are *in situ* or transported. These occur as sandy (desert) plains,

TABLE 6.1 (Contd...)

Sl. No.	Land use/ land cover category	Tone/ color	Size	Shape	Texture	Pattern	Location	Association	Season	Remarks
		vegetation						natural levees		in the form of sand dunes, beach sands and dune (wind blown) sands, etc. They support vegetation wherever moisture is available.
17.	Barren rocky/ stony waste/ sheet rock areas	Greenish blue to yellow	Varying in sizes	Irregular, discon-tinuous	Very coarse to coarse and medium	Linear to Contiguous and dispersed	Steep isolated hillocks, hill slopes/crest plateau and eroded plains	Barren and exposed rock/stony wastes, lateritic out-crops, mined areas and quarried sites, boulders	January to March	These are rock exposures of different rock types which occur as massive rocks, boulders, talus material; stony waste, etc. these are barren and are devoid of soil cover and vegetation.
18.	River/streams	Long blue to dark blue	Long narrow to wide	Irregular, sinuous	Smooth to medium	Contiguous, non-linear to dendritic/sub-dendritic, etc.	Natural rivers/ streams (perennial and non-perennial)	Drainage pattern on hill slopes, flood plains, uplands, etc. also with vegetation along the banks and in river bed	January to December	These are water courses in the channels of different dimensions and lengths. It may be perennial or non-perennial river or stream. These exhibit different drainage patterns depending on surface lithology, landform, and climate, etc. excess run-off of water results in overflow of river/stream banks causing floods.

TABLE 6.1 (Contd...)

Sl. No.	Land use/ land cover category	Tone/ color	Size	Shape	Texture	Pattern	Location	Association	Season	Remarks
19.	Lake/ reservoir/ tank/canal	Light blue to dark blue (subject to weeds/ vegetation)	Small medium to large	Regular to irregular	Smooth to mottled (subject to vegetation)	Non-contiguous dispersed, linear for canals	Tanks and lakes in low lands/ plains, reservoirs surrounded by hills and across rivers, canals in plains	Amidst cultivated lands, low lands, reservoirs with hilly terrain and rivers, canals with irrigated arable lands	January to December	These are impounded water bodies in low lands. Plains across river/ streams, unirrigated lands, etc. Surface water spread of reservoir/tank/ lake vary from season to season. These act as the source of irrigation, power generation and flood control.
20.	Shifting cultivation	Light yellow or greenish (subject to vegetation)	Small patches	Irregular	Coarse to mottled	Non-contiguous dispersed	On hill slopes/ hill tops	Mountains/hilly areas amidst forest cover and forest cleared areas	January to April	These are more often known as Jhum lands. Jhuming is a cyclic process where forests are slashed and burnt to bring cleared land under cultivation. Jhum lands contribute to loss of forest wealth and erosion.
21.	Grassland grazing land	Light red to light brown	Varying in size	Irregular	Coarse to mottled subject to vegetation	Contiguous to non-contiguous	Plains, uplands, hill slopes/foot slopes, close to river/ streams	Amidst agriculture lands, dry lands and lands fenced from cultivation	December to March	It includes lands where grasses occur naturally providing fodder to the cattle. It includes natural pasture and meadows. Excessive grazing contributes to loss of

TABLE 6.1 (Concld.)

Sl. No.	Land use/ land cover category	Tone/ color	Size	Shape	Texture	Pattern	Location	Association	Season	Remarks
										vegetative cover and accelerates erosion.
22.	Snow covered/ glacier areas	Bright to white (subject to moisture contents thickness)	Large and extensive	Irregular, disconti-nuous	Smooth	Contiguous	Mountain peaks and slopes	High relief and glaciers	December to March	It includes permanently snow covered areas such as Himalayas. Glacier is a mass of accumulated ice on the permanently snow covered areas. Snow accumulates and melts resulting in surface run-off of water to form river/stream.

Source: Manual of Nationwide Land Use/Land Cover mapping using Satellite Imagery, NRSA, Hyderabad.

TABLE 6.2: Coastal wetland mapping image interpretation key using remotely sensed data

Category	Tone	Shape	Texture	Location	Association	Remarks
Non-vegetated Wetland						
High tidal/supra tidal mud flat	Bluish green	Irregular	Smooth	Near high tide mark		Doesn't get inundated by daily tides. Inundated only during monsoon or highest of high tides.
Inter-tidal mud flat	Persian green	Irregular	Smooth	In inter-tidal area	Low energy coasts and quite depo-sitional areas, where tidal influence is not too strong	Made up of fine grained soft mud in intertidal zone. Acts as a suitable habitat for the growth of the mangrove marsh veg. and a variety of flora and fauna.
Subtidal mud flat	Brown/ black	Irregular	Smooth	Along lowest waterline	Low energy coasts	Usually clayey and silty. Vegetation may be present.
Muddy beach	Pale grayish brown	Linear smooth	Smooth	Near the subtidal area, on the sea-ward side and exposed to the sea	It is followed by sandy beach on the land ward side	Clayey and silty
Sandy beach	White	Linear crescent	Smooth	Adjacent to coast on the land-water boundary	Open coast	Made up of fine sand particles, broken molluscan shell, etc.
Spit	White	Linear may be curved at the end	Smooth	Part of beach extending in the sea	Open coast	Narrow embankment of land consisting of sand/gravel having one end attached to the mainland and other terminating in open water.
Bar	Whitish	Linear and narrow	Smooth	Running parallel to the breach	In shallow epiconti-nantal or shelf water.	Linear deposit of sand/gravel.

TABLE 6.2: (Contd...)

Category	Tone	Shape	Texture	Location	Association	Remarks
					Found singly or together and internally terminated	
Shoal	Brown	Irregular	Smooth	Near the coast or offshore, within estuary, delta, gulf, etc.		Either submerged ridge, bank or bar producing shoal. Consisting of sand, mud, gravel, etc.
Fringing reef	Greenish black	Irregular broad to narrow	Smooth except at the edge	Adjacent fringing coast	Warm waters	Deed reef gives greenish black tone. Forms a rocky eminence on sea floor.
Platform reef	Greenish black	Oval to round flat on the top	Smooth	Offshore	—	Forming islands without lagoon (large). Grows upward to the tide limit.
Patch reef	Greenish black	Ovoid to round. Flat on the top	Smooth	Offshore	—	Forming islands without lagoon (small). Grows upward to the tide limit.
Coral pinnacle	Greenish black	Ovoid	Smooth	Offshore	—	Forming islands without lagoon. Very small.
Rocky coast	—	—	—	Adjacent to the coast	Pocket beach and cliffs	Beach, mudflats are normally absent.
Vegetated Wetland						
Mangrove	Bright red if dense, pale red if sparse. Orange red	Irregular	Smooth	In the inter-tidal area/ along the creeks and low lying flats	Low energy coasts or quiet depositional areas. Warm waters	Grows on substrate mud/ mud-sand composition, where the percentage of mud is more. Evergreen.
Marsh	Dark brown if dense, brown if sparse	Irregular	Smooth	In the inter-tidal/ supra-tidal area	Low energy coasts or quiet depositional areas.Cold waters	Usually occur in temperate countries. Certain tolerant species occur within mangroves or after the mangrove belt. Evergreen.

TABLE 6.2 (Contd...)

Category	Tone	Shape	Texture	Location	Association	Remarks
Algae	Pink if, dense, light pink if sparse	Irregular	—	inter-tidal/ sub-tidal area	On reefs, rocks and mud flats	Distinct on reef edge, reef flats and rocky coast. Oct. to Feb./March.
High-tidal flat with vegetation	Pale red		Smooth	High-tidal area		Supports scanty vegetation of Xerophytes, *Salvadora, Prosopis, Acacia, Tamarix,* etc.
Mud flats with scattered vegetation	Pale red	Irregular	Smooth	Inter-tidal area	Mud flat	May be scattered mangroves, Marsh, algae, etc.
Beach vegetation	Pink	Irregular	Smooth	Adjacent to beach/ on the beach	Sand	Monsoonal. *Ipomoea, pescaprae, cenchrus* grass, *salvadora* may be present.
Water Bodies						
Bay	Dark blue	Semi-circular	Smooth	Adjacent to coast	—	—
Estuary	Dark	Funnel	Smooth	Connection with sea and land	—	Semi-enclosed body. Part of lower river course that is affected by mixing of salt water with fresh water.
Creek	Blue	Meandering	Smooth	inter-tidal/ supra-tidal area	Mud flat	Intricate network of narrow inlets of sea water in tidal flats.
Lagoon	Deep blue	Varying oval, linear	Smooth	Parallel to coast/ offshore	In between the reef or body or water lying parallel to coast	Semi-enclosed body connected to sea by one or more opening. Lagoons in between the reef have sandy beach on one side.
Lagoon with sandy bottom	Whitish blue	Ovoid/ large	Smooth	Central/ western side of island	Enclosing the island	Depth varies from 1-16 m. Central region of lagoon consists of fine sand. Shallow region consists of medium-coarse sand.

TABLE 6.2 (Contd...)

Category	Tone	Shape	Texture	Location	Association	Remarks
Shore Land						
Saline area	Creamish	Meander-ing / in patches	Smooth	On the main land above high waterline	—	On the land generally after the mangrove patch.
Deltaic plain	—	Tri-angular/ cuspate	Smooth	Relatively still stand-ing body of water	—	Rapid deposition of stream-borne sediments into still standing body of water.
Flood plain	—	Linear mean-dering	Rough	Near the river course	—	Lies parallel to the river course.
Older Flood plain	Bright red	Linear mean-dering	Rough	Near the river course	—	Lies parallel to the river course. Presence of vegetation.
Paleo-mud flat	Grayish	Irregular	Rough	Above high tide flat	—	They may have inherent salinity.
Coastal dune	Red/ white	Linear	Smooth	Behind the beach on land-ward side	Develops in any environ-ment in which loose particles of sand are with exposed to wind action	Detected only when vegetation present on it, otherwise the signature of dune gets mixed up beach. Casuarina species may be present.
Paleo channel	Whitish/ pink	Linear/ mean-dering	Smooth	Vicinity of river	—	Tonal discontinuity.
Oxbow lake	Dark blue	Crescent	Smooth	Near the river	—	Cut-off portion of the meander.
Reclaimed mud flat	Bluish white	Irregular	Rough	Above high waterline	Presence of earthen bund	Crops are grown on it.
Relict alluvium	Bluish reddish (veg.)	Irregular	Rough	Within inter-tidal mud flat	Lying high waterline	Indicates sea level changes.
Transitional zone	Whitish, red spot	Irregular	Rough	Above high tidal flat	Agricultural land and plaeo-mudflat	Lying between good agricultural land and paleo-mudflat.
Strand plain	White to dull	Linear	—	Parallel to the coast	Boarders estuaries	May have many strandlines.

TABLE 6.2 (Concld.)

Category	Tone	Shape	Texture	Location	Association	Remarks
Reclaimed land/ Chiraland	Dull White	—	Rough	May be connected to water bodies	Presence of earthen bund	Agriculture is prime land use.
Other landforms High and low water line	White (HWL), blue (LWL)	Linear	Smooth	HWL located upto highest sea water influx in land during high tide and LWL to lowest level to which water recedes during low tide	HWL in supra-tidal area, LWL in sub-tidal area	Maximum and minimum level of water reached during tidal cycle.
Tidal flood boundary	Creamish	Thin/ linear	Smooth	On the land	—	Sharp contrast on land up to which the flood water reaches during monsoon.
Strandline	White	Linear	Smooth	Out of reach of present wave action. Parallel to present coast	—	Ancient shoreline.
Cliff	—	—	—	Along coast	Pocket beaches	Detected by absence of beach/mud-flat on the coast.
Salt pan	Dark blue, light blue, white	Rectangular, square	Smooth	Near high tide limit	On the land. Tidal influx not daily	Dry salt pan appears white. Shallow/ depression in which water accumulates and evaporates leaving salt deposit.
Lineament	Whitish	Linear and narrow	—	—	Rocky area	—

TABLE 6.3: Urban cover mapping image interpretation key using remotely sensed data (False color composite made using green, red and infrared bands)

Sl. No.	Urban Cover	Image Characteristics						
		Tone	Texture	Size	Shape	Shadow	Pattern	Association
1.	Built-up Land	Dark cyan	Coarse	Variable	Variable	Shadow	Regular roads/ streets	Mixed with different multi rise buildings.
2.	Residential:							
	a. High rise	White	Coarse	Variable	Variable	Small shadow	Regular streets	A group of uniform houses.
	Low rise	White	Medium	Variable	Variable	Small shadow	Regular streets	A group of uniform houses.
	Detached	White	Coarse	Variable	Variable	Small shadow	Regular streets	A group of uniform houses.
	Slums	Light cyan	Smooth	—	—	—	Irregular	Surrounded by marsh and along roads, railways.
	Mixed	Mottled gray	Coarse	Variable	Variable	Small shadow	Unorganised	—
3.	Residential area without vegetation	White and cyan mixed with pinkish specks	Coarse	Variable	Variable	—	—	—
4.	Industrial	White	Medium	Variable	Rectangle	—	Regular	Interspersed with vegetation and open lands.
5.	Recreational:							
	a. Garden/Parks	Red	Medium	Variable	Variable	—	Dispersed	—
	b. Playgrounds	Pinkish white	Medium	Variable	Variable	—	Dispersed	—
	c. Race course/stadiums	*Reddish*	Medium	Variable	Definite	—	—	—
6.	Transportation							
	a. Roads	Dark gray	Medium	Variable	Straight with bends	—	Linear	Interspersed with pink tone.

TABLE 6.3 (Contd...)

Sl. No.	Urban Cover	Image Characteristics						
		Tone	Texture	Size	Shape	Shadow	Pattern	Association
	b. Railways	Light gray	Medium	Fixed	Straight	—	Linear	—
	c. Runways	Light blue	Smooth	Variable	Long straight strips	—	Linear	Interspersed with pink tone.
	d. Harbour	Light cyan	Smooth to medium	Variable	Irregular	—	—	Surrounded by blue tone.
	e. Jetties	White	Smooth	Variable	—	—	—	Surrounded by blue tone.
	f. Groynes	Pale cyan	Medium	Variable	—	—	—	—
7.	Agriculture:							
	a. cropped land	Pinkish red	Fine	Variable	Variable	—	Scattered	—
	b. Fallow land	Whitish blue	Coarse	Variable	Variable	—	Scattered	—
	c. plantations	Dark pink	Smooth	Variable	Variable	—	Scattered	—
8.	Forest land	Dark red	Medium	Variable	Variable	—	Scattered	—
9.	Scrub lands	Light pink to gray	Medium to Coarse	Variable	Variable	—	Scattered	—
10.	Forest blanks	Pale red	Medium	Variable	Variable	—	Scattered	—
11.	Barren hills with sparse vegetation	Gray mixed with pink	Medium	Variable	Variable	—	Scattered	—
12.	Salt pans	Bright white	Smooth	Uniform	Square/ rectangle	—	Contiguous	—
13.	Rock outcrops	Medium gray	Medium	Variable	Variable	—	Scattered	—
14.	Quarries	Whitish cyan	Coarse	Variable	Variable	—	Scattered	—

TABLE 6.3 (Concld.)

Sl. No.	Urban Cover	Image Characteristics: Tone	Texture	Size	Shape	Shadow	Pattern	Association
15.	Water bodies:							
	a. Clear water	Dark blue	Smooth	Variable	Variable	—	Scattered	—
	b. Turbid water	Light blue	Medium	Variable	Variable	—	Scattered	—
	Sandy areas	White	Coarse	Variable	Variable	—	Scattered	—
16.	Wetland:							
	a. Wetland with vegetation	Pale red	Smooth	Variable	Variable	—	Scattered	—
	b. Wetland without vegetation	Dark gray	Medium	Variable	Variable	—	Scattered	—
	c. Mudflats	Light cyan mixed with pink	Coarse	Variable	Variable	—	Scattered	—

Source: Pathan (1993).

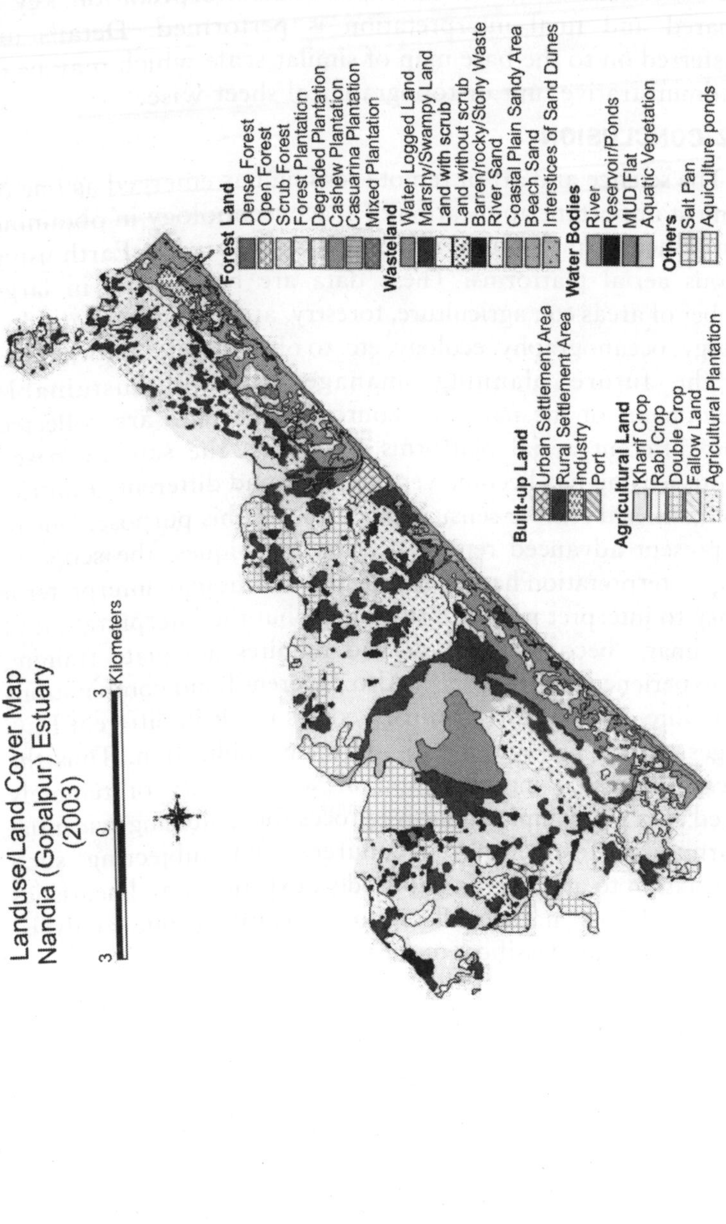
Landuse/Land Cover Map
Nandia (Gopalpur) Estuary
(2003)
3
0
3
Kilometers
Forest Land
Dense Forest
Open Forest
Scrub Forest
Forest Plantation
Degraded Plantation
Cashew Plantation
Casuarina Plantation
Mixed Plantation
Wasteland
Water Logged Land
Marshy/Swampy Land
Land with scrub
Land without scrub
Barren/rocky/Stony Waste
River Sand
Coastal Plain Sandy Area
Beach Sand
Interstices of Sand Dunes
Water Bodies
River
Reservoir/Ponds
MUD Flat
Aquatic Vegetation
Others
Salt Pan
Aquiculture ponds
Built-up Land
Urban Settlement Area
Rural Settlement Area
Industry
Port
Agricultural Land
Kharif Crop
Rabi Crop
Double Crop
Fallow Land
Agricultural Plantation

Figure 6.5

categories such as built-up land, agricultural land, forest, wastelands and waterbodies is shown in **Figure 6.5.**

The preliminary map thus prepared is used for ground truth data collection. Based on this data final interpretation key is prepared and final interpretation is performed. Details are transferred on to the base map of similar scale which may be of an administrative unit or topographical sheet-wise.

6.12 CONCLUSION

The science and art of remote sensing has emerged as one of the most important application of space technology in obtaining environmental and natural resources data related to Earth using various aerial platforms. These data are being used in large number of areas e.g. agriculture, forestry, archeology, geography, geology, oceanography, ecology, etc. to obtain useful information for the future planning, management and sustainable development of the natural resources. Such data are collected using different aerial platforms. However, the satellite based remote sensing has become very popular and different countries have launched remote-sensing satellites, for this purpose. Due to the present advanced remote sensing techniques, the scope of image interpretation has been widened. For human interpreter it is easy to interpret natural color image but the interpretation of FCC image becomes difficult and requires adequate training and experience (Short, 1999). Also different band combinations of satellite data for three primary colors result in different FCC images, which are suitable for different application. Thus the process of visual interpretation of wide variety of remotely sensed data is a complex intuitive process of combining evidential information from different sources and subjecting such information to an expert's knowledge, experience and heuristics at each levels namely detection, identification, analysis, recognition and classification of the process.

7

Digital Image Processing

7.1 INTRODUCTION

Image processing is primarily a set of techniques to make images more meaningful to humans. Human eye perceives and understands the happenings in the earth surrounding it. In analogous to it, computer using prescribed image processing algorithms create the same effect of human vision by electronically perceiving and understanding the event (image). In order to have the human perception in the computer, certain image processing techniques are necessarily applied. Digital image processing (DIP) is, therefore, said to be the use of computer algorithms to perform image processing on digital images. In other words, digital image processing refers to processing of digital image by means of a digital computer. Improvement of pictorial information for human interpretation and processing of digital image for storage, transmission and representation for autonomous machine perception led to interest in digital image processing methods.

In comparison to other disciplines, the digital image processing plays a crucial role in the areas of remote sensing. DIP is continuously enhanced by other areas such as neural nets, wavelet theory, mathematical morphology, data compression and recognition, and artificial intelligence. Lot of complex environment problems is being solved effectively by the DIP in combination with powerful human ability to visualize and interpret data. Remote sensing scientists throughout the world have recognized the capability of Digital Image Processing in expanding the unique human ability to visualize and interpret data. The scientists have delineated no boundary in application areas of DIP.

Remotely sensed images are recorded in digital form and then processed by the computers to produce images for interpretation purposes. These digital images contain large amount of data. For example, a Landsat scene contains about 30 million digital numbers. Computers with large data handling capability are used to process such images.

Digital image processing is a collection of techniques for the manipulation of digital data to enhance the image qualities with the help of computer. Thus, it is the computer manipulation of images to extract useful information from them. The techniques involved in digital image processing include noise removal, geometric and radiometric correction, image enhancement, information extraction, data compaction, image display and recording, and image data manipulation and management. Such techniques help to enhance the clarity, sharpness and details of features of interest towards information extraction and further analysis. The availability of digital multispectral remote sensing data from the Landsat mission accelerated the development and application of digital image processing in the mid 1970's.

7.1.1 Visual and Digital Interpretation

Interpretation of an image of a scene yields same result whether it is done by manual or digital. These are detection, identification, measurement and problem-solving. The key elements of image interpretation are also common to both manual and computer assisted procedure. These are tone, color, texture, shape, size, association, pattern, shadows. While a human analyst simultaneously use all these elements in interpretation, majority of digital interpretation techniques use only tone or color. Efficient digital procedures employ elements other than tone or color.

7.1.2 Advantages of Digital Techniques

With the help of key elements of image interpretation, although human analysts efficiently perform the task of image interpretation visually and manually they prefer to digital techniques because of the following advantages:

(a) A large number of newly developed sensors collect and transmit data in digital format.

(b) Visual interpretation is a qualitative technique, whereas the digital processing and interpretation is quantitative in nature.

(c) A digital image can be copied any number of times without any loss or degradation of information and can be stored for desired span of time in the form hard disk or compact disk. Digital image can also be manipulated according to analyst's desired application.

(d) Human eye can discriminate 8 to 16 shades of gray level in a continuous black and white image. However, computer can recognize each gray level separately (256 shades of gray level in an 8-bit image) according to the assigned radiometric resolution.

(e) There is always more or less biasness in human interpretations, while a given computer algorithm will yield the same result.

(f) Computers are capable of keeping track of vast amount of quantitative information simultaneously. Computers can be programmed to repeat a set of steps consistently. It is relatively fast and provides results which are easier to communicate to others.

It may be mentioned that a variety of user friendly digital image processing softwares are commercially available, which can give fast and unbiased, accurate results by analyzing each and every pixel, which is not possible in visual interpretation by human analysts. Computer can use any number of bands (the recent hyperspectral image) to improve the classification accuracy, whereas human discrimination and analysis is fixed.

It may be mentioned that a dichotomy of human interpretation *vis-à-vis* computer interpretation is not typical in contemporary practice. All digital interpretation techniques require human intervention. The actual level is varying with the procedure. Contemporary image interpretation involves a significant mix of human and computer efforts to best combine the optimal and unique performances of each, viz. (i) the rich spectrum of interpretative elements used by humans, and (ii) the repeatable image improvements and high rate interpretation performance of the computer. Ideally this combination is synergistic (Estes et al., 1983).

7.2 DIGITAL IMAGE

Remotely sensed data refer to data of the earth collected from sensors onboard satellites or aircrafts. Such data are of two types: (i) analog and (ii) digital. An image that contains a continuous gray tone or color, like a photograph, is called analog image.

On the other hand, a digital image is a representation of a sampled scene containing a group of divided small cells, with integer values of average intensity, in a column-row rectangular pattern. Due to the recent advancement of sensor technology, the sensors record many types of remote sensing images in the digital form. The sensors collect and measure radiant energy reflected or emitted from the objects on the earth's surface. Remote sensing images are recorded in digital form by sophisticated non-imaging sensors such as: electro-optical scanners, imaging radars, passive microwave radiometers, etc. They collect and measure radiant intensities or reflectance values at 'n' points in a rectangular pattern. The radiant intensities from the target when received at the sensor detector, the sampled intensities are first translated into proportional electronic signals and then encoded into numerical equivalents as integers, from which digital image can be constructed. The integers are called digital numbers or DN values.

Before we proceed further into the components of digital image processing, it is essential to understand the basic, what a digital image is and how a digital image is constructed?

To start with, the digital image is a digital picture or representation of an object on the earth surface. Remotely sensed image data are digital representations of the planet Earth. A digital image is composed of a finite number of elements, each of which has a particular location assigned with a value that is arranged in a grid of rows and columns. These elements are called picture elements or pixels. In computer the analysts are familiar with the word "pixel" means an element of a digital image. We can see individual pixel when we zoom into an image. Usually when we look at an image without zooming, these pixels are so small that we do not see the outline of pixels. The process of displaying an image creates a graphical representation of array of pixels where the pixel values are assigned a particular gray-level (monochromatic image) or a particular color. Pixels with low values are assigned dark gray-levels, and high pixel values are assigned bright gray-levels. Digital image is regarded as a matrix of elements. For an 8-bit image, there are 256 pixels in the width direction and 256 pixels in the height direction starting from (0, 0) at the top left corner ending with 255 in both the directions.

The pixel in the upper left corner of the image always begins at co-ordinate (0, 0). The size of the image gives the number of pixels in each direction: X or width (column) and Y or height (row) directions i.e. a pixel in X-direction represents column while in Y-direction, it represents the row. More clearly, they are arranged in a column-row pattern. For example, a pixel in a location (3, 4) implies the pixel is at third column and fourth row (**Figure 7.1**).

Row Y

0.0	1.0	2.0	3.0	4.0
0.1	1.1	2.1	3.1	4.1
0.2	1.2	2.2	3.2	4.2
0.3	1.3	2.3	3.3	4.3
0.4	1.4	2.4	3.4	4.4

Column X

Figure 7.1

To sum up, a digital image is a sampled and quantized numeric representation of a scene. A multi-spectral digital image I (**Figure 7.2**) consisting of K spectral bands can be represented as K arrays of M × N elements of non-negative values or

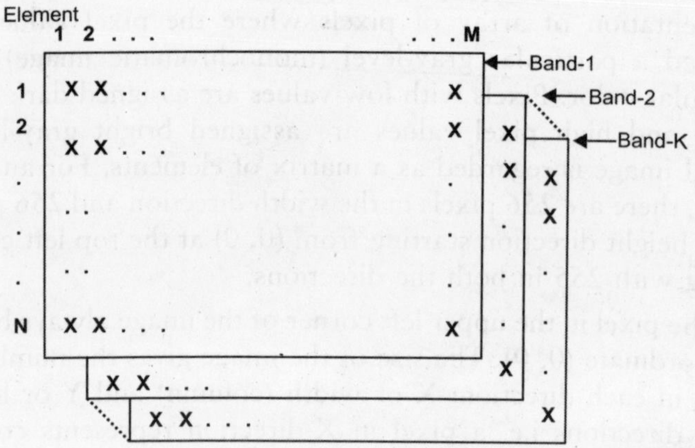

Figure 7.2

$$I^{(k)} = P_{ij}^{(k)}$$

Where k = band number: 1, 2, 3…K

i = element (column) number = 1, 2, 3….M

j = line (row) number = 1, 2, 3 … N

7.3 IMAGE DISPLAY

Digital Images can be stored as files in the computer. The image corresponding to each band is a set of data file values (radiant energy reflected or emitted from the specific portion of the electromagnetic spectrum), referred to as a layer of information when input into the GIS environment to create data base. The file contains not only the pixel values associated with each coordinate, but also an image header which provides additional information such as its size, the data type of the pixel elements, the color model, and other information.

An image may be considered to contain a finite number of sub-images sometimes referred to as areas of interest (AOIs) in ERDAS IMAGINE software. This concept reflects the fact that images frequently contain collections of objects each of which can be the basis for a region. Area of Interest in ERDAS

IMAGINE DIP software package is defined as a point, line, or polygon that is selected as a training sample or as the image area to be used in operation. AOIs can be stored in separate .aoi files in the computer.

7.3.1 BIT Concept for Image Plane

Remote sensing has been moving from analog or photographic methods to digital systems, which makes extensive use of digital technology. A bit in a digital system denotes binary digit. It is the smallest unit of data transfer on a computer network. It contains two binary values: 0 for off and 1 for on. Bits can also be "encoded" by electrical signals and pulses of light. Network performance has been measured in units of bits per second (bps). A set of bits can have many more values, depending upon the number of bits used. If we are using 8-bit data then we can have 2^8 i.e. 256 number of values. General expression for the number of values is 2^n, where n is the number of bits.

7.3.2 Display

Displays are referred to in terms of number of bits such as 6-bit or 8-bit. An 8-bit represents 256 number of brightness values in the display. For a 24-bit display color monitor, 24 bits per pixel breaks down to 8 bits for each of the three color guns (red, green and blue) per pixel. Thus each of the color guns of a pixel can have any one of 256 possible brightness values in the range of 0-255. The combination of three color guns thus yields $256 \times 256 \times 256$ i.e. 16,777,216 possible colors for each pixel on a 24-bit display color monitor (ERDAS Inc., 1999).

7.3.3 File Pixel vs. Screen Pixel

Pixel is the smallest part of a digital image in a computer or is called as the smallest unit of display in the computer monitor. In GIS, a satellite digital image is said to be in a raster format. Raster image is segmented into an array of grids arranged in column-row pattern like a chess board. Each pixel occupies each grid cell. File pixel is the data file value(s) i.e. remotely sensed energy for a pixel(s) in a digital image. Screen pixel is the corresponding brightness value for the unit of display in the color monitor.

For an image in a display monitor, each screen pixel corresponds to each file pixel. However, when we magnify or reduce an image, the one to one correspondence will be no longer. If we magnify an image zooming by 2, then one pixel can occupy 2 × 2 i.e. 4 grid cells (screen pixels) instead of previously occupied single grid cell.

7.4 INTRODUCTION TO COLOR

A reader must be familiar with the color concept before making an indepth understanding on display and mathematical manipulation of images in an RGB color monitor. In order to understand the attributes of commercially available image processing softwares, a brief summary of color is outlined below.

The human eye perceives color by means of light sensitive receptors in the retina called cones. The three types of color receptors or cones (sensors) in the eye measure the relative amount of red, green and blue light, as they are sensitive to wavelengths in the red, green and blue part of the electromagnetic spectrum respectively. Red, green and blue light can be added together to form a wide variety of colors. Therefore, red, green and blue are said to be the additive primary colors (**Figure 7.3**), because none of them can be produced by mixing the rest two. These additive primary colors are used in the screens of color monitors and the TV screen. The screens are coated with regularly spaced dots (screen pixels) of red, green, and blue phosphors. On display, the dots are bombarded with electrons from the red, green and blue color guns. The colored phosphors glow at certain frequencies to produce different colors. The human eye perceives different colors on integrating these dots together. Each pixel is represented by an equal number of red, green and blue phosphors.

All the colors and wavelengths of light that the human eye can see can be matched by mixtures of only three suitably chosen colors called Yellow, Magenta and Cyan. Each of these absorbs one third of all the wavelengths in white light while transmitting the other two-thirds. Yellow is called minus blue as it absorbs blue light, Magenta is called minus green as it absorbs green and Cyan is called minus red as it absorbs red light. The three colors Yellow, Magenta and Cyan are called secondary or complementary colors (**Figure 7.3**). The three complementary colors are used to match all of those in nature.

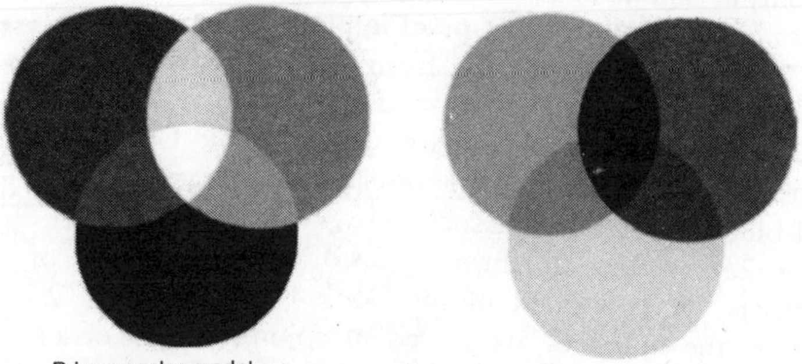

Figure 7.3

Colors can also be described in terms of three alternative attributes—intensity, hue and saturation (the IHS system).

Intensity (value): refers to the brightness of a color which for 8 bit display will vary from 0-255.

Hue: refers to relative amounts of two primary colors making up a color.

Saturation: the pureness of a primary color.

7.4.1 Brightness Values

The term brightness is a measure of the visual perception of how much light an area is emitting. Let us examine a case of a color monitor that we use. A color monitor is just a collection of thousands of pixels. As discussed earlier, each pixel emits red, green, and blue light. Brightness values are the quantities of each primary color to be output to each displayed pixel. The total brightness of an area on the monitor is the sum of brightness from each color. When an image is displayed, brightness values are calculated for all three color guns for every pixel.

7.4.2 Look Up Tables and Colormaps

In digital image processing the role of Look Up Table (LUT) in many aspects is very vital and fundamental. Look Up Table is a pre-set function to enable conversion from an input signal to an output signal in real time. In an image processing system, 8-bits color resolution per pixel implies each pixel may assume any one of 256 values i.e. from 0-255. However, the exact nature of the display is controlled by the size and characteristics of a separate block of memory called Look Up Table which contains the exact disposition of each combination of red, green and blue (RGB) values associated with each 8-bit pixel. It can store 256 values (0 through 255) as a sequence of 8-bit numbers. If you think of the Look Up Table as a 256 × 3 matrix, the numbers are stored in column-major order. This means that 256 R-values are followed by 256 G-values, followed by 256 B-values, for a total of 768 bytes exactly. The numbers can be either in a single column or in three lines of 256 values each.

Look Up Table acts as an interface, a table used to transform index numbers to output values. The pixel value of an image is used as an index into the intensity transformation array. For each pixel value of the image, there is one entry in the intensity transformation array. Thus if it is of 8-bit data, there are 2^8 i.e. 256 entries in the said array or table. Each entry in this table contains an index into the LUT, which gives the proportion of red, green and blue used to make the spot of color. If the number of entries in the Look-Up Table matches the number of entries in the intensity transformation table, then the default intensity transformation table contains a linear mapping, which means that the pixel values point directly into the look-up table. If the tables are of different length, then the default mapping will scale linearly between the two. In 24-bit RGB format, instead of storing a definite color for each pixel in an image, the value of each pixel is treated as an index number (input) into the LUT. Its use is to determine the colors and intensity values with which a particular image will be displayed. Typically, the output values stored in the LUT would be RGB color values.

The idea behind the colormap is that instead of storing a definite color for each pixel in an image, for instance in 24-bit RGB format, each pixel's value is instead treated as an index number into the colormap. When the image is to be displayed or otherwise processed, the colormap is used to look up the actual colors corresponding to each index number. The color map is a function of the display hardware.

7.4.3 Colorcells and Colormaps

We discussed earlier that, 24 bit color is 8 bits each of RGB, allowing 256 shades of each primary color, and 256 × 256 × 256 = 16.7 million color combinations. Such combinations create color on the screen of a 24-bit color monitor. A colormap is used to translate the 8-bit value into a color; however, it is not limited to 8-bit display. It is an ordered set of color cells which is used to perform a function on a set of input values. To display or print an image, the colormap translates data file values in memory into brightness values for each color gun.

For each data file, there is a colorcell in the colormap. The red, green, and blue values assigned to the colorcell control the brightness of the color guns for the displayed pixel. Number of bits in the display determines the number of colorcells in a colormap. There are array of 256 colorcells arranged in a rectangular manner to display an image in an 8-bit display. The following example explains how a pixel is assigned with the display color.

Let us consider a pixel of data file value 35, assigned with a display value of 12. This we mean that the said pixel uses the brightness values for the 12th colorcell in the colormap. In the colormap below (**Table 7.1**) this pixel is displayed as blue.

TABLE 7.1: Colorcell example

Colorcell index	Red	Green	Blue
1	255	0	0
2	0	160	100
3	0	0	255
12	0	0	255

The color that we see is a linear combination of the three primary colors: red, green and blue. These three colors form the basis for the RGB-colorspace. In an 8-bit display, the colormap contains 256 colorcells i.e. 256 colors can be displayed simultaneously on the display. In case of 24-bit display, the monitor displays 24-bit per pixel in their display memories. It offers up to 256 shades of red, 256 shades of green, and 256 shades of blue which is approximately 16 million of different colors. The data file values for each band are transformed into colorcell values. The colorcell that is specified by these values is used to define the color to be displayed. Thus, 24-bit display is capable of displaying a full range of colors. In such a case, each and every pixel is specified by the intensities of three color components to completely and accurately specify its color. Intensity of each component is stored as 8-bit integer and so each pixel requires 24 bits to completely and accurately specify its color.

7.4.4 24-bit True Color

In a true color graphic, image information is stored in the computer's memory where each pixel is represented by three or more bytes. One byte is used for each channel maintaining the order either RGB or BGR. The use of at least 24 bits (8 bits per channel) to represent each pixel in a graphic are referred to as true colors. True color is sometimes known as 24-bit color. A 24-bit true color display enables to view up to three continuous raster layers (bands) of data at one time, creating displayed pixels that represent the relationships between the bands by their colors (ERDAS Inc., 1999). The file pixels are transformed into screen pixels. The colors of the image so produced are due to the screen values (brightness values).

7.5 IMAGE PROCESSING OVERVIEW

Figure 7.4 depicts the general steps of digital image processing involved to get back the originality of a scene from its raw satellite digital image. Accordingly, image processing methods may be broadly organized into three functional categories:

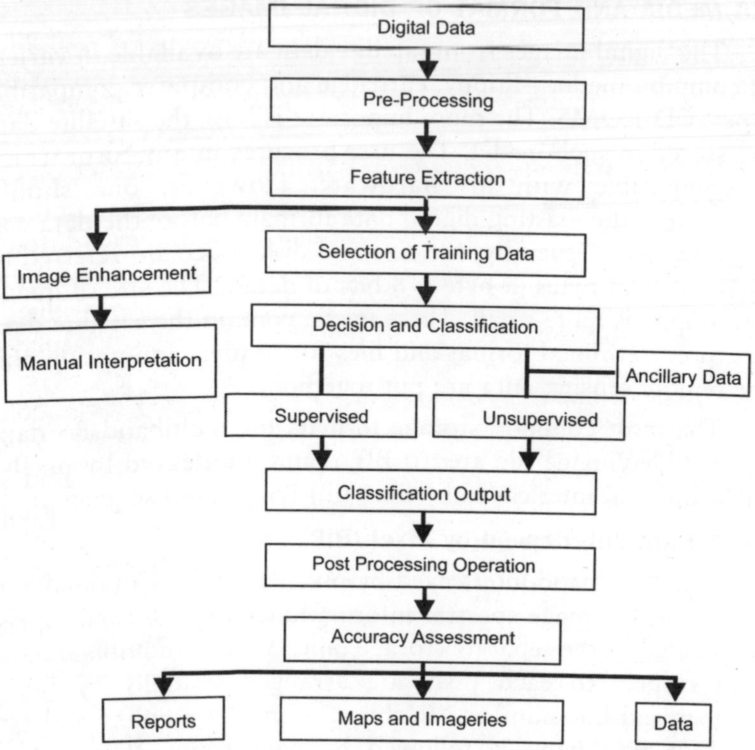

Figure 7.4

1. Image restoration:

(a) Restoring periodic line dropouts, (b) Restoring periodic line stripping, (c) Atmospheric correction, (d) Geometric correction.

2. Image enhancement:

(a) Contrast enhancement, (b) Intensity, hue and saturation transformation, (c) Density slicing, (d) Edge enhancement, (e) Digital mosaicing, (f) Producing synthetic stereo-images.

3. Information extraction:

(a) Principal Components Analysis, (b) Ratioing of images, (c) Multispectral classification—supervised and unsupervised, (d) Change detection analysis.

7.6 MEDIA AND FORMAT OF DIGITAL IMAGES

The digital images from satellite data are available in variety of common media—floppy, cartridge and computer compatible tapes, CD-ROMS. The more important is how the satellite data are stored in such media. The user procures in any form which is compatible with his hardware. However, one should understand the existing digital data formats before the data can be processed. Generally, file size and disk space are referred to by number of bytes (a byte is 8 bits of data). The size of image in a floppy is quite small. These media contain the satellite data in a predetermined format and files for header, annotation and the remote sensing data are put together.

The most common storage formats for multiband RS data in the video/image file are: (i) BIP (band interleaved by pixels) (ii) BIL (band interleaved by line) (iii) BSQ (band sequential).

7.6.1 Band Interleaved by Pixel (BIP)

The BIP (band-interleaved-by-pixel) format is optimal for accessing the image spectral information. In this format pixels are treated as the separate storage unit. A set of multiband data with respect to each pixel are arranged spatially by pixel number and line number. In this case the 1st scanline and 1st pixel of first band is followed by data from other bands, sequentially. Then the process is repeated for subsequent pixels as well as scanlines. It is practical to use if all bands in an image are to be used.

Record	Scanline	Pixels								
		1	1	1	1	2	2	2	2	.....
1	1	b1	b1	b1	b1	b1	b1	b1	b1	....
2	2	b2	b2	b2	b2	b2	b2	b2	b2	....

7.6.2 Band Interleaved by Line (BIL)

Each pixel of data is treated as the separate unit and the BIL format is stored by lines.

In this type of format, each record in the file contains a scanline (row) of data for all the pixels of one band. When all bands of the first scanline have been written, the method of storing shifts to the second scanline and so on.

Record	Scanline	Pixels								
		1	2	3	4	5	6	7	8	9.......
1	1	b1	b1	b1	b1	b1	b1	b1	b1	b1.......
2	1	b2	b2	b2	b2	b2	b2	b2	b2	b2.......
3	1	b3	b3	b3	b3	b3	b3	b3	b3	b3.......
4	2	b1	b1	b1	b1	b1	b1	b1	b1	b1.......
5	2	b2	b2	b2	b2	b2	b2	b2	b2	b2.......

Above example illustrates that each line is represented in all three bands before the next line is recorded. Like the BIP format, the format is useful if we make use of all the bands in our image analysis. This format is said to be inefficient, if we do not use some of its band since it is necessary to read serially even the unuseful band data.

7.6.3 Band Sequential (BSQ)

In BSQ format, each entire band is stored consecutively in the same file (Slater, 1980) i.e. first the entire image of 1st band is stored and is then followed by entire image of 2nd band and so on. The advantage in this format is that one band can easily be accessed and multiband data can easily be loaded irrespective of order.

Record	Scanline	Pixels								
		1	2	3	4	5	6	7	8	9.....
1	1	b1	b1	b1	b1	b1	b1	b1	b1	b1 ....
2	2	b1	b1	b1	b1	b1	b1	b1	b1	b1 ....
X		X	X	X	X	X	X	X	X	X
X	X	b2	b2	b2	b2	b2	b2	b2	b2	b2
X	X	X	X	X	X	X	X	X	X	X

7.7 IMAGE RESTORATION

When remotely sensed data are collected from the satellite sensors they are corrupted due to introduction of unwanted signals from image/system noise, atmospheric interference and sensor motion. The data corruption may be coined with the term distortions which are geometric, radiometric and atmospheric. Thus, due to geometric, radiometric and atmospheric distortions the real shape and geo-location of the object in the image is altered. Such image data are said to be the raw data i.e. unprocessed data. Radiometric distortion occurs due to the instrumentation error or the effect of the atmosphere on the pixel brightness value. Radiometric correction is usually carried out while preprocessing the satellite image and supplied to the user. Then it is necessary for the user to carry out geometric correction for georeferencing the satellite image (Richards and Xiaping, 1999). The geometric distortion is the distortion of the image caused due to Earth curvature, Earth rotation, variation in altitude and attitude's pitch, roll and yaw (details explained below). A satellite image needs restoration of its ground reality which requires geometric and radiometric correction especially if any interpretation, analysis or measurement has to be carried out. It would then only be possible to measure distances, areas and accurate geographic locations more correctly. Thus, image restoration is necessary and concerned with the correction or removal of distortion, degradation, and noise induced in the image process. The aim of image restoration is to produce from the degraded recorded image, a corrected image that is as close as possible, both geometrically and radiometrically, to the original object radiant energy distribution. Image corrections explained below are broadly considered under three heads: (i) geometric correction, (ii) radiometric correction, and (iii) atmospheric correction.

7.7.1 Geometric Errors and Correction

Remotely sensed data contain a number of geometric distortions and geographic locational errors due to data processing procedure and Earth's curvature and sensor motion. Such situations can never be avoided as the responsible processes are inevitable. Hence, a geometrically uncorrected image so obtained is not compatible with that of a map due to its different geometry and shift in geographic location. Some of these errors are commonly removed at the sensor's data processing center.

Image geometry refers to the projection, scale and orientation of the image, while geometric correction refers to the modification of the input geometry to achieve the desired geometry i.e. to assign the properties of a map to an image. This includes the removal of geometric errors due to sensor, orientation of thc platform and scene cffcct. When remote sensor data is collected it is not geo-rectified i.e. it is not free from geometrical errors. Therefore, geometric correction must be applied in order to place the data into a real-world coordinate system.

Raw level satellite image data are not true and accurate representation of the areas they image. Even images of seemingly flat areas are distorted by both the curvature of the Earth and the sensor being used. Unlike a cartographic map, distances between points on the earth's surface, relative positions of the points, shapes of areas, etc. are not accurately represented at a given scale in the raw satellite data. More or less, they represent irregular surface of the Earth. The raw images are not useful in the field as they do not have geographic co-ordinate systems but have row and column numbers. For the image to be useful in the field one needs to assign map coordinates to image coordinates. These coordinates may be latitudes and longitudes, or Universal Transverse Mercator (UTM) coordinates. The process of assigning map coordinates to image data is referred to as georeferencing.

7.7.1.1 Source of errors

Geometric errors introduced in the satellite images are from a number of sources. They are (i) systematic or predictable (ii) non-systematic or random.

Remotely sensed imagery potentially has many geometric and locational errors that are inherent within the data. These distortions are due to a variety of factors as follows:

(a) Scene related distortions:

1. The rotation of the Earth during image acquisition;
2. The curvature of the Earth;
3. Distortions due to differential scale;
4. Map projection.

(b) Platform related distortions:

1. The finite scan rate of the sensor;

2. The field of view of the scanner;
3. Panoramic distortion arising out of the tilt angles;
4. Variations in platform altitude, attitude and velocity.

Scene and platform related distortions are grouped together as geometric errors.

Systematic Errors

Systematic geometric errors may arise from the viewing geometry, scan geometry, earth rotation, curvature of the earth etc. (as explained above). Such errors are constant over time and they are predictable. A majority of these are removed as part of data products generation and analyst may not have to worry about these. The aim in removing these errors is to correct the pixel locational errors so that the ground features in the image are placed in their actual positions. This can be achieved by accurate modeling of the sensor and platform motion and the geometric relationship of the platform with the Earth.

In aerial photograph since the images are acquired for smaller areas that too in a very short time, they need not require following corrections as required by satellite imageries:

(i) Scan skew:

During the time required for the sensor mirror to complete an active scan, the platform moves forward along the ground track. This causes the ground swath slightly skew from its normalcy to the ground track producing cross-scan geometric distortion.

(ii) Panoramic distortion:

The image is a two dimensional representation of a ground object, which is a three dimensional spheroid representation (Earth's curved surface). Panoramic distortion is the greatest source of geometric error (map projection error) in imagery acquired from sensors with large angular field of view (**Figure 7.5**).

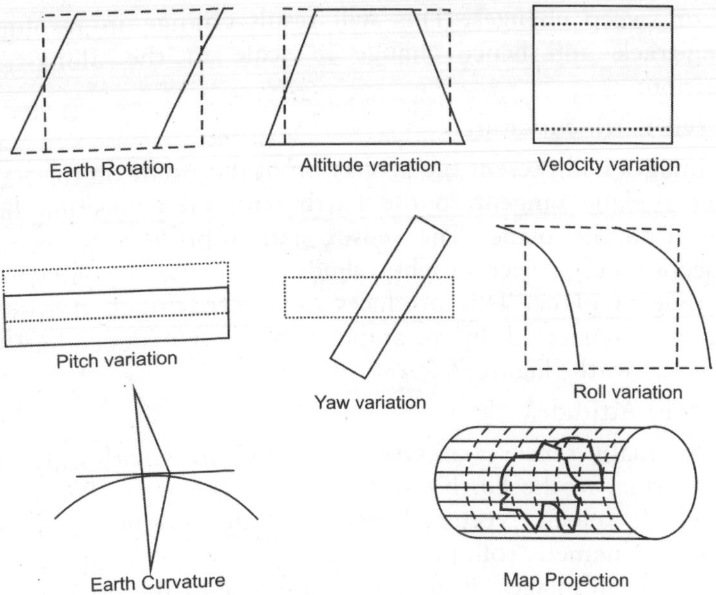

Figure 7.5

(iii) Earth's rotation:

Earth rotates beneath the sensor that scans the terrain. Thus there is a gradual westward shift of the ground swath being scanned. This causes along-scan distortion. This error is a function of latitude at a given time and table look up method is used for correction (Figure 7.5).

(iv) Altitude variation:

Variation in altitude of the satellite results in change in geometric resolution. Departure of the sensor platform from its normal altitude or the terrain increases in elevation produces scale distortions in the remote sensor data. This in turn will result in change in swath and hence in scale in the across-track direction. This error is compensated by shifting pixels in opposite direction (Figure 7.5).

(v) Spacecraft velocity:

Variation in altitude causes variation in velocity of the spacecraft. If the spacecraft velocity departs from its nominal, the ground track covered by a fixed number of successive mirror sweep changes. This will result change in resolution along-track and hence change in scale in the along-track direction.

(vi) Earth curvature:

Images will present the projection of the point on the Earth upon a plane tangent to the Earth with all projection lines normal to the plane. The sensor data represent perspective projection i.e. projection whose lines all meet at a point above the tangent plane. This produces only across-track distortion and it is corrected by resampling pixel size depending on latitude, swath (Figure 7.5).

(vii) Attitude:

Normally sensor axis system is maintained with only one axis normal to the Earth's surface and another parallel to the spacecraft velocity vector. Rotation of the satellite about its three axes namely, roll, pitch and yaw will result in substantial distortions in images. Roll and pitch will shift the positions of sample points in across-track and along-track distortions while yaw will have the skew effect in the image (Figure 7.5).

Roll error is due to the rotation of the spacecraft about the axis in the direction of motion. This will shift pixels in across-track direction. The positive roll will shift the pixels towards right (Figure 7.5).

Pitch error is due to rotation of the spacecraft about the line perpendicular to the direction of motion. This will shift the pixels in the along-track direction. Positive pitch angle will shift the image in the direction of motion of satellite (Figure 7.5).

Yaw error is due to rotation of spacecraft about the local vertical vector. This will have skew effect in the picture. Positive yaw will shift the pixels at the end of the scan line in the direction opposite to the direction of motion of satellite (Figure 7.5).

(viii) Scale variation:

Differential scale errors are due to departure of the sensor platform from its normal altitude or the terrain increases in elevation. This happens due to continuous in the distance between Earth and sensor.

(ix) Map projection:

A map projection is the method of representation of a three dimensional solid surface onto a two dimensional planar surface (piece of paper or on a computer screen). Remote sensors providing images represent projection of the Earth's curved surface onto a planar surface. Hence they suffer distortions due to map projection errors.

Unsystematic Errors

Unsystematic (random) errors are due to altitude and attitude variations (roll, pitch, and yaw) and topographic elevation differences. Removal of these errors can be achieved by the process of georectification. Rectification is a geometric process by which imagery is 'corrected, from distorted or degraded to create a more faithful representation of the original scene' (Lillesand and Kiefer, 1987). It is the process of projecting the data (three dimensional) onto a plane (two dimensional) and making it conforms to a map projection system. Assigning map coordinates to the distorted image data is called georeferencing. Since all map projection systems are associated with map coordinates, rectification involves georeferencing (ERDAS Inc., 1999). In other words, it is the process of picking up an image in its raw format and linking it to the land that it covers. This process depends on the scene and makes use of well distributed geographic features on the image called Ground Control Points (GCPs) (e.g. highway intersections, airport runway, and other such landmarks) that are clearly identifiable in both the distorted image (called the input or source image) and the reference map. More clearly, GCP is the point on the surface of the Earth which is identifiable on the image, in terms of rows and columns and map coordinates i.e. latitudes and longitudes. If there is a DEM of the study area, then, a process called orthorectification, a form of rectification that corrects for terrain displacement can be used. In relatively flat areas, orthorectification is not necessary, but in mountainous areas (or on aerial photographs of buildings) where a high degree of accuracy is required, one can make use of orthorectification. Ground Control Points should be collected at locations spaced evenly throughout the remote sensor image. The more scattered the GCPs are, the more accurate the rectification is. In fact, the relative location of the control points is at least as important as the number of points. Each GCP is identified by the coordinates in both the row-column image and latitude-longitude map. Likewise, after selecting several well distributed GCP pairs, the coordinate information is processed by the computer to determine the proper transformation equations to apply to the original (row and column) image coordinates to map them into their new ground coordinates. This process corrects for the distortion in the input image and is termed rectification. Depending upon the distortion in the imagery, the number of GCPs used, and their locations relative to one another, complex polynomial equations may be required to express the needed transformation.

Rectification of the source image (distorted image) removes much of the geometric errors adhered to it, but not all. During the process of transforming the data from one grid system into another grid system, use of selected polynomial equations tend to minimise the error but practically an error free coordinate conversion is not possible. There is an every possibility of influence of each GCP on the value of the coefficients in the transformation matrix. Total error introduced can be calculated using a transformation matrix. The matrix is computed from a particular set of GCPs and contains coefficients that are used in polynomial equations to convert the coordinates. The size of the matrix depends upon the order of transformation. The coefficients of the transformation matrix help to derive the polynomial equations for which there is least possible amount of error when they are used to transform the reference coordinates of the GCPs into the source coordinates. A first-order transformation will probably handle most of georeferencing requirements. It is a linear transformation. With a minimum of three links, the mathematical equation used with a first-order transformation can exactly map each image point to the target location. Any more than three links introduces errors, or residuals, that are distributed throughout all the links. However, one should add more than three links because if one link is positionally wrong, it has a much greater impact on the transformation. Thus, even though the mathematical transformation error may increase as you create more links; the overall accuracy of the transformation will increase as well. The higher the transformation order, the more complex is the distortion that can be corrected. However, transformations higher than third order are rarely needed. Higher-order transformations require more links and thus will involve progressively more processing time. It is the responsibility of the analyst to decide how much error is acceptable. Accurate rectification needs selection of accurate GCPs. The GCPs should be (i) reliably matched between source and reference image, and (ii) widely dispersed throughout the source image.

7.7.1.2 Resampling

From the above discussion it is clear that the GCPs are first edited, and then a transformation matrix is derived using the GCPs to geometrically correct the image to a suitable level of accuracy. Then an approach of resampling (original pixels are resampled to match the geometric coordinates) is adopted which is a method that produce analyst's required geo-coded image for further digital image processing.

Resampling is a process of determining the DN value of a pixel shifted to a new, more accurate position from its initial after the various geometric corrections and translations have been applied to the distorted image. In other words, the process of resampling estimates in a better way the DN values for individual pixels in the output image from manipulating the pixels of the input image.

After georectification, a matrix of "empty" cells is computed using the map coordinates. Then, resampling calculates and assigns each empty cell a DN value by examining the cells in the ungeoreferenced (input) image. The corrections have led to a pixel that at the time of sampling covered ground 'A' being shifted to a position that have 'A' values but should if properly located represent ground 'B'. Resampling actually geometrically corrected the original distorted image by determining the digital values to place in the new pixel locations of the corrected output image.

The method of resampling can easily be understood if one closely examines the schematic diagram presented in **Figure 7.6.**

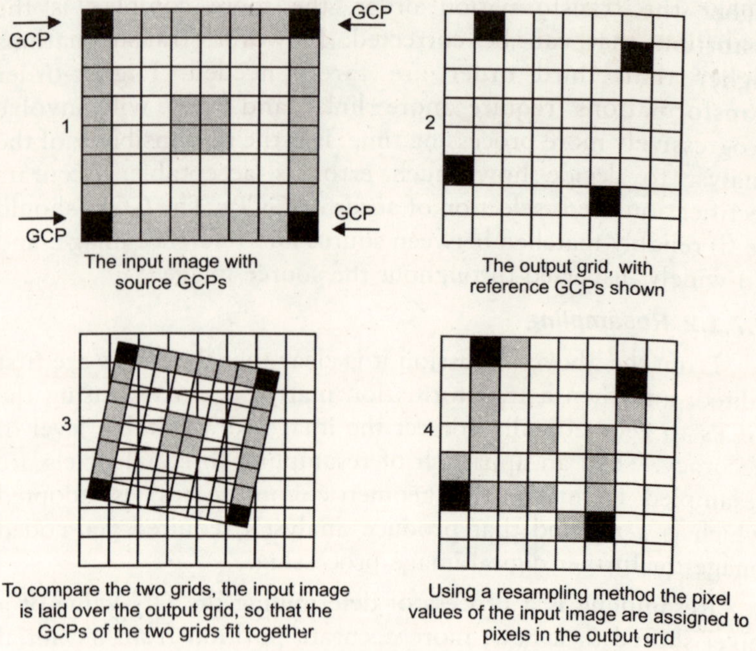

Figure 7.6

The three most common resampling techniques are nearest neighbor assignment, bilinear interpolation, and cubic convolution. Each resampling method employs a different strategy to estimate values at output grid for given known values for the input grid.

Nearest Neighbor Assignment

The most basic type of resampling method is nearest neighbor assignment or point sampling. The method is simply to assign the DN value of the nearest pixel in the input image (uncorrected) to the pixel in the output image (corrected)

(**Figure 7.7**). That is the DN value of the pixel closest to the retransformed coordinates becomes the new data value of the pixel in the output image. Figure shows the output image is overlaid on the input image. Each output cell (empty) is assigned with the DN of the input which is closest to the center of the output cell. This method is chosen for use with classified image data. There is no loss in the original color and tone values and no new averaged colors are created and hence, the edge details remain sharp.

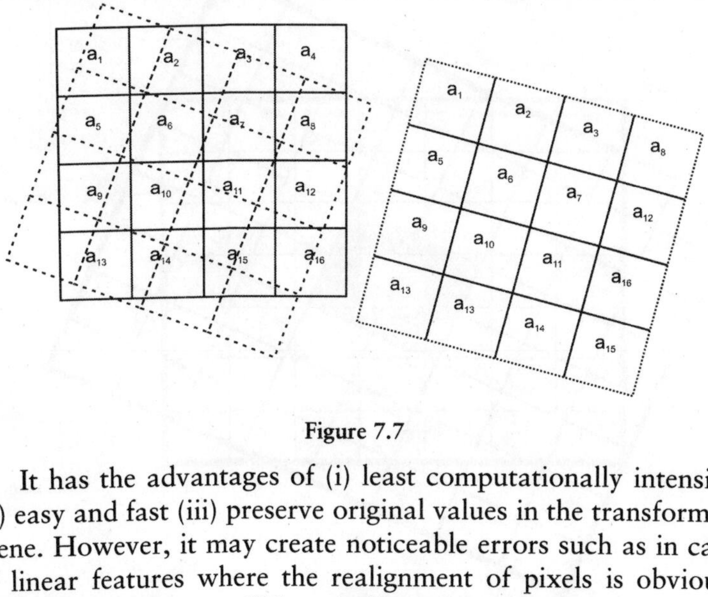

Figure 7.7

It has the advantages of (i) least computationally intensive (ii) easy and fast (iii) preserve original values in the transformed scene. However, it may create noticeable errors such as in case of linear features where the realignment of pixels is obvious. The geometric error will be a half pixel at maximum.

The disadvantage of the nearest neighbor resampling is that the rectified image tends to have stairstep shaped linear features.

Bilinear Interpolation

The principle of this resampling method is to calculate the DN value of each output pixel based on a weighted average of the four nearest input pixels. That is, the data value of the pixel in the output image is interpolated from the four nearest pixels to the transformed coordinates, using a linear function. The advantage here is that since the output value is created from several input values, the resultant image will give a natural look. However, some changes can be observed if it creates new pixel value (**Figure 7.8**). The disadvantages of this method are (i) the DN values in the input image are lost, and (ii) the spatial resolution of the output image decreases as the average is taken over areas while resampling the output image. The spectral data will be smoothed after the interpolation.

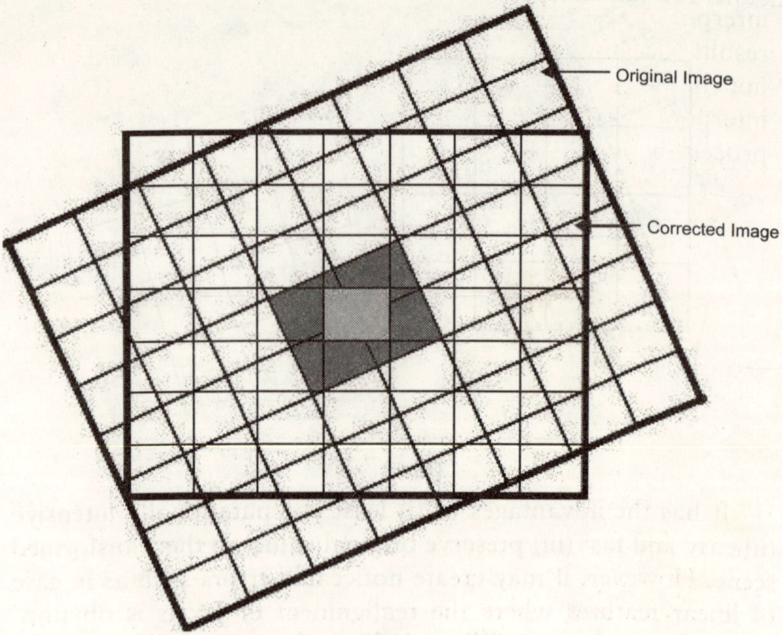

Figure 7.8

Resampling by bilinear interpolation is several times slower than resampling by nearest-neighbor and due to the smoothing effect of this method, the image resolution is not as good as nearest-neighbor.

Cubic Convolution

It is the most sophisticated and computationally intensive method of resampling. Cubic convolution generalises bilinear interpolation and determines the DN value of the pixel in the output image from a 4 × 4 grid compromising thc 16 surrounding pixels, using a linear function. First, four cubic "convolutions" are performed in one direction (horizontally or vertically) and then one more cubic convolution is performed in the perpendicular direction. A "convolution" replaces each neighboring value by a standard curve centered at the neighbor's location and scaled according to the neighbor's value. Cubic convolution requires a longer time for the computation when compared with the other methods. In comparison to bilinear interpolation, the image so produced by cubic convolution results in sharpening as well as smoothing (**Figure 7.9**). It does not have the aliasing effect produced by nearest neighbor interpolation at sharp intensity due to the inherent smoothing process.

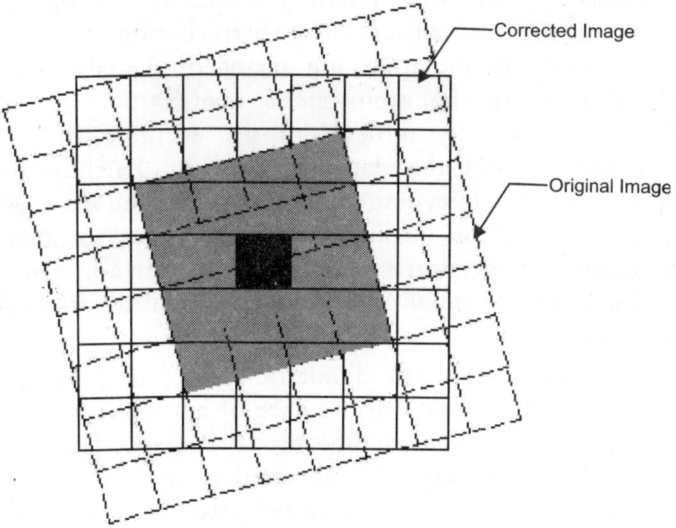

Figure 7.9

7.7.2 Radiometric Errors and Correction

Radiometric errors may arise due to detectors, imaging system noise, sensor degradation, transmission and atmosphere. The last two processes are non-systematic error sources and may require scene to scene analysis. Radiometric calibration, destriping, periodic noise removal and synthetic line generation for transmission losses are some of the techniques used in radiometric correction.

Let us consider an ideal Earth observation system. If it is equipped with perfect radiometric measurement instrument then it can record accurately and uniformly the reflected energy emanating from the object on the Earth surface. However, the solar radiation that reaches the object on the Earth's surface traverses through the atmosphere and hence is perturbed due to interference with the atmosphere. The result is: speed, direction, intensity, wavelength and spectral distribution of the electromagnetic radiation are affected. Similar is the case with the light reflected by the objects and travels through the atmosphere to reach the sensor. In its journey through the atmosphere the sensor signal again becomes perturbed due to interference of the atmospheric constituents. More clearly, because of the intensification of perturbation, the spectral properties of the radiation are changed. Signals from the target thus carry the atmospheric contribution (unwanted signal). It is actually difficult to get accurate radiometric measurements from the data recorded by Earth observing satellites. It is, therefore, desired to have meaningful measurement of reflectance at the Earth's surface. In order for that, the atmospheric contribution must be removed from the recorded sensor's signal. This process is called atmospheric correction.

The entire process of radiometric correction involves three steps (**Figure 7.10**). Spectral radiance of an object is governed by certain factors such as time of day, season, latitude, etc. Object's spectral radiance is converted to spectral reflectance, which is the amount of radiance reflected out of the incident solar radiation. The spectral reflectance provides a standardised measure that is directly comparable between images by different satellites or recorded by the same satellite at different times and several solutions do exist to try to overcome these errors. Some of them are based on complex mathematical models that describe the main interactions involved. These models are effective.

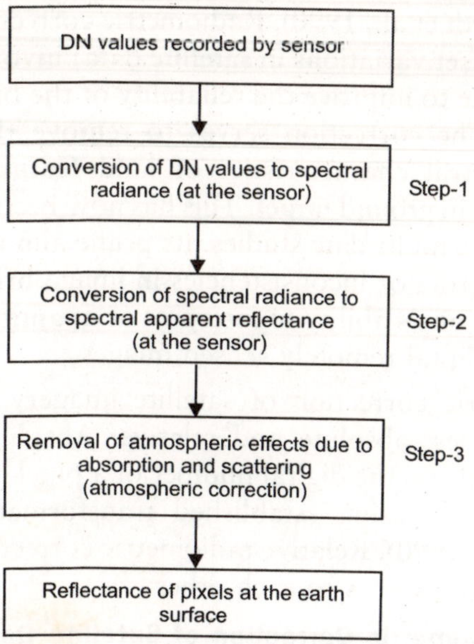

Figure 7.10

Other radiometric correction methods are based on the observation of reference targets whose radiometry is known. Examples are the surfaces of bodies of water, glacial ice caps, and expanses of desert sand. In the visible part of the solar spectrum the reflectance of natural surfaces spans the range from very dark oceans and dense forests to very bright surfaces such as snow, ice and clouds. Particularly high reflectances should also be expected in sun glint areas. A high radiometric performance, together with an accurate calibration procedure is therefore required. Actually making such corrections often is not that easy.

Remotely sensed satellite data play a key role especially in environmental monitoring and modeling. Multidate satellite data of a given area provides valuable and potential information in many forms of change detection analysis. These repeated observations are confounded in terms of radiometric consistency due to changes in sensor calibration over time, differences in illumination and observation angles, and variation in atmospheric effects (Eckhardt et al., 1990). Radiometric correction (correction of gain and offset variations in satellite data) involves processing of digital image to improve the reliability of the brightness value magnitudes. The correction serves to remove the effects that alter the spectral characteristics of land features, except for actual changes in ground target. This has now become mandatory in multi-sensor, multi-date studies. Its prime aim is to reduce the influence of errors or inconsistencies in image brightness values that may limit one's ability to interpret or quantitatively process and analyze digital remotely sensed images.

Radiometric correction of satellite imagery falls into two broad categories, absolute and relative. Absolute radiometric correction converts the digital number of a pixel to a percentage reflectance value using established transformation equations (Richter et al., 1990). Relative radiometric correction normalizes multiple satellite scenes to each other.

Steps in Radiometric Correction of Satellite Data

Step 1. Conversion of DN to spectral radiance:

The radiometric correction in satellite image processing is critical, since the digital radiometric data measured by the sensors have to be converted to albedo values (visible and near infrared channels) or brightness temperature values (TB, middle and thermal infrared channels) for the subsequent processes.

Pixel values in satellite imagery represent the radiances of the surface in the form of Digital Numbers (DN) which are calibrated to fit a certain range of values. The first step in radiometric correction is to convert the pixel values into "at-satellite radiance". Conversion of DN value to spectral radiance requires information on the gain and bias of the sensor in each band (**Figure 7.11**). It is necessary to make conversion of DN to absolute radiance values for comparative analysis of several images taken by different sensors. Each sensor has its own calibration parameters used in recording the DN values. This implies the same DN values in two images recorded by two different sensors will have two different radiance values.

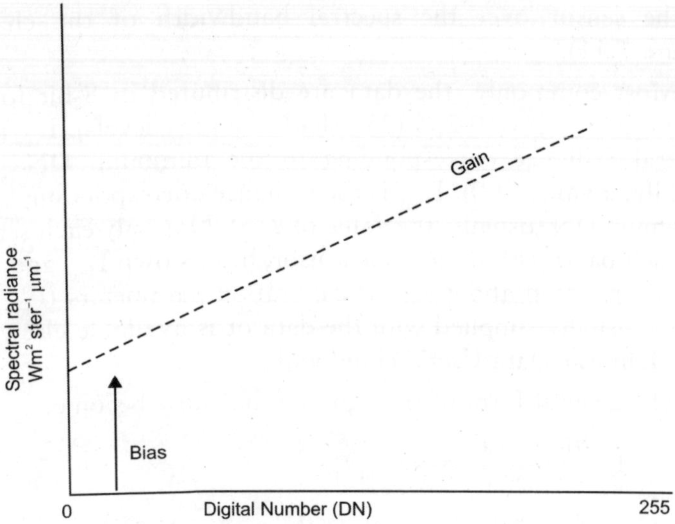

Figure 7.11

Transformation of DN to radiance value depends on calibration curve of the sensor. Sensors are calibrated before the satellite is launched and the accuracy declines as the sensitivity of the sensor changes over time. Therefore, recalibration of the sensor is attempted periodically. The principle behind the sensor's calibration is that there exists linearity between DN and spectral radiance.

The linear equation for calibration is expressed by the following equation:

$$L_\lambda = \text{Bias} + (\text{Gain} \times \text{DN}),$$

where, L_λ is satellite spectral radiance.

Using the simple linear equation above one can convert each DN value to corresponding spectral radiance after knowing the values of Bias and Gain. Gains and Biases are calculated from the lower (L_{min}) and upper (L_{max}) limits of the post calibration spectral radiance range for each band.

In this linear equation, the parameters involved are the range of DN values in the image (in X-axis), and the lowest (L_{min}) and highest (L_{max}) radiances (in Y-axis) measured by the sensor over the spectral bandwidth of the channel (Figure 7.11).

Most commonly, the data are distributed in 8-bit format within the range of 0-255 (256 shades of gray levels). L_{min} is the spectral radiance corresponding to the minimum DN value (usually, a value of 0). L_{max} is the radiance corresponding to the maximum DN (usually, the value of 255). Not only each sensor, but each band within the same sensor, has its own L_{min} and L_{max}. The information about sensor calibration parameters (L_{min} and L_{max}) is usually supplied with the data or is available elsewhere (e.g., Landsat Data User's Handbook).

The general form of the equation has now become:

$$L = \frac{(L_{max} - L_{min})}{255} \times L_{min}$$

where Bias = L_{min}, and $G_{ain} = \frac{(L_{max} - L_{min})}{255}$ and,

L is the radiance expressed in $Wm^{-2}\ sr^{-1}$.

Step 2. Conversion of spectral radiance to apparent reflectance:

The second step involves correction for atmospheric propagation effects, to obtain "at-surface radiances". The apparent reflectance, which is for satellite images sometimes termed as exoatmospheric reflectance (E_{exo}).

For many quantitative applications whenever multispectral images are generated from visible-NIR spectral regions, they are affected unequally by atmospheric scattering. The physical processes of Rayleigh or molecular scattering, aerosol scattering, multiple scattering due to Rayleigh-aerosol interaction, sun glitter radiance are the principal components that contribute path radiance to the real signal from the scene on the Earth. Because, (i) Rayleigh scattering is inversely proportional to the fourth power of the wavelength, and (ii) Mie scattering is also inversely related to the wavelength, which takes place when the particle size is comparable to the incident wavelength; the images representing the shorter wavelength regions are affected the most. In case of thermal imageries, the sensor records the brightness temperature of the radiation reaching the sensor, in stead the brightness temperature of the radiation leaving the surface. So there is an atmospheric contribution to every pixel value. Therefore, it is desirable to correct the data for the effects of atmospheric propagation (Cracknell and Hayes, 1991). If the data are accurately calibrated, the variable that is measured is the radiance reaching the sensor, but the variable that is wanted is the reflectance of the surface. If there were no atmosphere, the calculation would be a straightforward one, requiring only knowledge of the radiance at the Earth's surface due to solar illumination, and of the geometry of the sun-target-sensor arrangement. However, the presence of atmosphere significantly complicates the relationship between the solar radiance and the radiance detected at the sensor.

Spectral reflectance is the ratio of incident-to-reflected radiant flux measured from an object or area over specified wavelengths. Unlike radiance and irradiance values, reflectance is an inherent property of an object and is independent of time, location, illumination intensity, atmospheric conditions and weather. Although reflectance is a key unit of measure in remote sensing, it is not measured directly and instead must be derived. For a Lambertian surface, the incident solar radiation is in units of irradiance (radiant power per area, within a specified spectral band) and the outgoing radiation is in units of radiance (radiant power per area per solid angle, within the same band). Calibration of spectral reflectance involves at a given solar elevation, (i) measurement of the irradiance from the sun (incoming radiance) and the upwelling radiance from the Lambertian surface to reach at the sensor (outgoing radiance), and (ii) taking the ratio of the two quantities. Calibration converts pixel values from radiance to apparent reflectance which is the radiance received by the sensor at the top of the atmosphere (TOA) at a given wavelength.

In the above context, the effective planetary reflectance of the surface (apparent) is expressed as:

$$\rho = \frac{(\pi . L . d^2)}{[E_{exo} . \cos(SZ)]}$$

where, ρ = unitless planetary reflectance of the surface (on a scale of 0-1)

π = 3.141593

L = Reflected at-surface radiance (the radiance at the surface heading towards the sensor) in mW cm^{-2} $ster^{-1}\mu m^{-1}$

d = Actual distance from the sun to the area being sensed

E_{exo} = Mean exoatmospheric irradiance of the Sun at a distance of one astronomical unit (1 AU = 1.496×10^{11} m)

SZ = Sun zenith angle in degrees when the scene was recorded

Step 3. Removal of atmospheric effects due to absorption and scattering:

The effects of the atmosphere upon remotely sensed data are not considered errors, since they are part of the signal received by the sensing device (Bernstein, 1983). Before any interpretation of remotely sensed data about the objects on the Earth's surface is made, it is however crucial to make accurate atmospheric correction. It is because of the fact that backscattered radiation by air molecules and aerosols is predominant (80% or more) in forming the radiance detected at the top of the atmosphere (TOA) as per as the visible part of the spectrum is concerned. Removal of atmospheric effects from remotely sensed data is, therefore, considered to be one of the most challenging problems in remote sensing. However, precise corrections are extremely difficult because of the dynamic nature of the atmosphere. The knowledge of the following, apart from the characteristics of the sensor system is essential for correcting spectral reflectance signature:

(i) Irradiance at the ground feature of interest with reflectance.

(ii) Atmospheric spectral transmittance between the ground and the sensor.

(iii) The upwelling atmospheric radiance between ground and sensor, called as path radiance.

However, if we are considering radiation emitted by the object, then, of course, we need no irradiance.

Atmospheric correction procedure can be broadly divided into three groups:

(A) Removal of path radiance (B) Direct calibration using field radiometric measurements (C) Atmospheric modeling.

(A) Removal of Path Radiance

"In most cases the atmosphere is perceived as a hostile entity whose adverse impact must be neutralized or eliminated before remotely sensed data can be properly analysed" (Schott, 1997). The radiometry of the satellite imageries is influenced by various effects from the outside of the sensor. Path radiance from the atmospherically scattered sunlight is one of the prominent effects. The atmosphere introduces two forms of path radiance into the signal, radiance from Rayleigh or molecular scattering, and radiance from aerosols or haze. This especially occurs at shorter wavelengths and long atmospherical path lengths, which gives rise to brightening of the images. These can be removed simultaneously using dark object subtractions. However, an improvement in atmospheric correction offers advantages in treating atmospheric and Rayleigh corrections separately. If no dark water exists in the scene, the Rayleigh correction is critical. Rayleigh radiance is removed before the dark object subtraction. While not critical, when dark water is present, removal of the Rayleigh path radiance permits pixel by pixel correction of aerosols for water pixels and allows better control on the adjustment for aerosols.

A1. Dark pixel subtraction (Chavez et al., 1977)

The sophisticated methods of atmospheric correction require auxiliary data; however, the simple methods are based only on the image itself and are served to be suitable for operational use. One of the most widely used and well-known simple atmospheric correction methods is the dark pixel subtraction. Dark pixel subtraction is a technique which determines the pixel in the image with the lowest brightness value. This pixel is assumed to have a zero ground reflectance such that its radiometric value represents the additive effect of the atmosphere (Crippen, 1987).

This method is also known as the histogram minimum method, based on the assumption that in the image there exists a pixel whose DN value is essentially zero, so that the radiance recorded by the sensor is solely due to path radiance. That is the pixel of the lowest DN in each band in reality should be zero, and hence its radiometric value (DN) is the result of the atmosphere induced additive errors. In order to remove the path radiance, an area of shadow or a very dark object (such as clear deep sea water) is selected and the minimum pixel value is determined. The signatures are corrected by subtracting the minimum observed value, determined for each specific band, from all pixel values in each respective band.

Despite the considerable number of available atmospheric correction algorithms, there is little literature concerning their relative merits. Over water bodies, atmospheric effects account for the majority of the at-satellite measured radiance in the visible bands, and therefore targets of this type provide an opportunity for assessing the effectiveness of the different methods available.

Dark pixel subtraction methods are inclined to suffer greatly from noise within the dataset and the technique requires suitable dark pixels to exist somewhere on the image. However, the assumptions are said to be very vague, and application of this method may degrade the image rather than improve.

Mather (1987) applied dark object subtraction method of correction to an image in bands 1-4 acquired from Landsat MSS. Bi-plots of band 4 against bands 1-3 are generated for pixels of the dark regions. Regression techniques are then used to calculate the Y-intercept which represents the path radiance in bands 1-3. This value is then subtracted from all pixels in the imagery.

A2. Regression method

Potter and Mendlowitz (1975) described two methods for determining haze levels (specified by haze optical depth at a wavelength of 0.5 micron) from Landsat multispectral scanner data. The method (channel correlation) is developed based on selecting an area of homogeneity of an image and constructing a series of bi-plots between each band pairs. The haze level is related to the y-intercept of the regression line through a plot of the data in the plane of the multispectral channels MSS 4 and MSS 5. This is due to the fact that the slope of the plot represents the relative reflectance of the homogeneous area in the two channels. However, in the absence of the atmospheric haze the slope must pass through the origin. **Figure 7.12** illustrates two instances of measured radiance on ground and space for a homogeneous area (does not have low reflectance). It is found that in case of ground (handheld spectrometer measured radiance), the slope of the bi-plot passes through the origin, implies there is no atmospheric haze, which is a fact. However, in the other case (satellite measured radiance), the slope shows offset from the origin. This is due to calibration offsets of the sensor and the addition of the atmospheric path radiance. The other method, minimum value relates haze level to the minimum value of individual lines in the MSS 4 data set.

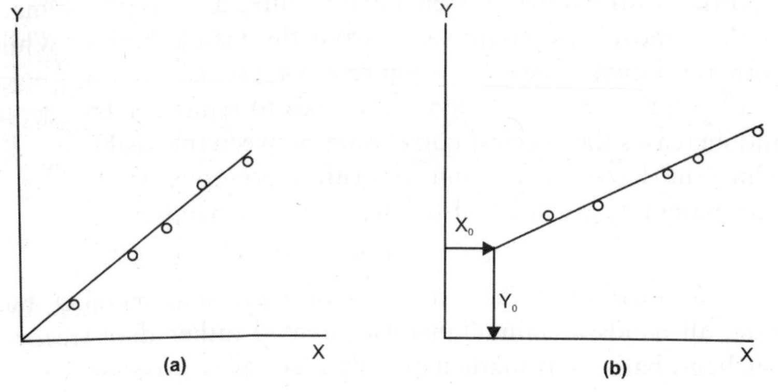

Figure 7.12

Using the regression method as a base, Crippen (1987) has developed an atmospheric correction method that is simple to compute, easily performed on many image analysis systems and can be achieved without measurements of the prevailing meteorological conditions. The regression based technique which determines a 'best fit' line for multispectral plots of pixels within homogenous cover types. The slope of the plot is proportional to the ratio of the reflective material (Crippen, 1987).

Carlotto (1999) described new method for reducing the effects of space-varying, wavelength-dependent scattering in multispectral imagery caused by smoke and haze. It is intended for use in situations where atmospheric scattering affects the shorter wavelengths and varies in space. The method converts an image in which space-varying scattering is present into an image where the scattering has been equalized over the entire image, so that previously developed techniques for removing constant scattering effects can be used. The spectral measurement space is viewed as consisting of two subspaces: one spanned by the bands that are affected by scattering, the other by the bands that are not. A correspondence between the two subspaces is established and used to predict the values of the former bands (i.e., what their values would be without scattering) from the latter. Carlotto compared haze equalization algorithm to an earlier de-hazing algorithm developed by Lavreau that subtracts a portion of the fourth tasseled cap feature, used as an estimate of the atmospheric component, from the visible bands. While both are shown to be effective in removing space-varying smoke and haze, the de-hazing algorithm tends to remove subtle detail and increases the spectral correlation between the visible bands, while the haze-equalization algorithm preserves subtle detail and maintains the spectral balance between bands.

A3. Covariance matrix method (Switzer et al., 1981)

This method is the extension of regression method, but here, all bands are simultaneously treated rather than treating per band basis. It is mathematically a complex process.

A4. Regression intersection method (Crippen, 1987)

The method is complex mathematically. It minimises the effect of the atmosphere and is more familiar as it provides absolute results from the image data without the use of ancillary information. The method itself does not require (i) any knowledge or assumptions about the scene and the target themselves, (ii) any knowledge or assumptions regarding atmospheric conditions or sensor calibrations, any assumptions about the results (Crippen, 1987). It is based on a similar regression principle to those above but generally involves the calculation of regression lines for a number of surface materials of contrasting spectral properties. The regression lines must intersect at a point, the point of zero ground reflectance where the radiometric values of two spectrally different materials can be equal. In practice however, the path radiance will offset this intersection. The desired path radiance correction is, therefore, the measured displacement of the hypothetical zero ground radiance pixels from the zero data number position on the graph. Crippen (1987) recommends the collection of a series of training areas resulting in many regression lines intersecting in two dimensional space. The relative values of path radiance between bands are more accurate in the covariance matrix method. If this methodology is used, it is best to use it in conjunction with the covariance matrix method (Crippen, 1987).

(B) Direct Calibration Using Field Radiometric Measurements

It is a second, simple method referred to as empirical line calibration. This method attempts an entire radiometric correction rather than just removing atmospheric effects. In this method, several *in situ* radiometric measurements are taken over various objects concurrently with the acquisition of the remote sensor data. The instrument used for these reflectance measurements is called a radiometer. Unlike the remote sensor system, the radiometer is used to take measurements *in situ* with almost no atmosphere. The data collected using the radiometer is used to develop a simple linear mathematical function to predict what the radiometric values should be over the entire image. In order to make accurate measurements of surface reflectance properties of various vicarious calibration target sites; the Remote Sensing scientists make use of field spectral radiometers. The users are required to do some field work to measure the reflectance of two or more regions of the area covered by the image.

(C) Atmospheric Modeling

A third method is computationally more complex than either of the previous two. It relies on either assumptions or acquisition of explicit information on the atmospheric conditions at the time of imaging by the remote sensor.

This method involves theoretical determination of the path radiance contribution of the atmosphere with the conjunction of satellite overpass. The spaceborne sensor detected radiance is a mixture of the radiance emerging from the target on the Earth's surface and the solar radiation backscattered by the air molecules (Rayleigh scattering) and the aerosols (principally, Mie scattering) in the atmosphere. The Rayleigh and aerosols contribution together constitute atmospheric path radiance. To calculate the contribution of the scattering on the reflected radiance requires that many atmospheric variables at the time of the satellite overpass be recorded and inputted into theoretically derived equations to determine the effect of the atmosphere on each spectral band (Forester, 1984; Crippen, 1987). The variables might include a detailed profile of temperature and humidity within the atmospheric column, the Sun-Earth geometry, and the position of the sensor with respect to each pixel. These data are first processed to retrieve Aerosol Optical Depth (related to total amount of absorbers and scatterers along the solar path). The aerosol optical depths are used in an inversion scheme (e.g. Principal Component Inversion Scheme) to estimate the aerosol particle size distribution given by Junge (1958) and columnar amounts of ozone and water vapor. All of these results are used as input to a radiative transfer code to predict the radiance at the top of the atmosphere.

There are number of radiative transfer models available to predict the atmospheric effects on recorded electromagnetic radiation. These can be used to produce a radiometrically corrected image.

How to decide on atmospheric correction?

1. Examine summary statistics for each scene—means, variances and frequency histograms for each band.
2. Look for absence of dark values, particularly if image contains large water bodies.
3. Loss of resolution and low contrast indicate poor atmospheric conditions.

4. Date of image-summer time implies high humidity, haze and poor visibility; winter, spring and autumn dates indicate clearer atmospheric conditions.

7.7.3 Noise Removal

One of the most important considerations in processing of digital image is noise. As per www.dictionary.com, noise is irrelevant or meaningless data. Noise in a remotely sensed image is referred to introduction of any unwanted signal or disturbance due to the sensor response and data recording process, which is independent of the scene signal. Despite, noise can also be said as random or repetitive events that obscure or interfere with the desired information. Noise creates inconsistencies in image brightness values or DNs that may limit one's ability to interpret or quantitatively process and analyze digital remotely sensed images. It can either degrade or totally mask the true radiometric information content of a digital image. In all image processing systems, one must consider how much of the detected signal can be regarded as true and how much is associated with random background events resulting from either the detection or transmission process.

Noise can result from various potential sources such as: (i) Differences in sensitivity of the elementary detectors of the viewing instrument may result in the acquisition of spurious data, (ii) Data transmission from the satellite to the ground based receiving Earth stations may cause in loss of data, (iii) Data recording error during recording of data onto the Computer Compatible Tapes (CCTs) at the receiving stations, (iv) There may be scale distortions of the images, (v) Errors due to atmospheric component, and (vi) Image quantization errors.

When a digital image is recorded by the sensor on a satellite or aircraft, it may contain noise introduced errors in the measured brightness values of the pixels. Additive noise can be of basically two types in image data: random and non-random noise. Random noise typically occurs as individual pixels with digital numbers that are much higher and lower that the surrounding pixels (Sabins, 1987). In the image these pixels produce inconsistencies in brightness value result in bright and dark spots in regions of homogeneity that can degrade the image. These spots also interfere with information extraction. It can be removed from the image data by applying image processing techniques like low pass filtering or smoothing, unfortunately at the expense of some high frequency information in the image (Richards, 1986).

Random Noise Removal

Image denoising still remains a challenge for researchers because noise removal introduces artifacts and causes blurring of the images. Hence before proceeding for any subsequent enhancement or classification of the image data it is mandatory to make the digital image free of noise so that it can be a closure approximation of the original scene. Random noise is not due to system equipment and hence is said to be the natural noise. Sophisticated restoration methods are needed to remove the random noise effect from the digital data.

Random noise is characterized by the statistical variation in DNs from pixel to pixel called bit errors. Normally it originates in the atmosphere or in electronic components and is, therefore, considerably more difficult to separate from valid image data. Removal of random noise implies removal of the bit errors. Addition of bit errors unsystematically onto original pixel values in a digital image brings about an abrupt change in DNs than true image values. Abrupt change in values one must recognize, in order to handle the bit errors. In order for that, the analyst compares each pixel with its neighbors. If the difference between a candidate pixel and its neighbors is exceedingly more than a threshold value defined by the analyst, then the pixel value for the said candidate is considered to contain the noise component.

Most of the methods for noise removal in a digital image are template based techniques in which a template or a filter or box or window is defined and moved over the image row by row and column by column. Noise removal can be accomplished by low pass filtering either using an average operator or a median operator. In case of average operator, the noisy pixel is corrected by assigning it with the value, averaging the surrounding pixels covered by the template at a particular position. When this is done for every pixel in the image, a radiometrically modified image is produced that enhances or smoothens geometric features according to the specific numbers loaded into the template. If the noise is uncorrelated from pixel to pixel, its variance will be reduced by low pass filter. In the median filtering, in stead of averaging the surrounding pixel values, the median value of all the surrounding pixels is assigned to the central noisy pixel.

Normally to remove bit errors due to noise from the true gray values in a digital image, windows of 3 × 3 or 5 × 5 are recommended (Lillesand and Kiefer, 1987). The concept of a moving widow of 3 × 3 pixel neighborhood adapted from Schowengerdt (1983) is explained in **Figure 7.13** and a typical noise suppression algorithm adapted from Lillesand and Kiefer (1987) is illustrated in **Figure 7.14.**

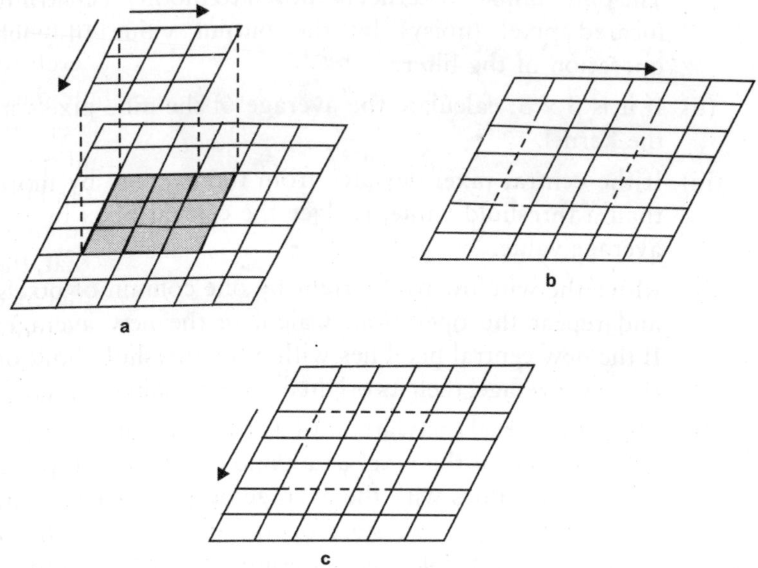

Figure 7.13

DN_1	DN_2	DN_3
DN_4	DN_5	DN_6
DN_7	DN_8	DN_9

$AVE_A = (DN_1 + DN_3 + DN_7 + DN_9)/4$

$AVE_B = (DN_2 + DN_4 + DN_5 + DN_8)/4$

$DIFF = | AVE_A - AVE_B |$

$THRESH = DIFF \times WEIGHT$

IF : $| DN_5 - AVE_A |$ OR $| DN_5 - AVE_B | > THRESH$

THEN: $DN_5' = AVE_B$ OTHER WISE $DN_5' = DN_5$

Figure 7.14

Steps to be followed while moving a window (filter kernel)

The following steps are normally followed while moving an average filter in the image:

(i) A filter kernel is a matrix of odd number of pixels (e.g. 3 × 3 or 5 × 5) in both the column and row dimension. The odd numbered kernel is chosen to modify a centrally located pixel (noisy) by the output value of each operation of the filter.

(ii) If it is 3 × 3, calculate the average of the nine pixels in the kernel.

(iii) If the central pixel deviates from the average by more than a threshold value, replace the central pixel by the average value.

(iv) Move the window to the right by one column of pixels and repeat the operation. Calculate the new average. If the new central pixel lies within the threshold limit of the new average, then its original value remains unaltered.

(v) Move the kernel eastward until its right margin coincides the right end of the pixel array and continue to replace the central values with the average outputs. Then again drop down one row, start operating the kernel from the extreme left. Likewise, continue the operation until the entire image is modified by the application of the moving average filter.

Non-random Noise Removal

Non-random noise (periodic) such as periodic line drop-outs and stripping is caused due to malfunction of the detectors. It is more easily predictable, and hence easier to remove.

Periodic line drop-out

It is caused due to sensor related effects. While scanning, if one of the detectors ceases functioning, then it will result in missing or defective data along a scan line. This type of error is detected on Landsat MSS or TM images in which no data are recorded for every sixth or sixteenth scan line, causing a black line on the image. These are called periodic line drop-outs. Such missing lines are repaired and image line replacement is a simple procedure that allows the operator to fill-in missing lines. This involves two steps calculation: (i) identify the defective lines, and (ii) replace the defective lines. In order to identify, the average DN value for each scan line is compared with the scene average. A scan line is identified as defective if it deviates from the average by more than a threshold value. Then, the pixel values on defective lines (missing lines) are restored by replacing the line with the pixel values of the preceding or succeeding scan line, or multiplying a correction factor with an average of the two. Although the repaired line is a line of artificial data, the output image is a major improvement over the original one.

Periodic line stripping

Scan line banding or stripping effect is one of the flaws or deficiencies noticed in a remotely sensed data in raw form generally in both push broom and whisk broom scanners. The presence of a systematic horizontal banding pattern is due to sensitivity of one detector being higher or lower than the others. This banding effect interferes with the visual appreciation of the patterns and features present on the image. In case of Landsat MSS or TM images every sixth or sixteenth scan line is brighter or darker than the others because of banding effect of a detector. Sensitivity of a detector varies slowly over time. This causes some detectors more or less sensitive to radiance than its neighbors giving rise to stripping or banding effect in images. This effect is more significant with Landsat MSS than with TM or SPOT, and affects dark objects such as water in particular. It is, therefore, necessary to make correction of the image particularly if a linear pattern is prominent in the image. The destripping procedure on scenes should be applied in the initial phase of pre-processing, before rectification to eliminate or reduce the effect by modifying all the data values in the image.

Removal of this effect can be possible by the detector's master calibration curve. A simple method which is widely used in the digital technique is as follows (Bernstein and Ferneyhough, 1975; Bernstein, 1983; Jenson, 1983):

$$Y_k(i,j) = \frac{\sigma}{\sigma_k} \times [X_k(i,j) - M_k] + M$$

where, Y_k (i, j) = Output pixel gray value

X_k (i, j) = Input pixel gray value

M = Mean of the full image

M_k = Mean of the kth detector

σ_k = Standard deviation of the kth detector

σ = Standard deviation of the full image

7.8 IMAGE ENHANCEMENT

7.8.1 Introduction

Once the image has been restored after treatment with required geometric, radiometric and atmospheric corrections, the image can be subsequently used for display and enhancement. The goal of image enhancement is to improve the quality of an image so that a human analyst can easily detect and discriminate objects or patterns in a digital image for visual interpretation. In image enhancement certain transformations are applied to an input image to derive a visually more clear and detailed, more pleasant, and less noisy output image. The transformation usually requires interpretation and feedback from a human evaluator of the output result image. Therefore, image enhancement is considered a difficult task when attempting to automate the analysis process and eliminate the human intervention.

Faust (1989) opined that image enhancement is the process of making an image more interpretable for a particular application. While enhancing, computer modifies the radiometric nature of an image to improve its visual impact. This process makes important features of raw, remotely sensed data more interpretable to the human eye. Enhancement techniques are often used instead of classification techniques for feature extraction—studying and locating areas and objects on the ground and deriving useful information about them from images (ERDAS Inc., 1999).

Enhancement generally distorts the original digital values; therefore enhancement is not done until the restoration processes are completed. Once restoration of a digital satellite image is completed, it can be spatially, spectrally and radiometrically enhanced to visually and mentally identify and interpret obscure or subtle features in the image. In other words image enhancement techniques are developed to improve the visual appearance of the objects on the image. A great variety of digital image enhancement techniques are available. Commonly used image enhancement techniques include image reduction, image magnification; transect extraction, contrast adjustments (linear and non-linear), band ratioing, spatial filtering, Fourier transformations, principal components analysis, and texture transformation (Jensen, 1996). The choice of particular technique(s) depends on the objectives of interpretation, data available, and experience and preferences of the image analysts. This makes improve the perception of information in image, and thus, helps the analyst easier for visual interpretation, and understanding of imagery. In this section we will highlight some procedures which are in general usage in analyzing/interpreting remote sensing images.

To improve the visual contrast of an image on screen, it is not necessary to modify the data file values i.e. the radiance values of the image itself. While performing on screen enhancement, a function called LUT transforms the original DN values (of the original pixel) into brightness values (of the screen pixel) without modifying the original data file values. The function LUT associates the on-screen brightness values corresponding to red, green and blue color guns to the pixels' digital values in an image band. Operating on the LUTs the original image values are preserved; also, the contrast enhancement is quicker and allows further modifications.

Different displays have different ranges of possible brightness values. The range of most displays is 0 to 255 for each color gun. Each screen primary color i.e. red, green and blue has its own look up table or LUT. When three image bands are visualized on screen, each LUT is modified separately to optimize the representation of the corresponding image band. In a B/W visualization, the same band is loaded on all three primary colors on screen; thus the three LUTs are modified in the same way.

Improvement to an image can be divided into two categories: (i) Contrast modification, those that operate on individual pixels without reference to their spatial context, and (ii) Spatial filtering those that also make use of spatial information (Rees, 2001). In order to aid visual interpretation, visual appearance of the objects in the image can be improved by image enhancement techniques such as gray level stretching to improve the contrast and spatial filtering for enhancing the edges. Therefore, there is no selective or particular enhancement technique considered being the best as the human analyst ultimately evaluate and judge the results of the techniques as to whether an applied image enhancement technique is suitable or not.

7.8.2 Contrast Enhancement

Contrast enhancement techniques expand the range of brightness values in an image so that the image can be efficiently displayed in a manner desired by the analyst. In other words, the original input brightness values are expanded to make use of the total dynamic range of the output device. The DN values in a scene are literally pulled farther apart i.e., expanded over a greater range. The effect is to increase the visual contrast between two areas of different uniform densities. This enables the analyst to discriminate easily between areas initially having a small difference in DN value.

In most of the digital image display devices (particularly for 24 bits display), each primary color gun of a pixel will have brightness values ranging from 0-255 (8 bits/pixel) i.e. 256 shades of gray levels. Thus it contains the full dynamic range from minimum value to maximum value. If any image uses this entire intensity scale, from minimum at 0 to maximum at 255, then the image is said to have excellent contrast, covering fully saturated colors extending from black to white. On the other hand, if an image uses a narrow range of numerical values, then the image is said to have poor contrast. Because, the image does not use the full dynamic range from minimum value to the maximum value. To get the contrast, radiometric enhancements are performed on the display device, where data file values are transformed into brightness values.

The planet Earth is gifted with a big variety of spatial features ranging from features of very low radiance (oceans, low solar elevation angles, high latitudes) to that of very high radiance (snow, sand, high solar elevation angles, low latitudes). The remote sensor that collects the spectral signatures must be capable of accommodating such a huge number of scene radiances and it is desirable to have as many bits/pixel possible over this range for precise measurements. This dynamic range is set for the sensor at the design stage, so that while performing the contrast enhancement visual quality of the displayed image can be improved.

The radiance values of any particular scene generally do not cover the full dynamic range but much less than that. Therefore, when a sensor records the radiance values of such a scene, it records the low contrast from the scene itself. When it is displayed in a color monitor (typically 8 bits/pixel in each color or more), the displayed image uses only a fraction of the full dynamic quantisation range (i.e. 0-255 for 8 bits/pixel). Since the displayed image is unable to utilize the full dynamic range, it will result in possessing low contrast. The range of brightness values present on an image is referred to as contrast (C).

Contrast in an image can also be defined as a ratio of maximum DN to minimum DN.

C can also be expressed as, $C = \frac{DN_{max}}{DN_{min}}$

To improve the contrast, the role of image enhancement techniques is tremendous, as such techniques make an image easier to analyze and interpret. Contrast enhancement is a process that makes the image features stand out more clearly by making optimal use of the colors available on the display or output device.

Reasons for low contrast in an image

Most of the satellite images lack adequate contrast and require contrast improvement. Low contrast may result from the following reasons:

1. The individual materials and the background that make up the terrain often reflect similar amounts of energy throughout the visible and near-infrared portion of the electromagnetic spectrum that is recorded by the remote sensing system. In other words, the scene itself has a low contrast ratio.
2. Scattering of the electromagnetic energy by the atmosphere can reduce the contrast of a scene. This effect is most pronounced in the shorter wavelength portions.
3. Besides, cultural factors also result in low contrast image. For example, people in developing countries often use natural building materials (e.g., wood and soil) in the construction of urban areas (Haack et al., 1995). This results in remotely sensed imagery with a much lower contrast as opposed to urban areas in developed countries where concrete, asphalt and fertilized green vegetation may be more prevalent (Jensen, 1996).
4. In addition to above, sensitivity of the detectors is also a factor creating low contrast image. The detectors are designed to record a relatively wide range of scene brightness values (e.g., 0 to 127 or 0 to 255) without getting saturated. Saturation occurs when radiometric sensitivity of a detector is insufficient to record the full

range of intensities of reflected or emitted energy emanating from the scene.

Thus, it is important to consider such factors while enhancing an image for maximum contrast. Images with low contrast ratio are commonly referred to as "washed out" with nearly uniform tones of gray.

There are a number of approaches available in most software packages. Linear and nonlinear digital techniques are two widely practiced methods of increasing the contrast of an image.

7.8.2.1 Linear contrast stretch

The terms linear and non-linear, when describing types of spectral enhancement, refer to the function that is applied to the data to perform enhancement. A linear contrast stretch is a simple way to improve the visual contrast of an image. It is often necessary to enhance the contrast in a raw image, so that it can be seen on the display utilizing the total range of the display device.

Increasing the contrast by expanding the narrow range of brightness values typically present in an input image over a wider range of DN values can be accomplished by linear contrast stretch procedure. It is merely the simplest of a range of possible contrast modification procedures. This uses a simple algorithm, where, the DN values in the original image and the modified image follow a linear relationship. The minimum DN in the original histogram is stretched to extremely black (assigning DN = 0) and the maximum DN in the original histogram is stretched to extremely white (assigning DN = 255), and the remaining intermediate pixel values are distributed linearly between the extremes (0 and 255). The features or details that were obscure or subtle in the original image will stand out clearly in the output contrast stretched image.

The linear contrast stretch or enhancement algorithm greatly improves the contrast of most of the original brightness values (DN), but there is a loss of contrast of the extreme high and low ends i.e. at the tail of the histogram (Sabins, 1986). However, when compared to the overall contrast improvement, the contrast losses in the extremes can be bearable unless the analyst is specifically interested for the information at the extremes.

The linear equation used for linear contrast stretching is given by:

$$Y = a X + b \quad ...(1)$$

where X = input pixel value,

Y = output pixel value

and, a & b are the slope and intercept values respectively. The values of a & b are computed from the equations.

Let X_{min} and X_{max} be the minimum and maximum input DN values and Y_{min} and Y_{max} be the corresponding minimum and maximum output DN values respectively.

Then,

$$Y_{min} = a X_{min} + b \quad ...(2)$$

$$Y_{max} = a X_{max} + b \quad ...(3)$$

by solving the above two equations, we get

$$a = \frac{(Y_{max} - Y_{min})}{(X_{max} - X_{min})}$$

$$b = \frac{(X_{max} . Y_{min} - Y_{max} . X_{min})}{(X_{max} - X_{min})}$$

Thus, putting the values of a & b in the linear equation as above, we will have:

$$Y = \frac{(Y_{max} - Y_{min})}{(X_{max} - X_{min})}(X) + \frac{(Y_{max} \cdot Y_{min} - Y_{max} \cdot X_{min})}{(X_{max} - X_{min})} \quad ...(4)$$

For Y_{min} = 0 and Y_{max} = 255 (as these are usually fixed at 0 and 255 for Landsat data), equation (4) becomes:

$$Y = \frac{(X - X_{min})}{X_{max} - X_{min}} \times 255 \quad ...(5)$$

Linear contrast stretch operation can be represented graphically as shown in **Figure 7.15.**

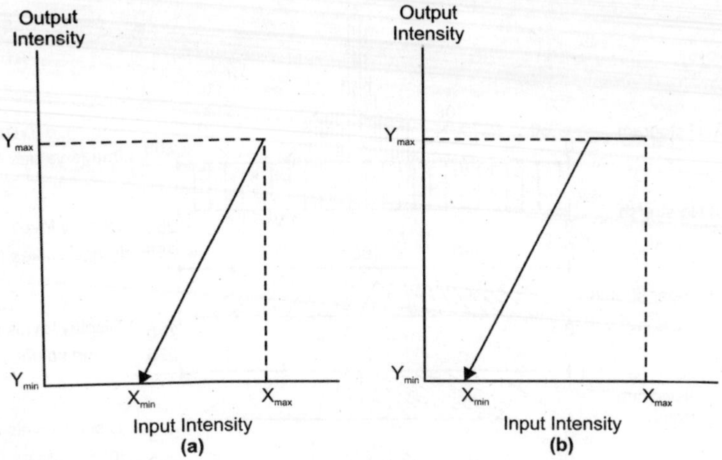

Figure 7.15

The intermediate values will be extrapolated as per equation (5).

The advantage of this digital enhancement technique is that the analyst can enhance any portion of the dynamic range of the histogram to achieve the desired results, since the digital image is flexible. It is generally preferred to expand the digital values ranging from the 5th to the 95th percentile of the distribution, i.e. the tails of the histogram that may contain anomalous values are not used to build the linear stretch.

Linear contrast stretch operation on a digital image has been well explained by Lillesand and Kiefer (1987). The diagram (**Figure 7.16**) explains the underlying fundamental concepts how and why contrast stretching is carried out. Figure 7.16 shows how the contrast stretch manipulates the histogram of the data increasing contrast in some areas and decreasing it in others. The outlines are explained below.

Figure 7.16(a) shows a histogram plot of DN values ranging from 60 to 158 (out of the available dynamic range 0–255) of a georectified digital image. The image is displayed in a hypothetical output device capable of displaying 256 shades of gray levels (0-255). The effective DN range as seen from the histogram is limited to 108-158, as below 108, towards left, there are few pixels of DNs extending up to 68.

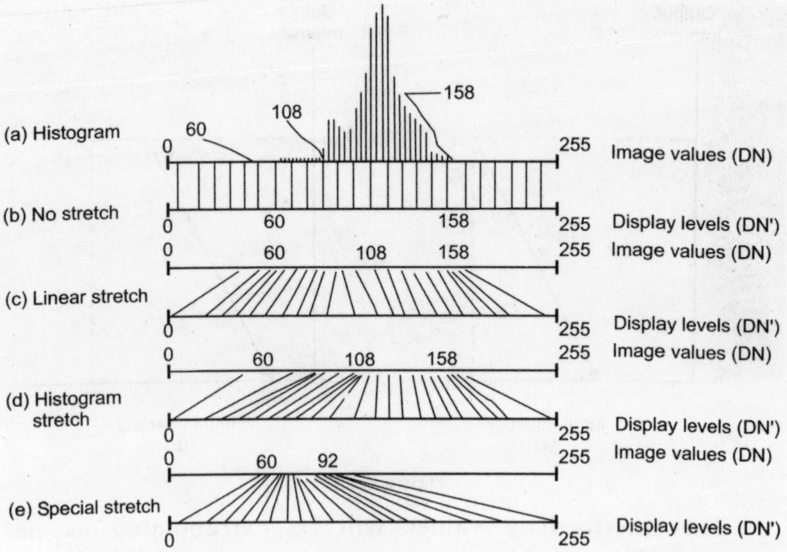

Figure 7.16

Figure 7.16(b) illustrates, display of the image in the device without application of any stretch algorithm. This results in one to one correspondence between the input DN and display brightness values. The range of gray levels is mostly limited to 40 (158 – 118 = 40) DN values and the resulting image is of low contrast as is only using a small portion of the full range of the display levels (0-255). The interpreter is not able to distinguish the radiometric details as the tonal information in the scene is squeezed into a small portion, masking the information content in the ranges 0-59 and 159-255.

For ease understanding, let us examine **Figure 7.17** of Richards (1993), a typical histogram corresponding to a Landsat band 4 image. It shows a peak of slightly scattered, low DN values. A few pixels may have very high values; they correspond to highly reflective surfaces, clouds or sensor errors. In the graph, the histogram image is overlapped by the LUT dotted line, facilitating the interpreter to determine which portion of the image needs to be enhanced by modifying the LUT. The image histogram is generally used as reference to modify the image contrast.

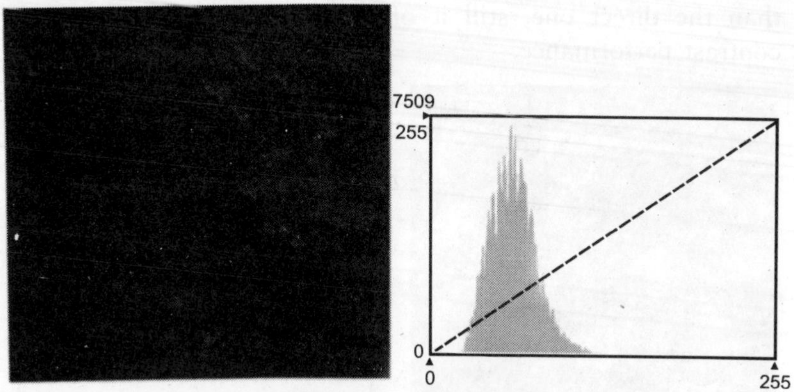

Figure 7.17

The graph that appears to the right is the default look up table and shows that the pixel brightness values in the image (x-axis) are mapped linearly onto the display scale (y-axis) with no change. Thus the image pixel with 0 value is displayed with brightness 0 and digital value of 125 say, will be represented by display value of 125.

Figure 7.16(c) is the case where the standard linear stretch algorithm (explained above) is applied to the image (also refer **Figure 7.18**). The image would able to use the full range of display values (0-255). The image DN 60 is assigned to 0 and the DN 158 is assigned to 255 of the display device. The remaining intermediate image values are distributed proportionately (following equation-1) to fill the total range of the display device. This uniform distribution of the original pixels is called a standard linear stretch. The interpreter is now able to discriminate the subtle variations in input image more readily in the output device. The extreme pixels are highlighted with the expense of the central values. One drawback of this algorithm is that it does not take into account the pixel frequency distribution. It assigns half of the dynamic range of the output device (0-127) to a scarce to absent pixels with image values (60-108). On the other hand, most of the pixies in the input image are concentrated within DNs of 109 to 158. The 109-158 DN range is given the other half stretch of the display values from 128 to 255. The resulting image although is better than the direct one; still it often does not provide the best contrast performance.

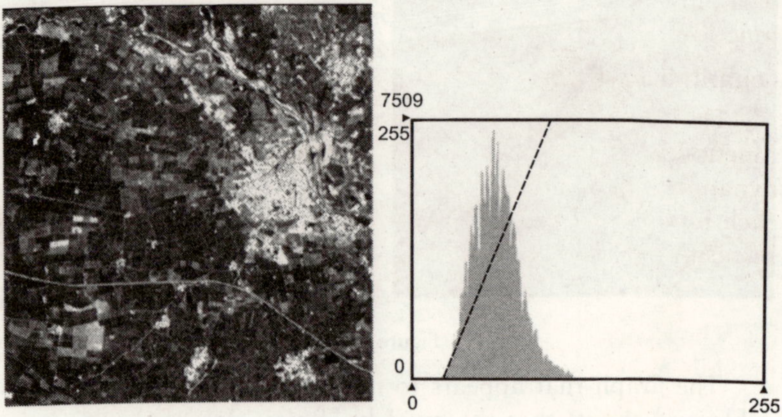

Figure 7.18

7.8.2.2 Non-linear contrast stretch

The linear contrast stretch although is widely used for the improvement of contrast, it is found that there is a loss of contrast of the extreme high and low ends at the tail of the histogram (Sabins, 1986). The non-linear contrast enhancement procedures are therefore thought to be of much useful.

In these methods, the inputs DNs are not mapped linearly onto the output DNs, but are exhibiting nonlinear relationship. Non-linear contrast enhancement is represented generally by y = f(x), a non-linear equation, where x is the input data value and y is the output data value. The non-linear contrast enhancement techniques have been found to be useful for enhancing the color contrast between the nearly classes and subclasses of a main class.

Though there are several non-linear contrast enhancement algorithms available in literature, the use of non-linear contrast enhancement is restricted by the type of application. Good judgment by the analyst and several iterations through the computer are usually required to produce the desired results. Logarithmic and exponential transformation algorithms are used with Landsat data. This enhancement has greatest impact on the brightness values found in the darker part of the histogram. It could be reversed to enhance values in brighter part of histogram by scaling the input data using the inverse log function.

Logarithmic Contrast Enhancement

The image enhancement due to logarithmic transformation function is a non-linear contrast enhancement technique. The dynamic range of an image can be compressed by replacing each input value with its logarithm function. The effect is low intensity pixel values are enhanced. It uses the formula

$$Y = a \log bX$$

The constants "a" and "b" are determined by taking the maximum and minimum gray values of the input image and the corresponding maximum and minimum values in the output image. These are called stretch parameters. The detail solution is very much similar to the linear contrast stretch. This transformation only highlights the features lying in the darker region of the histogram (Hall, 1979; Jenson, 1986).

The constants "a" and "b" are computed from the equations as discussed.

$$a = \frac{(Y_{max} - Y_{min})}{\log X_{max} - \log X_{min}}$$

$$b = \frac{(Y_{max} . \log X_{min} - Y_{min} . \log X_{max})}{(Y_{max} - Y_{min})}$$

where, X = input data value

Y = output data value

X_{max} and X_{min} are the maximum and minimum input data values obtained from the histogram of the scene.

Y_{max} and Y_{min} are the maximum and minimum output data values.

The following are advantages and important characteristics of logarithmic contrast enhancement (Hall, 1978):

(i) It makes the low contrast details more visible by enhancing low contrast edges.

(ii) It provides a contrast signal to noise ratio.

(iii) It somewhat matches the responses of the human visual system.

(iv) It usually provides a more equal distribution of gray values.

(v) It transforms multiplicative noise into additive noise.

Exponential Contrast Enhancement

Both exponential and logarithmic contrasts are useful for contrast enhancement in data that clusters in a small range of a distribution that covers several orders of magnitude. Exponential contrast enhancement function helps to darken the images which are too bright.

$$Y = a\,e^{bX} + c$$

where X represents the input value of a pixel in the raw image, Y is the modified value and a, b and c are constants. The value of b is arbitrarily chosen from 0.01 to 0.1 in order to avoid higher value of e. The constants can be computed from the equations stated above in logarithmic contrast enhancement. The result is in the output image subtle distinctions between shades of gray have become emphasized and, therefore, easier to distinguish by eye. The effect of application of the exponential transformation procedure on the edges in an image is to compress low contrast edges, while expanding high contrast edges. Its effect is just the reverse of the logarithmic transformation procedure.

This technique generally produces an image with less visible detail than the original and is thus of limited use for image enhancement (Hall, 1978).

Gaussian Stretch

Image enhancement techniques are used in maximizing clarity, sharpness and details of features of interest towards information extraction and further analysis. There is no single image enhancement procedure that can be considered for optimal use. The best one is the one that displays the best the features of greatest interest to the analyst.

Gaussian is a statistical term that refers to the normal distribution of values. The discrete variable (DN value) has become continuous and the frequencies have merged into each other without a break (i.e. continuous), which is the characteristic of the Gaussian curve. The distribution of gray values is determined by the mean (μ) and standard deviation (σ), the two constant parameters. The mean (μ) locates the center of the distribution and standard deviation (σ), measures the spread or variation of the individual measurements; in fact, σ is the scale of the DN which is normally distributed. From the **Figure 7.19**, it is seen that within one σ, the frequency is decreasing even more rapidly but beyond that point it decreases at a continuously lesser rate. By the time the pixel value (DN) has reached $\pm 3\sigma$ the percentage frequencies $f_g(X)$ are negligibly small. The frequency of occurrence never vanishes entirely (theoretically), but it approaches zero as X (here, DN value) increases indefinitely. The 2/3 of the observations (DNs) i.e. around 68% lie in the interval $\mu \pm \sigma$, while some 95% of them are in the interval $\mu \pm 2\sigma$. The 99.7% of the data elements are within three standards deviation of the mean ($\mu \pm 3\sigma$). In this procedure, the DNs of the original histogram are to be distributed in the normal curve or Gaussian curve as explained, following an algorithm.

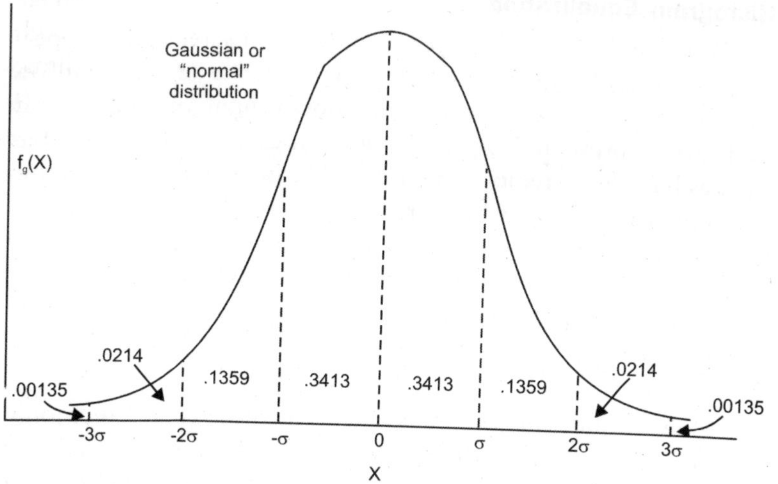

Figure 7.19

A Gaussian stretch is a non-linear stretch that enhances contrast within the tails of the histogram i.e. in the light and dark ranges of the image. It assumes distribution of the pixel values in a bell-shaped curve (Figure 7.19). This method of contrast enhancement fits the observed histogram to a normal or Gaussian histogram between the 0 and 255 limits. Such enhancement occurs at the expense of contrast in the middle gray range.

The algorithm used here is:

$$f_g(X) = \left(\frac{\sigma}{\Pi}\right)^{0.5} e^{-\sigma x^2}$$

where, σ is the standard deviation range of the X for which $f_g(X)$ drops by a factor of $e^{-0.5}$ or 0.607 of its maximum value

and the maximum value $= \dfrac{1}{(2a)^{0.5}}$.

In this method of enhancement each pixel value of input image can be converted to the LUT value based on the probability of each pixel value with respect to a class following the Gaussian law.

Histogram Equalization

Raw satellite images when visualized on screen they appear to be dark and poorly contrasted. As a matter of fact that low contrast images are mostly dark, mostly light or mostly gray and are represented by narrow histograms. High contrast images have large regions of dark and large regions of white and are represented by flat histograms. An interpreter needs a good contrast image that exhibits a wide range of pixel values i.e. no single gray level dominates the image. Histogram equalization provides good contrast images based on equalizing occurrence of gray levels in which there is a good representation of detail at all intensity values.

One of the image enhancement methods that commonly used is "histogram equalization". It is a technique of non-linear contrast stretch used for automatically increasing contrast in an image. Histogram equalization is the process whereby the computer attempts to redistribute the DNs to produce a more easily interpreted image. In this approach, image values are assigned to display levels on the basis of their frequency of occurrence, so that more display levels are assigned to the image values that occur frequently. In this case, contrast is increased at the peaks of the histogram and lessened at the tails.

When two or more images are acquired under different conditions and one wishes to compare those on a specific basis, the first and common approach is to normalize their histograms to a "standard" histogram. The normalization of histograms is done by histogram equalization. In this technique, one attempts to change the histogram through the use of an algorithm into a histogram that is constant for all brightness values. This would correspond to a brightness distribution where all values are equally probable.

The nonlinear contrast stretch method has one major disadvantage. Each value in the input image can have several values in the output image, so that objects in the original scene lose their correct relative brightness values.

Figure 7.16(d) explains histogram equalized stretch in which DN values are transformed into the brightness values of the display device based on their frequency of occurrence. The frequently occurring portion of the histogram, the 109-158 DN range, is stretched to 39 to 255, which is a large portion of the display level. However, the infrequently occurring DNs from 60 to 108 are spread over differently to a smaller portion of the display level (0-38). The result is a histogram-equalization stretch.

Figure 7.20 illustrates a transformed histogram and the corresponding stretched image. While comparing the original histogram with the transformed one, it is observed that the transformed histogram is quite uniform in the average and flatter than the original.

Computation of transformation function

Gonzalez and Wintz (1977) demonstrated the technique for modification of original pixels using cumulative frequency distribution of the original histogram.

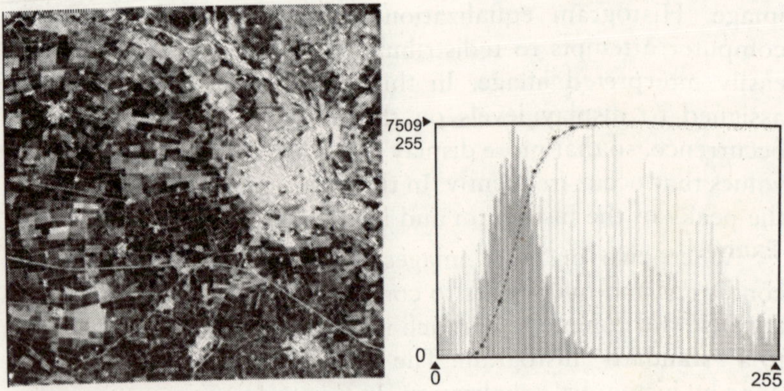

Figure 7.20

Steps involved in Computation

(i) Determine the probability distribution of gray levels i in the input image by dividing n_i the number of occurrences of gray level (frequency) i by n being the total number of pixels in the image:

$$p(x_i) = \frac{n_i}{n}, \; i \, \varepsilon \, o, \, ..., \, ...L - 1$$

where, L = total number of gray levels in the image

p = image's histogram, normalized to 0.1.

(ii) Compute the transformation function (cumulative probability function corresponding to (p) for each of the individual DN value defined by:

$$c(i) = \sum_{j=0}^{i} P(x_j)$$

where c is the image's accumulated normalized histogram.

(iii) Normalize the DNs between 0-1 by dividing the minimum value by the maximum in each item.

(iv) The values of c(i) will have to be scaled up by 255 and rounded to the nearest integer so that the output values of this transformation will range from 0 to 255. To do that, compare each of the transformation function c with each of the scaled DN (0-1) and assign each c a value of DN close to it. Thus the discretization and rounding of c(i) to the nearest integer will mean that the transformed image will not have a perfectly uniform histogram.

(v) Stretch the transformation function linearly to obtain the stretched/new DN value.

Example-1

Demonstrate histogram equalization technique to enhance a hypothetical digital image of dimension 64 × 64 pixels having DN values 0,1,2,3,4,5,6 and 7.

Follow the steps as mentioned above.

Probability function and transformation function are calculated following steps (i) and (ii) and are given in **Table 7.2.** After following (iv) through (iii), it is found that only five distinct histogram equalized gray values are prominent and are presented in **Table 7.3.**

TABLE 7.2: Results of Histogram Equalization Mapping

Original Gray values	Scaled value	Frequency n	Probability P(xi)=ni/n	Transformation Function c(i)
0	0	790	0.19	0.19
1	0.14	1023	0.25	0.44
2	0.28	850	0.21	0.65
3	0.42	656	0.16	0.81
4	0.57	329	0.08	0.89
5	0.71	245	0.06	0.95
6	0.85	122	0.03	0.98
7	1.00	81	0.02	1.00

TABLE 7.3: Output of Histogram Equalization Mapping

New Gray value	New Frequency	Probability Function	Scaled Gray Between 0-255
0.14	790	0.19	0
0.42	1023	0.25	83
0.71	850	0.21	169
0.85	985	0.24	210
1.00	448	0.11	255

Source: Hall, 1978.

Example-2

Figure 7.21 represents the histogram of the original image. It contains 240 pixels and 10 bins or bars (0-9). The pixel values will be reassigned to certain number of bins. The pixels are then given new values, based upon the bins to which they are assigned. Then,

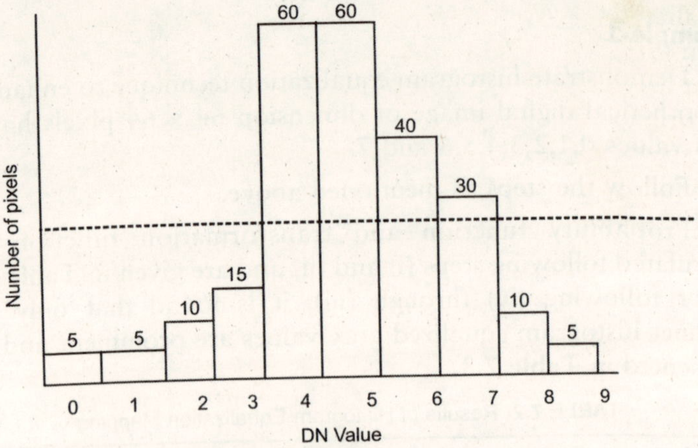

Figure 7.21

A = equalized number of pixels per bin = 240/10 = 24

To assign pixels to bins, the following equation is used:

$$B_i = \text{int}\left[\frac{\left(\sum_{k=1}^{i-1} H_k\right) + \frac{H_i}{2}}{A}\right]$$

where:

A = equalized number of pixels per bin

H_i = the number of values with the value i (histogram)

int = integer function (truncating real number to integer)

B_i = bin number for pixels with value i

Source: Modified from Gonzalez and Wintz (1977).

The input DN values range from 0-9. Therefore, the 10 bins are rescaled to 0-9 in order to compare the equalized histogram with the original.

The enhanced image corresponding to the equalized histogram (**Figure** 7.22) will gain in contrast in the peak area of the original histogram. Figures depict the closely packed bins from 3 to 7 in the original image is stretched to 1 to 8 in the equalized histogram. The data file values in the original histogram from 0 to 2 in the extreme left and from 8 to 9 in the extreme right are grouped together in Figure 7.22. It is observed that in the original image DN values of 5, 5, and 10 pixels are 0, 1 and 2 respectively. However, in the equalized histogram, because of grouping, those 20 pixels are grouped together and are assigned with the output value of 0. Similarly, input values 8 and 9 both have the output value of 9.

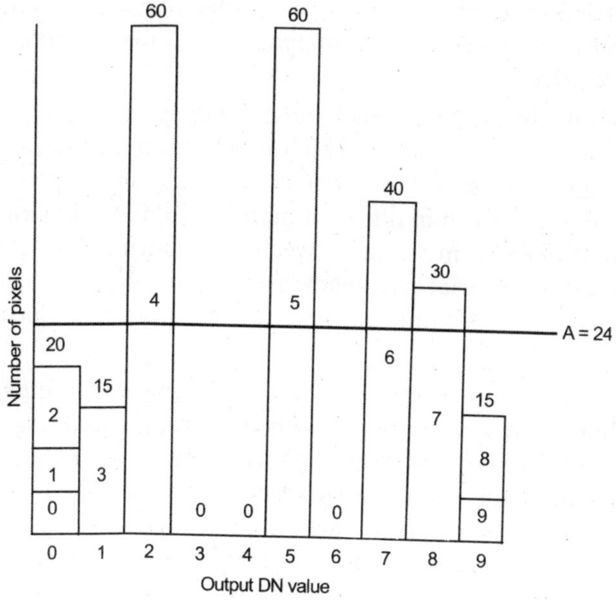

Figure 7.22

From the above contrast manipulation, it is on the other hand seen that the contrast in the darkest and the brightest regions at the tails of the input image is lost.

7.9 SPATIAL FILTERING

7.9.1 Introduction

An analyst interacts with the satellite image using different image processing techniques principally to identify different objects and their associationships to make an assessment about a phenomenon or process. In order to meet the specific requirements, there are large numbers of functions performed on a digital data. While radiometric enhancement operations are based on each individual pixel (called point operation), there are spatial enhancement techniques that modify pixel values based on the neighboring pixels (i.e. area operation).

Spatial enhancement is the mathematical processing of image pixel data to emphasize spatial relationships. It modifies pixel values based on the DN values of the surrounding pixels unlike spectral enhancement based on modifying pixel values in an image to enhance certain feature over others. Spatial enhancement is, therefore, defined as "enhancing DN values of pixels in an image relative to surrounding pixels".

Spatial enhancement technique deals largely with spatial frequency within an image, which is the difference between the highest and lowest values of a contiguous set of pixels. The definition is put forth in different manners in different situations. It is the difference in value between neighboring pixels. Jensen (1986) defines spatial frequency as "the number of changes in brightness value per unit distance for any particular part of an image".

Lillesand and Kiefer (1987) refer spatial frequency as "roughness" of the tonal variations occurring in an image. If the frequency of brightness values in a segment of an image is zero, then it will result in no change in contrast due to similar value of all pixels [**Figure 7.23 (a)**]. If an image contains smooth variation in gray tone due to low spatial frequency, then it will result in low contrast [**Figure 7.23(b)**]. However, an image is said to have high contrast (high spatial frequency) if it contains an abrupt change in pixels over a relatively small number of pixels (very short distances) [**Figure 7.23(c)**]. High frequency areas in an image are regarded as rough whereas, low frequency areas are smooth as gray values change only gradually over a relatively large number of pixels [(Figure 7.23(b)]. It is a common experience that the spatial objects when imaged, found that it contains a combination of both high and low spatial frequencies, i.e. it contains generally features of rapid change in DN such as, water body, urban, vegetation, etc.

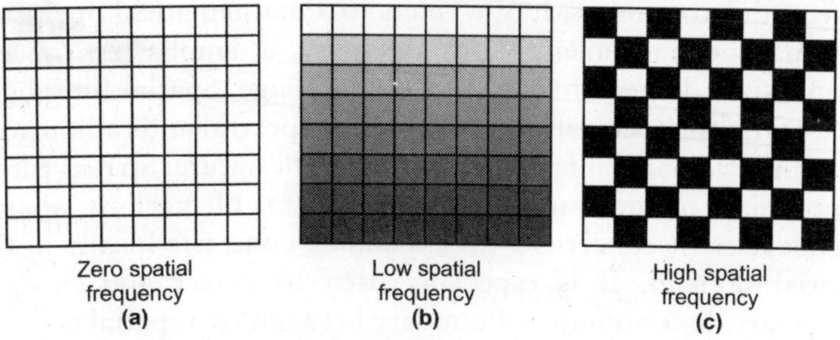

Figure 7.23

The concept behind spatial filtering is that an image is constructed from spatial components at different scales. To understand the concept, let us consider an image and filter it with a small window of size 3 × 3. Replace each pixel with an output value taking the average of the neighboring pixels under the window. The resulting output image is a blurred image of the original. Such a blurred image we call it as Low Pass version of the original. A simple high pass filter may be implemented by subtracting a low pass filtered image from the original. Now if we subtract each output pixel value (of blurred image) from the corresponding input value (of original image), the resulting image will be an array of pixels, each representing the difference between the original and the average values of its neighboring pixels. Such a resulting image we call it as High Pass version of the original image. From the above concept, mathematically we can relate the both:

$$\text{Image}(x, y) = \text{Low Pass Image}\ (x, y) + \text{High Pass Image}\ (x, y)$$

From the above it is clear that if we increase the size of the window say from 3 × 3 to 5 × 5, 7 × 7 and so on, then Low pass image isolates successively larger and larger structures, while the High pass image picks up the smaller structures lost in the low-pass image in order to maintain the above relation. This decomposition of an image into a sum of components at different scales is the basis of all spatial filtering (Schowengerdt, 1997).

Filtering is a process meant to enhance certain features or to remove unwanted signal or disturbance (noise) or to remove the effects due to sensors. It also refers to a mathematical operation of image enhancement when operates, it emphasizes or de-emphasizes degree of roughness on an image. Spatial filtering is one of the enhancement procedures like spectral and radiometric revealing valuable information of different nature. Spatial filters emphasize or de-emphasize image data of various spatial frequencies. It concerns with expanding contrasts locally in the spatial domain. It is especially used to detect and sharpen boundary discontinuities. Boundary between two spatial features in the real world is regarded as the separation of abrupt changes in DN value on either side of it. Such a boundary can be emphasized in performing any such filtering techniques. The output image appears quite distinctive.

Spatial filtering technique like convolution operates only within a relatively small number of neighboring pixels to modify or transform the candidate pixel. However, technique like the Fourier transformation can process the full image for global spatial structure. The objective of spatial filtering is to enhance areas of high or low spatial frequency. It involves pixel by pixel transformation moving a window of few pixels in dimension (e.g. 3 × 3, 5 × 5, etc.) over each pixel in the image. The DN value of the central pixel is replaced with the new value calculated mathematically using the pixels under that window. The window starts moving from the extreme top left corner of the input image. It first moves along the line one pixel at a time. Soon it completes the line; it restarts again for the next line. This process continues till it filters the entire image to generate the output image. Therefore, spatial filtering is regarded as "context dependent" operation as it alters the gray level of a pixel according to its relationship with the gray levels of the other pixels in the immediate neighborhood.

7.9.2 Convolution

A linear spatial filter is a filter for which the DN at any location in output image is a function of some weighted average of DN located around that location in the original image. The process of evaluating the weighted neighboring pixel values is called a two dimensional convolution.

Convolution is a simple mathematical operator which takes two functions, modifying one by another to produce a third (output) function. In spatial filtering this implies the operation of a filter (one function) on an input image (another function) to produce a filtered image or convolved image (the output). It is fundamental to many common image processing operators. Statistically, two matrices of numbers having equal dimensions and generally of different sizes are multiplied together to produce a third matrix of same dimension. This can be used in image processing to implement operators whose output pixel values are simple linear combinations of certain input pixel values. Under this process the image is divided into its constituent spatial frequencies, and selectively altering certain spatial frequencies to emphasize some image features. Convolution filtering is used to change the spatial frequency characteristics of an image (Jensen, 1996).

An Example

A brief account of how a convolution modifies an input image into an output image is explained below:

(i) The input image can be fragmented into a set spatial components or compartments, each of which can be viewed as a scaled and shifted function.

(ii) The output resulting from each component is a scaled and shifted version of the convolution kernel or filter kernel or simply kernel.

(iii) The overall output image can be found by adding these scaled and shifted kernels. In other words, if we are given with a filter kernel, then we can calculate what the output will be for any possible input image. In digital image processing, the kernel is called the point spread function.

Convolution is nothing but a mathematical operation. It is a combination of operations namely multiplication, addition, and integration. It takes two functions such as input and kernel and produces a third function, the resulting output image. Convolution explains the relationship between the input image, the kernel or window, and the output image in linear spatial filtering.

Convolution Formula

The following formula is used to derive an output data file value for the pixel being convolved (in the center):

$$V = \left[\frac{\sum_{i=1}^{q} \left(\sum_{j=1}^{q} f_{ij} \, d_{ij} \right)}{F} \right]$$

where:

f_{ij} = the coefficient of a convolution kernel at position i, j (in the kernel)

d_{ij} = the data value of the pixel that corresponds to f_{ij}

q = the dimension of the kernel, assuming a square kernel (if q = 3, the kernel is 3 × 3)

F = either the sum of the coefficients of the kernel, or 1 if the sum of coefficients is 0

V = the output pixel value

In cases where V is less than 0, V is clipped to 0.

Source: Modified from Jensen, 1996; Schowengerdt, 1983.

Using the above formula, the input image can be convolved with the kernel to provide the output image.

Convolution can be explained based on the followings viewpoints:

(i) Each pixel in the input image contributes many points in the output image.

(ii) Each pixel in the output image has received information from many points in the input image.

Suppose that we are given some original image F (i, j) and kernel K, and want to find the convolution of the two. The output image can be given by:

$$Y(m,n) = \sum\sum F(i,j) \times K$$

where, F (i, j) = Input pixel array of an image

K = Convolution kernel

Y (m, n) = Output image pixel array

The above equation tells that, the output image is equal to some combination of the many values in the input image and kernel. Each pixel in the output image is equal to a sum of weighted inputs. Each output pixel is influenced by a region of pixels in the input image, as determined by what the weighing coefficients are chosen to be. For example, if there are nine weighing coefficients, each with a value of one-ninth, this makes each pixel in the output image the average of nine pixels from the input.

The operation of a kernel over an input image involves the following steps:

(i) A kernel or a moving window is an array of numbers; each called a coefficient or weighing factor. Kernels are normally pixels of odd dimensions e.g. 3 × 3, 5 × 5, 7 × 7. Initially a kernel is fixed.

(ii) The kernel is laid over the input image and moved throughout. Each DN of the output can be obtained by multiplying each kernel element by its corresponding input element and adding all the resulting products.

1/9	1/9	1/9
1/9	1/9	1/9
1/9	1/9	1/9

(a) Kernel

67	67	72
70	68	71
72	71	72

(b) Original image DN

	70	

(c) Convolved image DN

Figure 7.24

Concept of convolution. Shown is a 3 × 3 pixel kernel with all coefficients equal to 1/9. The central pixel in the convolved image contains the average of the DNs within the kernel (Lillesand and Kiefer, 1987).

Convolution as calculated:

1/9 * (67) + 1/9 * (67) + 1/9 * (72) + 1/9 * (70) + 1/9 * (68) + 1/9 * (71) + 1/9 * (72) + 1/9 * (71) + 1/9 * (72) = 630/9 = 70 (the central value)

If an original image contains M rows and N columns, and that of a kernel has m rows and n columns, then the resulting output image will be a matrix of size given by (M – N + 1) rows, and (N – n + 1) columns.

Convolution can be used to implement many different operators, particularly spatial filters and feature detectors. Examples include Gaussian smoothing and the Sobel edge detector.

7.9.3 Types of Filters

A digital image contains both low and high frequency spatial information. Sum of both constitutes the original image. Basically there are three different kinds of filters. They are:

(i) Low pass filters, (ii) High pass filters, and (iii) Band pass filters.

Low Pass Filters

Different filtering techniques are used to increase subjective image quality by sharpening certain image features viz. edges and boundaries and by attenuating noise. Filtering provides a means of improving images by suppressing or enhancing certain spatial frequencies, directions and textures (Rosenfeld and Kak, 1976). Low pass filters usually enhance a degraded image because the image signal tends to be concentrated in the low frequencies while background noise is spread across all frequencies.

A low pass filter is transparent to low frequency data i.e. it retains low frequencies unaltered but suppresses high frequency details. It allows low tonal variation data i.e. the DN values does not change much from pixel to neighboring pixel, to pass through, removing the high tonal variation data, i.e. data that changes rapidly from pixel to neighboring pixel. Since it blocks high frequency details correspond to sharp intensity changes (e.g. noise), and defines smooth areas like forest cover, lava field or water of an image it is regarded as smoothing filter. This technique brings out gradual changes in brightness values over a large number of pixels. In this case high frequency details are suppressed while preserving the low frequency information. While removing noise, this technique, reduces gray level range, which has to be taken care of by rescaling. The effect of high pass and low pass filtering can be well understood when a drainage network is enhanced by filtering. Low pass filtering enhances major tributaries, while high pass filtering emphasizes high frequency brightness values resulting enhancement of minor tributaries and drainage pattern. An image containing lot of noise, a low pass filter would smooth out the image and reduce the noise with minimal affect on large features in the image.

The mathematical operation the low pass filter performs in order to achieve these effects is to replace each pixel in the image with the average of all the neighboring pixels.

Low pass filters are used in remote sensing for suppressing or eliminating and enhancing large features. This can be accomplished by choosing a filter size based on the size of the details to be eliminated. The degree of smoothing is a function of the size of the kernel and the weighting factors used in collecting statistics within the kernel. As the filter size increases, smoothing increases and the image becomes more homogeneous or blurred. In image processing, low pass filters can serve to remove noise, reduce resolution (without reducing the image size), or subdue details in one step of a procedure.

There are number of low pass filters: those are mean filter, median filter, and adaptive filter. Low pass filtering is used by Crippen (1989), Eliason and Mc Ewen (1990) to remove banding effects on remotely sensed images. Dale et al. (1996), used a low pass filter to smooth the image in the process of registration. They used the technique of mean filtering where each pixel is sequentially examined and if the DN value of a pixel is greater than the average DN of its surrounding pixels by some threshold; it is replaced by the mean of the 3 × 3 kernel.

The kernel can be of any size such as 5 × 5, 7 × 7 and so on. Time required for computation will be more for larger sized kernels.

A three by three kernel substitutes the central pixel of the kernel with a simple measure of digital number variance (range, kurtosis, standard deviation) as it passes over the image (Haralick et al., 1973; Logan et al., 1979). These have proved to be particularly useful for a number of operations like the smoothing out of over enlarged discrete images to make them appear continuous; the suppression of banding that is common to both Landsat MSS and TM images and the suppression of "speckle" in radar images (Lee, 1981; Bruniquel et al., 1997; Gineste, 1999; Frulla et al., 2000; NdiNyoungui et al., 2002; and Dekker, 1998). Likewise, filters have been used for image sharpening or edge enhancement (Short, 1982), geological mapping (Thomas et al., 1981), terrain analysis (Weszka et al., 1976; Shi and Schowengerdt, 1983), land cover mapping (Hsu, 1978; Irons and Petersen, 1981), forest mapping (Logan et al., 1979), urban classification (Jensen, 1981) and the differentiation of sea ice (Gersen and Rosenfeld, 1975) and clouds (Harris, 1977).

The low pass filter calculates the average value for each 3 × 3 neighborhood. The effect is that the high and low values within each neighborhood will be averaged out, reducing the extreme values in the data. An example of simple low pass filter of 3 × 3 which decreases spatial frequency is given below:

1	1	1
1	1	1
1	1	1

The value of 1 in each cell of the neighborhood indicates the use of each cell in the calculation. This kernel simply averages the values of the pixels causing them to be more homogeneous. The resulting image looks either smoother or more blurred.

Calculation

Let us consider an example of how a kernel operates on a spatial component of a digital image. In this case, the processing cell let it be a_5, which is at the center.

$$\begin{matrix} a_1 & a_2 & a_3 \\ a_4 & a_5 & a_6 \\ a_7 & a_8 & a_9 \end{matrix}$$

The calculation for the processing cell (central pixel) is given by the average of the input cells. It is the sum of all the values in the input contained by the neighborhood, divided by the number of cells in the neighborhood (3 × 3 = 9):

$$\text{Output value for } a_5 = (a_1{*}1 + a_2{*}1 + a_3{*}1 + a_4{*}1 + a_5{*}1 + a_6{*}1 + a_7{*}1 + a_8{*}1 + a_9{*}1) \div (3 \times 3 = 9)$$

The central value a_5 will be replaced by the output (average) value. From this simple calculation, note that, since the average is being calculated, a_5, the highest value at the center will be the average value.

High Pass Filters

High pass filters are defined as the spatial filters that pass high frequencies emphasizing the details of an image. That is, these filters keep high frequency information, which corresponds to areas of rapid change in brightness, such as edges or fine textures. The functioning of a high pass filter is essentially exactly the same as that of a low pass filter; however, in this case, it allows high frequencies to pass through while attenuating or eliminating low spatial frequencies. That is, it enhances areas that have high spatial frequencies (rapid changes in DN over a short distance). This is on the other hand, a method of detecting rapid changes in gray levels enhancing high frequency component of a digital image. These filters are useful for detecting and enhancing lines and edges (e.g. roads, land water boundaries), or sharpening an image. Since they enhance edges without affecting the low frequency areas of the image, they are sometimes referred to as sharpening filters or edge enhancement filters. Small kernels will enhance small features and details. Large kernels will allow large features to pass through, suppressing or eliminating smaller features.

The advantage in using these filters is that we are not losing the overall image information.

A high pass filter can be defined by the kernel as:

-1	-1	-1
-1	c	-1
-1	-1	-1

where $c \geq 8$

An increase in the value of c will decrease the high pass effect, as the central coefficient will dominate.

An Example

Below given a high frequency kernel of 3×3, where $c = 16$ operating on an array of pixels in which (i) a relatively low value at the center surrounded by its neighborhood of higher values, and (ii) relatively higher value at the center surrounded by the lower values at the neighborhood.

Case (i): After kernel operation, the central value which is lower still gets lower.

(Before)			(After)		
204	200	197	204	200	197
201	106	209	201	9	209
198	200	210	198	200	210

Case (ii): After kernel operation, the higher value gets higher value.

(Before)			(After)		
64	60	57	64	60	57
61	125	69	61	187	69
58	60	70	58	60	70

Source: ERDAS Inc., 1999.

Average Filter

The Convolution Filters are used for sharpening as well as blurring of an image, and detecting edges within an image, or other kernel-based enhancements. There are five methods of convolving an image. Averaging is one such convolution methods of filtering. This technique is based on the fact that a candidate pixel (pixel under consideration) is replaced with the average (mean) value of the neighboring pixels within the kernel if its DN value is greater than its neighboring average by a threshold value. The template or window may be of any size such as 5 × 5, 7 × 7 and so on.

Below given a kernel sample, where the central pixel is assigned with the value X, which has no correlation with the neighboring pixels. It will be then replaced by the neighboring average.

$$\begin{matrix} a_1 & a_2 & a_3 \\ a_4 & X & a_5 \\ a_6 & a_7 & a_8 \end{matrix}$$

$$X = \frac{1}{8} \times \sum_{a_{i=1}}^{8} a_i$$

$$\text{If } x - \frac{1}{8}\sum a_i > t,$$

where x = DN value of X and t = threshold value

The general equation for averaging filter is given by:

$$y^{(i)} = \frac{1}{M}\sum_{j=0}^{j-1} x^{(i+j)}$$

where, $x^{(i+j)}$ is the input pixel value, $y^{(i)}$ is the output pixel value, and M is the number of pixels used in the moving average.

In spite of its simplicity, the moving average filter is optimal for a common task: reducing random noise while retaining the sharpness. However, it is the worst filter for frequency domain encoded signals, with little ability to separate one band of frequencies from another.

Median Filter

The median filter is a non-linear sliding-window spatial filter. Median filter may be used when the objective is to achieve noise reduction with a minimum amount of blurring. These are non-linear filters. This is performed using a window of odd number of pixels such as 3 × 3, 5 × 5, etc.

The central value in the window is replaced with the median value of all the pixel values in the window. This is applicable when the central value differs more than a certain value called a threshold value that determines whether or not to modify a pixel. Median value is calculated by arranging the DNs in the kernel in an ascending or descending order using the simple statistical formula:

$$M = \text{size of } \frac{(N+1)}{2} \text{ th item}$$

where M = Median, N = Number of items in the series.

The calculation repeats for all the values in the input image to get the kerneled output image. In the Figure median value of a pixel neighbor is calculated. The central pixel seems to be abnormal when compared to its neighboring pixels in the window 3 × 3. This central pixel (160) is not the representative of the surrounding pixels and therefore is replaced with the median value 126.

125	124	124	135	140
120	**128**	**126**	**124**	133
120	**125**	**160**	**122**	130
140	**120**	**135**	**130**	128
130	120	125	124	122

Neighborhood values arranged in the order are:

120, 122, 124, 125, **126**, 128, 130, 135, 160

In the kernel sample, the central value 160 will be replaced by the median value of its neighborhood i.e. 126.

Median filtering is used to remove the impulsive noise from an image (Baxes, 1994; Haralick, 1992; Yin et al., 1996). It is also useful in cases where edge blurring is not required as it claims sharp edge preserving.

Edge Enhancement Filters

An edge we mean the boundary between objects or parts of objects in images. Change in contrast usually occurs between features. An edge can be a boundary that separates two different features or can be a line that differs from the feature on both its sides. Sudden change in DN value from one pixel to another, usually indicate a boundary or edge between the two adjacent objects in an image. It is a transition in the gray value of an image. A digital image is composed of array of pixels or resolution elements which have finite number of dimensions tends to blur edges such as faults, roads, land use boundaries, etc., the transition of sudden change in DN value. Therefore, in order to overcome such blurring effect in the image, edge sharpening is necessary.

Edge enhancement is a technique used to delineate edges surrounding objects such as rock joints and faults, field boundaries, and street patterns which are characterized by abrupt discontinuities. It is used to map lineaments and drainage pattern. It makes a digital image looks sharper than its original. However, it actually makes the image less detailed. High pass filtering can enhance these edges or boundaries. The edge crispening filtering technique is typically employed to exaggerate the edges between contrasting cover type or bring out linear trend of geologic significance (Pratt, 1978; Sabins, 1986). In this technique, high pass filter is used across the image. Sharp change in the gray value at a particular pixel point indicates features of the built environment, roads or field boundaries.

Template Operators

Enhancement of edges or lines of separation between features in an image leads to image sharpening. It is accomplished by defining a template or a window and operating over the image row by row and column by column. The input image can be modified radiometrically by replacing its array of pixel values with the calculated brightness values due to template operating over the image. The new brightness values are calculated by summation of the products of the brightness values covered by the template at a particular location and the corresponding convolution entries using the following equation:

$$r(i,j) = \sum_{1}^{M}\sum_{1}^{N} \phi(m,n)\,t(m,n)$$

where φ (m, n) = Input pixel brightness value addressed according to the template position

r (i, j) = Output value for image pixel (i, j) for an M × N pixel sized template

t (m, n) = Template entry at that location.

An Example

The templates commonly used are:

t =

0	-1	0
-1	5	-1
0	-1	0

(or)

t =

1	-2	1
-2	5	-2
1	-2	1

(or)

t =

-1	-1	-1
-1	9	-1
-1	-1	-1

If φ (m, n) and t (m, n) are defined as follows, then the output r (i, j) can be calculated using the above formula:

φ =

a_1	a_2	a_3
a_4	a_5	a_6
a_7	a_8	a_9

t =

0	-1	0
-1	5	-1
0	-1	0

x	x	x
x	r	x
x	x	x

Then, $r = 5\ a_5 - (a_2 + a_4 + a_6 + a_8)$ and input pixel a_5 will be replaced by the calculated brightness value r. It is important to note that the sum of the elements in the template or window used here is 1, whereas in other high pass filters like Laplacian, the sum is zero. Here since the sum of the entire coefficient in the mask matrix is equal to unity, the digital number would have been retained at the same value as in the original, while the exchange takes place at the edges (Galbaiti, 1990).

Laplacian Filter

The Laplacian filter differs from the other edge enhancement filters as it uses second derivative information about the intensity changes in an image through a difference equation. It measures the rate at which the first derivatives change. Therefore, it is a second order or second derivative method of enhancement. This helps to determine if a change in adjacent pixel values is an edge or a continuous progression. It can emphasize fine detail in an image. Any feature with a sharp discontinuity can be enhanced implementing such a filter. Thus, one application of a Laplacian operator is to restore fine detail to an image which has been smoothed to remove noise.

The Laplacian operator usually contains negative values in a cross pattern with the central value either negative or positive. It is implemented by a number of different filter masks. Three commonly used discrete approximations to the Laplacian filter are:

0	-1	0
-1	4	-1
0	-1	0

-1	-1	-1
-1	8	-1
-1	-1	-1

-1	0	-1
0	4	0
1	0	1

Input image

a_1	a_2	a_3
a_4	X	a_5
a_6	a_7	a_8

Filter or mask

0	-1	0
-1	4	-1
0	-1	0

Output image

a_1	a_2	a_3
a_4	Y	a_5
a_6	a_7	a_8

In the process of convolving an input image (as shown above), the 3 × 3 kernel is placed at the center on each pixel (X). The central pixel value (X) is replaced by new pixel value Y as follows:

The central pixel (Y) of the output image = $4X - (a_2 + a_4 + a_5 + a_7)$.

It is observed that the algebraic sum of the kernel coefficients in each of the three Laplacians is zero. Therefore, it is a type of high pass filtering. If this mask is operated over an image containing pixels of same gray values, then the resulting image will be black as the sum of product of each pixel and its mask value will yield zero.

Sobel Filter

Sobel filter is a standard non-linear edge enhancement operator described by two templates (X and Y). This modifier emphasizes pixels that are closer to the center of the mask. It is a discrete differentiation operator that creates an image through edge detection. Only a 3 × 3 filter size can be used with this filter. The operator calculates approximate rate of change in DN values at each point of the image (intensity gradient). This demonstrates the direction of the largest possible increase from light to dark and the rate of change in that direction. That is it shows the part of the image that represents the edge as well as how the edge is likely to be oriented. If the Sobel operator applies to an image point in a region of constant DN value, then, the gradient vector will be zero. However, at a point on an edge, the vector direction will be across the edge from darker to brighter values.

Each pixel of a digital image represents a discrete intensity point. However, there is no continuity in intensity function as the intensity values in a digital image are available only at the discrete points. Therefore, to calculate the intensity gradient and its derivatives, it is assumed that there is an underlying continuous intensity function which has been sampled at the image points. Although it is inaccurate approximation of the image gradient still is of sufficient quality to be of practical use in many applications.

This filter uses two 3 × 3 templates to calculate the Sobel gradient value as shown below:

-1	0	1
-2	0	2
-1	0	1

X

1	2	1
0	0	0
-1	-2	-1

Y

a_1	a_2	a_3
a_4	a_5	a_6
a_7	a_8	a_9

Input image

$$X = -1^* a_1 + 1^* a_3 - 2^* a_4 + 2^* a_6 - 1^* a_7 + 1^* a_9$$
$$Y = 1^* a_1 + 2^* a_2 + 1^* a_3 - 1^* a_7 - 2^* a_8 - 1^* a_9$$

At each point in the image, the resulting gradient approximations can be combined to give the gradient magnitude, using:

$$\text{Sobel Gradient} = \sqrt{(X^2 + Y^2)}$$

Edge Detection Filters

Edge detection is a filtering technique of highlighting intensity points in a digital image at which there is a marked change in gray values. It is a type of digital image analysis wherein a report in the form of an edge map is extracted. Edge detection operator preserves the important structural properties of an image by filtering out significantly irrelevant information.

Edge detection concerns the localization of significant variations of the gray level image and the identification of the physical phenomena that originated them. This information is very useful for applications in 3D reconstruction, motion, recognition, image enhancement and restoration, image registration, image compression, and so on. Usually, edge detection requires smoothing and differentiation of the image. Differentiation is an ill-conditioned problem and smoothing results in a loss of information. It is difficult to design a general edge detection algorithm which performs well in many contexts and captures the requirements of subsequent processing stages. Consequently, over the history of digital image processing a variety of edge detectors have been devised which differ in their mathematical and algorithmic properties (Ziou and Tabbone, 1998).

There are many ways to perform edge detection. They are mostly grouped into two categories: (i) Gradient and (ii) Laplacian. The gradient method is a search-based method detects edges by looking for maxima and minima in the first derivative of the image. The Laplacian method is a zero-crossing based method searches for zero crossings in the second derivative of the image in order to find edges.

Chavez Filter

Filter Chavez belongs to high pass filtering category. This modifier when operates over an image it highlights or emphasizes (passes) high frequency features suppressing or filtering out low frequency changes in data values. Chavez filter can be used to extract or emphasize areas of rapid change in brightness values from a pixel to its neighbors, such as edges or fine textures and linear features. If the image contains noise, this edge detection filter will emphasize the noise. However, it is advised to use a low pass filter to remove the noise effects before convolving edge enhancement filters.

The default Chavez filter is given by:

-1	-1	-1
-1	17	-1
-1	-1	-1

Chavez Filter of 3 × 3

Each cell in the default filter carries 1 as weightage. The new cell value is calculated as 1/9 times the average of the weighted values in the window.

$$x = \frac{1}{9} \times \left(\frac{\sum (x_i \times W_i)}{n} \right)$$

where x_i = cells in the window

W_i = weight assigned to cell by the window

n = number of cells in the window

Roberts Filter

Roberts filter is one of the basic edge detection routines. This operator can highlight the edges between the features. It focuses on the diagonal pixel differentials, which emphasizes corners more clearly but can blur together small horizontal or vertical features.

a_1	a_2	a_3
a_8	X	a_4
a_7	a_6	a_5

Input image

1	1	1
1	0	-1
-1	-1	-1

Filter or mask

a_1	a_2	a_3
a_8	Y	a_4
a_7	a_6	a_5

Output image

For Robert, the output central pixel (Y) = $(a_1 - a_5) + (a_2 - a_6) + (a_3 - a_7) + (a_8 - a_4)$

Directional Filters

Many edge detection methods are based on directional filtering. An edge in an image is characterized by a steep gradient, which is the area of sharp change in gray values between adjacent pixels. The gradient or slope, which is the change in gray value with respect to distance, represents the first derivatives of the image. Directional filters can be used to compute the first derivatives of an image defining in any direction within a given space. However, X- and Y-directional filters are generally used to compute derivatives of an image in their respective directions.

A 3 × 3 kernel operating in X-direction of an image can be represented as follows:

-1	1	1
-1	0	1
-1	0	1

The kernel of the same dimension in Y-axis is the transpose of the above and given by:

-1	-1	-1
0	0	0
1	1	1

Apart from the above, there are other filters which may include weighting in the center of the non-zero columns.

Prewitt Edge Detector Filter

This filter creates an image where edges are shown. It uses two 3 × 3 templates. Prewitt gradient is calculated similar to the Sobel.

-1	0	1
-1	0	1
-1	0	1

X

1	1	1
0	0	0
-1	-1	-1

Y

a_1	a_2	a_3
a_4	a_5	a_6
a_7	a_8	a_9

Input image

where a_1 - a_9 are gray levels of each pixel in the filter window.

$X = -1^* a_1 + 1^* a_3 - 1^* a_4 + 1^* a_6 - 1^* a_7 + 1^* a_9$

$Y = 1^* a_1 + 1^* a_2 + 1^* a_3 - 1^* a_7 - 1^* a_8 - 1^* a_9$

$$\text{Prewitt Gradient} = \sqrt{\left(X^2 + Y^2\right)}$$

Kirsch Filter

Kirsch (1971) suggested Kirsch filter, a kind of non-linear edge crispening algorithm. Similar to the Prewitt and Sobel operators, the Kirsch filter approximates the first derivative of an image function operating as a compass operator to determine gradient directions within the image. It provides the maximal compass gradient magnitude about an image point ignoring the central pixel. It is one of a family of template matching filters in which the filter output is the maximum response from a set of linear filters which are sensitive to edges at different orientations.

a_1	a_2	a_3
a_8	Y	a_4
a_7	a_6	a_5

Input image

The central output pixel Y is given by:

$Y = 5S_i - 3T_i$

Where:

$S_i = a_1 + a_2 + a_3$

$T_i = a_4 + a_5 + a_6 + a_7 + a_8$

Typical 3 × 3 Kirsch filters are:

3	3	3
3	0	3
-5	-5	-5

3	3	3
-5	0	3
-5	-5	3

-5	3	3
-5	0	3
-5	3	3

7.10 FREQUENCY DOMAIN

There are two types of image modification techniques. They are: (i) image domain techniques, and (ii) spatial frequency domain techniques. The techniques which are applied directly to an image to modify and enhance its geometric detail could be called image domain techniques. All the filtering and enhancement techniques that we discussed in the previous section include in this category. They are either point based or neighborhood based. Point based techniques enhance the pixel, based on *its DN only* without depending on the neighboring pixels. They include contrast stretching, classification and density slicing. In case of neighborhood based techniques, the pixel is enhanced based *on its surrounding pixel values* through a widow operation. This involves enormous number of calculations due to processing of a possibly large number of pixels for each output pixel. To avoid such cumbersome to make calculations somewhat much faster, the enhancement due to convolution operation in the spatial domain can be performed as a simple multiplication in frequency space. To accomplish, a digital image is converted from the spatial domain into a frequency domain using Fast Fourier Transform (FFT) function.

Spatial frequency domain techniques are alternative to image domain techniques. In these techniques, an image is beforehand subjected to certain transformation functions and is separated into its various spatial frequency components. These transformation functions are Fourier transformation, Hadamard transformation, sine and cosine transformation (Jain, 1986; Pratt, 1978; Gonzalez and Wintz, 1987; Hall, 1979).

Fourier Transformation

The Fourier transform is a certain linear operator that maps functions to other functions. The operation Fourier Transformation fits a continuous function through the discrete DN values if they were plotted along each row and column in an image. In stead of representing an image as an array of pixels in columns and rows, alternatively it can be represented as the sum of many sine and cosine waves of different frequencies, amplitudes and directions. This is referred to as the frequency domain or Fourier representation. The parameters specifying the sine and cosine waves are termed Fourier coefficients. A Fourier transform results from the calculation of the parameters such as amplitude and phase for each possible spatial frequency in an image. Fourier analysis is the image decomposition technique as it separates the image into its various components of spatial frequencies. It describes the underlying structure of an image as a combination of simpler elements and displays the spatial frequency values in a two dimensional scatter plot known as Fourier spectrum. Fourier analysis mathematically attempts to describe the crests and troughs along any given row or column as a composite of sine and cosine waves with various amplitudes, frequencies and phases.

Concept of Fourier Analysis

There are ample treatments of Fourier analysis from a mathematical point of view. However, the description offered here is basically a conceptual treatment.

Fourier analysis is based on the concept that real world signals can be approximated by a sum of sinusoids, each at a different frequency. The more sinusoids included in the sum, the better the approximation.

The image is a record of array of reflectance or emittance values from the Earth's spatial features. Remote sensors record electromagnetic radiation in one or more regions of the electromagnetic spectrum that is reflected or emitted from the surface of the Earth. They are treated as discrete pixel values. If we plot a graph taking the pixel values of a row or a column, we will observe the character of a complex wave form as the EM radiations travel in the form of electromagnetic wave. Thus an image can be treated as a complex wave form. For a 2-dimensional image this would represent a surface may be land surface or waterbody. However, the complex wave pattern of the image was ultimately created by a set of very simple wave forms.

This logic is treated in reverse in Fourier analysis. It starts with a complex wave form and assumes that this is the result of the additive effects of some combination of sine and cosine waves of varying frequency, amplitude and phase. A digital image is composed of low spatial frequencies as well as high spatial frequencies. Low spatial frequency areas are those of infrequent gray scale changes that occur gradually over a relatively large number of pixel distances. High spatial frequencies represent areas of frequent change in gray level over a short pixel distance. A digital image is, therefore, thought to be of complex in nature as it is represented by many sine and cosine terms with their associated coefficients. It is decomposed into a finite set of simple waves with varying wave parameters ranging from the lowest frequency with single wavelength upto the highest frequency with a wavelength equal to twice the pixel resolution over the spatial extent of the image. Fourier analysis is then simply calculating the coefficients necessary for the sine and cosine terms to adequately represent the image.

Fourier transformation is being used for enhancement, compression, textural classification, compression, noise removal, or image restoration, quality assessment cross correlation, etc. (Hall, 1983; Jain, 1989; Lillesand and Kiefer, 1987).

Fourier filtering operation in the frequency domain of a digital image is carried out in three steps and shown in schematic diagram (**Figure 7.25**).

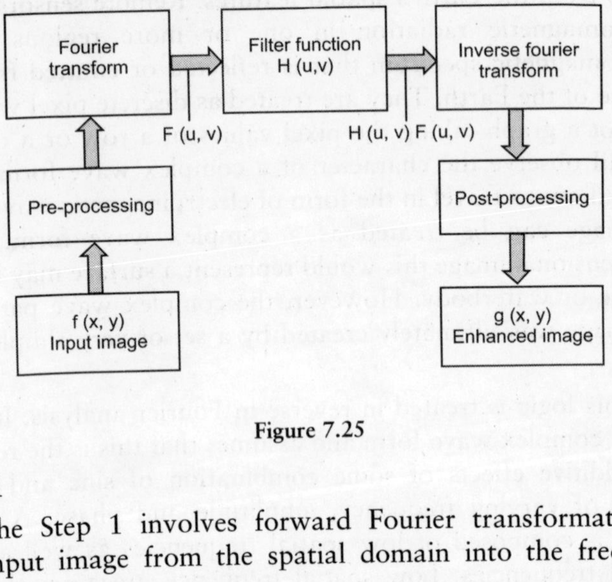

Figure 7.25

Step 1

The Step 1 involves forward Fourier transformation of the input image from the spatial domain into the frequency domain.

Transformation of pixel locations from the original image from the spatial domain into the destination (frequency domain) is called forward Fourier transformation or mapping. In mapping a frame of reference is defined in which each pixel in the input image is transformed to its new location in the frequency domain.

$$F(u,v) \leftarrow \sum_{x=0}^{M-1} \sum_{y=0}^{N-1} \left[f(x,y) e^{\frac{j2\Pi ux}{M} - \frac{j2\Pi vy}{N}} \right]$$

where:

M = the number of pixels horizontally

N = the number of pixels vertically

(u,v) = new coordinate = spatial frequencies that represents gray value change with respect to spatial distance (x,y)

F(u,v) = gray values after transformation

J = the imaginary component of a complex number

f(x,y) = input gray value

The number of pixels horizontally and vertically must each be a power of 2. If the pixels in the input image are not as a power of 2, they are padded up to the next higher power of 2. (*Source*: Oppenheim and Ronald, 1975; Press et al., 1988.)

The implementation of Fourier transform uses complex-valued inputs and outputs. A complex number contains a real number part and imaginary number part which is expressed in the form a + ib, where a, b, are real numbers and i is an imaginary number square of which is equal to -1 ($i^2 = -1$). For input numbers in the Fourier transform, the imaginary part is set to zero, since it does not exist. Thus the input to the Fourier transform is a single image.

The transformed Fourier image is not for optimal viewing and editing. It consists of two images—each corresponds to real and imaginary parts. Each pixel of the Fourier image contains (i) real component and (ii) imaginary component. The real part is attributed to the phase whereas the imaginary part is amplitude of the system at spatial frequency (u,v).

$$\text{Phase} = \tan^{-1}\left(\frac{b}{a}\right)$$

$$\text{Amplitude} = \sqrt{(a^2 + b^2)}$$

Together the real and imaginary parts express the frequency spectrum of an image.

Step 2

The transformed Fourier image which is in the frequency domain is manipulated by applying the familiar convolution operation in order for optimum viewing and editing. In this step both the operator and the input image must be in the frequency domain.

Image enhancement in the frequency domain is straightforward. In step 1 the Fourier transform of the input image is computed. The Fourier image in this step is multiplied by an operator rather than convolving the image in the spatial domain.

Mathematically, the multiplication procedure can be written as:

$$g(x,y) = h(x,y) \times f(x,y) = G(u,v) = H(u,v) \times F(u,v)$$

where:

$f(x,y)$ = input image

$h(x,y)$ = convolution operator

$g(x,y)$ = output image

G, F, H = Fourier transforms of g, f, h

The Fourier image is symmetry about the origin (u,v = 0,0). If the orgin lies in the upper left corner then the symmetry is more difficult to locate. However, if it is plotted at the center, then the symmetry can more readily be seen. Therefore, in the Fourier magnitude image, the origin is shifted to the center and the concentric circles separate the spectrum of the Fourier frequency plane. Nearer to the center are the low frequencies while the outer circles farther away from the center signify higher frequencies. The majority of the information in an image is confined to the region of lower frequencies.

The superimposed circles which have the radii of 5, 11, 22, 26, 36 and 48 enclose 90, 95, 98, 99.5, and 99.9 per cent information respectively. About 90% of the information of the original scene can be explained by the smallest radius of 5 units (Gonzalez and Wintz, 1987).

It is to be remembered that a position in a Fourier image (u,v) is a variable. It does not always represent the same frequency as it varies according to the size of the input image. If the spatial domain of the input image is large it will contain components of lower frequency compared to the small spatial domain of the input image. The lower frequencies are plotted near the origin (u,v = 0,0) which explain the majority of informations contained in the input image.

Relationship between sampling increment in spatial and frequency domain:

Spatial frequency, as we know, is expressed as inverse length (m^{-1}).

Let Δu and Δv be the horizontal and vertical sampling increments in the frequency domain respectively.

In the similar way, Δx and Δy may be defined for increment in pixel size in the spatial domain. Then the relationships can be expressed as follows:

$$\Delta u = \frac{1}{M\Delta x}$$

$$\Delta v = \frac{1}{N\Delta y}$$

where:

M = Horizontal image size in pixels

N = Vertical image size in pixels

In IRS-P6, LISS-III image (resolution = 23.5 m), a 512 × 512 image can be converted into frequency domain as follows:

$$\Delta u = \Delta v = \frac{1}{512 \times 23.5} = 8.31 \times 10^{-5} m^{-1}$$

u or v	Frequency
0	0
1	$8.31 \times 10^{-5}\ m^{-1}$
2	$16.62 \times 10^{-5}\ m^{-1}$

The above example demonstrates that the frequency depends upon the size of the image.

In the Fourier domain the simplest low pass filtering is accomplished using an ideal window. An ideal window defined for low pass filtering is:

$$H(u,v) = 1 \text{ if } D(u,v) \leq D_0$$
$$= 0 \text{ if } D(u,v) > D_0$$

where D_0 is the radius of the circle, called cutoff frequency.

The condition for attenuating the pixel frequencies in order to implement the low pass filters in the Fourtier domain is:

$$u^2 + v^2 > D_0^2$$

All frequencies inside the circle (radius = D_0) are retained, while the frequencies outside the circle are completely attenuated.

A low pass filter that satisfies the above condition is said to be an ideal low pass filter.

High pass filters are used for sharpening and edge enhancement by attenuating the low frequency details in the input image.

An ideal window defined for high pass filtering is:

$$H(u,v) = 0 \text{ if } D(u,v) \leq D_0$$
$$= 1 \text{ if } D(u,v) > D_0$$

Similar to the above, the condition for attenuating the pixel frequencies in order to implement the high pass filters in the Fourier domain is:

$$u^2 + v^2 < D_0^2$$

From the above it may be noted that in case of high pass filtering, all the frequencies outside the circle having radius D_0 will be preserved while all the frequencies inside the circle will be completely attenuated (Hall, 1983; VIPS-32, 1986; Gonzalez and Wintz, 1987; Pratt, 1991).

Step 3

In this step, the inverse function of the Fourier transform is carried out to get back the filtered Fourier complex image in the frequency domain into the spatial domain. Each pixel in the destination image is transformed to find where it was in the source image, which is the inverse transform:

$$f(x,y) \rightarrow \frac{1}{N_1 N_2} \sum_{u=0}^{M-1} \sum_{v=0}^{N-1} \left[F(u,v) e^{\frac{j2\Pi ux}{m} + \frac{j2\Pi vy}{n}} \right]$$

where:

$0 \leq x \leq M\text{-}1,\ 0 \leq x \leq N\text{-}1$

M = the number of pixels horizontally

N = the number of pixels vertically

(u,v) = spatial frequency variables

(*Source*: Oppenheim and Schafer, 1975; Press et al., 1988.)

The output image needs rescaling for display as the dynamic range of any transformation in the frequency domain is very large (Jain, 1987; VIPS-32, 1986). The technique for rescaling suggested by Gonzalez and Wintz (1987) is explained below:

$$D(m,n) = \log\,[1 + F(m,n)]$$

where,

F(m,n) = inverse output of Fourier transformation

D(m,n) = Output gray values for display

The generalized logarithmic can be written as:

$$D(m,n) = k \log \frac{[a + b\{F(m,n)\}]}{F(m,n)}$$

The above logarithmic transformation processing scaled down the dynamic range and provides a stretching of the resultant dynamic range in the final image display (Nag and Kudrat, 1998).

Conclusion

A filter is a device used to block certain objects or substances which are unwanted while letting others to pass through. In remotely sensed data, noise is an unwanted disturbance that adds certain intensity values to the real signal from the Earth's surface. Thus it creates hindrance while extracting meaningful information from the satellite imagery. Most images are affected to some extent by the noise. It is an unexplained variation in data. It can never be used to extract meaningful information. To make the image analysis simplifier, the noise should be filtered out. In signal processing, filtering is an operation that accentuates features of interest in data. However, in remote sensing the experts used the term in a broader sense. It refers to a mathematical operation of image enhancement, when operates it emphasizes or de-emphasizes degree of roughness on an image. It is used to emphasize edges—boundaries between objects or parts of objects in images. Filters provide an aid to visual interpretation of images.

The filters are considered under two categories: (i) linear and, (ii) non-linear.

Filters are said to be linear if their output values are the linear combinations of the pixels in the original image. They involve simple and understandable statistical calculations like, average, median, etc. Linear filters are studied in two ways both in Spatial and Frequency domains.

All nonlinear filters modify the value of each pixel in an image based on the value returned by a nonlinear function calculation that analyzes the neighboring pixels. Nonlinear filters are mostly used for noise removal and edge detection. Nonlinear filtering techniques have been one of the key research topics in digital image processing. In contrast to the linear filters, the problem of finding optimal nonlinear filter parameters is often slow and difficult since no simple analytical solution exists. Nonlinear filters are potentially more powerful as they can reduce noise without simultaneously blurring edges. They can detect edges at all orientations simultaneously.

7.11 BAND COMBINATION

Band combination is generally a combination of multiband (e.g. multispectral) images. It is defined as an output of multiband functions or operations. In a normal band combination, the same operation is carried out on each pixel in the image. The output of the operation is an image of new pixels generated due to some mathematical combination of the pixel values of the various bands of the input image.

An image recorded in a single band may not be sufficient to provide desired information of the analyst. This is due to record of the imperceptible signatures from the terrain features and visual display of minimal shades of gray levels. For example, in a panchromatic image, 8 to 9 shades of gray levels can be discriminated visually. However, in color print, it may be possible to distinguish 10 to 18 shades of colors. Therefore, in order to extract and enhance the analyst's features of interest, spectral enhancement techniques using more than one band of data is required. Multiband combination to generate one or more transformed combined bands is possible following certain mathematical operations. Band combination results in (i) minimizing the spectral redundancy in data, (ii) extracting new bands of data, which are not readily visible in their raw forms and are visually interpretable by the analyst, (iii) application of mathematical transforms and algorithms, and (iv) display of wide variety of information or emphasizing the features of interest in the three available color guns of the monitor: red, green and blue. Band combination includes mathematical transformation techniques such as addition, subtraction, ratioing, principal components analysis (PCA) and canonical analysis, IHS (intensity, hue and saturation) transformation, etc.

In this section, the concept, algorithms and advantage of various transformations are discussed. Special emphasis is given on more familiar and important transformations like vegetation indices, principal components analysis and fusion techniques in remote sensing.

Before focusing into those specific band transformations, it is desired to highlight the concept of false color composites which the remote sensing specialists are talking about.

False Color Composites (FCCs)

The individual bands can be combined to produce a color composite, which is useful to highlight information in the imagery. Color composites are made by utilizing the three different colors of light that a monitor can display (red, green and blue).

Color composites are considered under two categories: (i) true color composite and (ii) false color composite.

While displaying a multispectral digital image in a color monitor it is possible to assign which bands are to be displayed with each of the three color guns. The assignment of colors that allows for an easy interpretation of the image is the desire of the analyst. True colors are those which approximate the colors that would appear to a human observer of the scene. A true color image is formed when red, green, and blue band data are displayed in the corresponding red color gun, green color gun and blue color gun of the monitor respectively. To produce a false color composite image, any of the bands can be displayed in any of the color guns of the display device. The band combination which is used can be changed to highlight the desired feature or phenomena.

Remote sensing satellites record every material and synthetic object on the Earth's surface in several spectral wave bands. In other words, spaceborne sensors sample and record every material (natural and man-made) and natural phenomenon in several band widths in the optical and microwave regions of the electromagnetic spectrum. Several band combinations are, therefore, possible to construct with the help of red, green and blue which we call the primary colors. The resultant colors signify the characteristic of the objects in different parts of the electromagnetic spectrum.

Figure 7.26 illustrates 7 spectral bands of Landsat TM designed to record the spectral reflectance values in different wavelength regions to discriminate various objects on the Earth's surface. The figure also shows RGB combination of certain bands leads to images with different information content.

Band combination of red (3), green (2) and blue (1), can yield an image in natural color. The landscape in this combination looks exactly the same that we see with our naked eye. Low reflectance in the red region corresponds to chlorophyll absorption at around 0.65 µm which is detected on band 3 (0.61 - 0.69 µm). A relative lack of absorption in the green region allows band 2 (0.51 - 0.60 µm) to detect the green reflectance from vegetation. Blue light penetrates water best, even upto 25 m in clear water and is scattered back to the

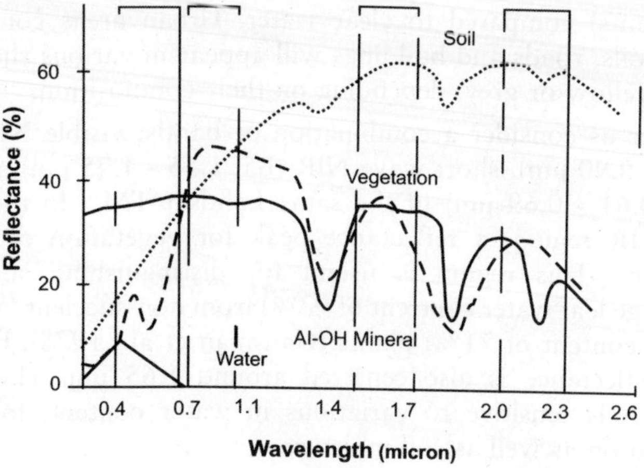

Figure 7.26

sensor whereas red light is rapidly absorbed in the top few meters. To derive information on water body and sediment distribution pattern band 1 (0.45 - 0.51 μm) is chosen to be more suitable. The capability of band 1 is that it can also discriminate soil from vegetation and distinguishes forest types.

A typical band combination of visible Near Infrared (4), red (3) and green (2) yields a conventional false color composite image. In near infrared region there is high reflectance (45%) from the vegetation which is detected on band 4 (0.76 - 0.90 μm) of Landsat TM. This band is suitable to discriminate different types of vegetation. Band 4 is familiarly being used to distinguish land parcel from water body.

In visual interpretation of a false color composite image, vegetation appears in different shades of red depending on the types and conditions of the vegetation, since it has a high reflectance in the NIR band. Brighter reds indicate more the growing vegetation. Soils with no or sparse vegetation appear in color varying from white (sand, salt) to greens or browns depending on moisture and organic matter content. Clear water appears dark-bluish (higher green band reflectance) to black, while turbid water appears cyan (higher red reflectance due to sediments) compared to clear water. Urban areas containing bare soils, roads and buildings will appear in various shades of blue, yellow or grey, depending on their composition.

Let us consider a combination of bands: visible NIR (B4: 0.76 - 0.90 µm), short wave NIR (B5: 1.55 - 1.75 µm) and red (B3: 0.61 - 0.69 µm) of the same Landsat TM. In the short wave IR region, a reflectance peak for vegetation occurs at 1.6 µm. This region is useful for distinguishing succulent (average leaf water contrent of 92%) from non-succlent (average water content of 71%) plants (Gausman et al., 1978). Highest soil reflectance is also centered around 1.65 µm. Therefore, Band 5 is sensitive to variations in water content, for leafy vegetation as well as soil moisture.

In this combination vegetation appears in different shades of red. When crops become disease affected their structure changes and hence their reflectance characteristics also changes. Due to their low moisture content, reflectance in band 5 will be relatively higher. This is due to more contribution of green and thus resulting in a more orange color. Lower reflectance value of vegetation in VNIR and higher reflectance in the SWIR allow green color to dominate in this combination. Bare soils and urban areas will appear in blue towards gray colors.

Let us consider a color composite image formed in combination of Short Wave Infrared (B7: 2.08 - 2.35 µm), Visible Near Infrared (B4: 0.76 - 0.90 µm) and Green (0.51 - 0.60 µm) bands. The VNIR band is confined to the high reflectance flat topped region of the vegetation curve. Vegetation, therefore, in this combination shows in various shades of green. The spectral reflectance curve for Al-OH minerals shows the absorption peaks at 2.17 and 2.20 µm, which is well within the spectral band width of SWIR. Band 7, which is confined to 2.08-2.35 µm is sensitive to variation in moisture content and therefore is used to identify clay minerals. Various rocks and minerals can also be discriminated using this band.

In this combination, rocks and minerals are visualised in shades of red to orange, soils in the brighter shades of blue and vegetation in bright green spots. Water appears in dark blue or black and urban areas will appear in dark blue or pink.

Algebraic Operation

Algebraic operations produce a single output image as a result of pixel by pixel sum, difference, product or ratioing of two or more input images. When this operation is performed on two or more input images it brings about the maximum possible thematic information than any of the original images.

Addition of images tends to extract information that is positively correlated from image to image. It on the other hand removes information that is uncorrelated from image to image. The output image gray value Y is given by:

$$Y = a_0 + a_1x_1 + a_2x_2 + a_3x_3 + \ldots\ldots\ldots\ldots\ldots\ldots + a_nx_n$$

where:

x_i, for i = 1, 2, 3...n, are the gray values of a pixel in different bands

a_i, for i = 0, 1, 2, 3...n, are the weighting coefficients of a pixel in different bands, where $a_I > 0$.

The coefficients a_i are the inputs of the analyst based on his knowledge and work experience. Huete et al. (1984) and Ezra et al. (1984) documented the importance of using coefficients based on site specific rather than using the global coefficients. Jackson (1983) described the determination of these coefficients.

Image subtraction tends to extract information that is uncorrelated or negatively correlated from image to image. Two images are taken as the inputs to produce a third one, the pixel values of which are simply the pixel values of the first minus the corresponding pixel values of the second. The pixel subtraction is also possible sometimes in a single image in which a constant value is deducted from all of its pixel values. Accordingly the output pixel values are given by:

$$Y(i, j) = x_1(i, j) - x_2(i, j) \text{ (in case of two images)}$$

Y: Output value of the pixel

$x_1(i, j)$, $x_2(i, j)$: Input pixel values corresponding to image 1 and image 2 respectively

$Y(i, j) = x_1(i, j) - C$ (for single image), where C is a constant value.

If there is any uncorrelated data exists from band to band, then the algebraic operation—subtraction is often used. For two different spectral bands, uncorrelated information is the spectral difference between the two images. If the images are from different time periods, the uncorrelated data represents the temporal difference between the two images.

Similar to the above, image division is performed by taking two input images to produce a third one, the pixel values of which are the pixel values of the first image divided by the corresponding pixel values of the second. It is also often possible to perform the division operation on a single image with respect to a specified constant.

Mathematically output can be written as:

$Y(i, j) = x_1(i, j) \div x_2(i, j)$ (in case of two images)

$Y(i, j) = x_1(i, j) \div C$ (for single image), where C is a constant value.

Tassel Cap Transformation

In remote sensing analysis, multispectral sensors provide a key tool in understanding and characterizing imagery. Different surface types have distinct spectral signatures that can be distinguished by analyzing their signals in the various bands of the sensor. The sensor bands often respond in a strongly correlated manner to surface features, so analyzing imagery using natural or false color may not distinguish the surface features in an optimal fashion. To overcome this difficulty, various spectral enhancement techniques have been addressed over the years to highlight certain surface types.

Spectral enhancement is the process of creating new spectral data from available bands to highlight the underlying features of the scene. New data are created on a pixel by pixel basis by applying operations to corresponding pixels in existing bands. Spectral enhancement only applies to multiband images. Tasseled Cap transformation is one of the available methods for enhancing spectral information content of an image to maximize the separation of different surface types. The concept of Tasseled Cap transformation is a useful tool for compressing spectral data into a few bands associated with physical scene characteristics (Crist and Cicone, 1984). This enhancement uses mathematical equations to transform the original *n* multispectral bands into a new *n*-dimensional space. The principle behind the transformation is that it rotates the data space to view the greatest amount of variation of scene class characteristic in the plane. It is the most important linear combination widely being used by the remote sensing community (Kauth and Thomas, 1976; Crist and Cicone, 1984b).

Kauth and Thomas (1976) and, Crist and Cicone (1984c) demonstrated that the Tasseled Cap transformation of Landsat Multispectral Scanner (MSS) and Thematic Mapper (TM) data represent examples of linear combination of features. This has been accepted by the remote sensing community. Crist and Kauth (1986) realized that the basic concepts and degree of understanding behind these transformations are not clear to many of the researchers. They opined that the users are either (a) using the Tasseled Cap transformation without really understanding them, (b) misapplying the transformations, (c) incorrectly applying the basic concepts to new sensors, or (d) hesitating to use the transformations because of their apparent mystery. Therefore, the basic concept behind the transformation is outlined below.

A sensor samples and records electromagnetic radiation in one or more regions of the electromagnetic spectrum in a specified number of bands. The signals recorded in different bands define the multidimensional space, where each sensor band corresponds to one dimension. If N is the number of bands in a multispectral image, then the data in N bands can be represented in N-dimensional space. Each dimensional axis is represented by the intensity values of a band from minimum to maximum.

Position of each pixel in the dimensional space is determined according to its DN value in each band. The scene class reflectance/absorptance spectra determine the clustering of pixels in the dimensional space. The clusters of data being formed at different portions of the space are termed as data structures.

The Tasseled Cap concept simply involves identifying the existing data structures for a particular sensor and application (i.e., set of scene classes), changing the viewing perspective (i.e., rotating the axes) such that those data structures can be viewed most directly, and defining feature directions (new *x*-, y-, and z-axes in the cube example) which correspond to spectral variation primarily or exclusively associated with a particular physical scene class characteristic.

The Tasseled Cap transformation provides excellent information for agricultural applications because it allows the separation of barren (bright) soils from vegetated and wet soils. In the case of the four-band Landsat Multispectral Scanner (MSS), vegetation and soils data were found to primarily occupy a single plane-like structure, which typically contains 95 per cent or more of the total data variation. In that plane, a feature named Brightness was defined in the direction of soil reflectance variation, and a feature named Greenness was defined in a perpendicular direction associated with the reflectance characteristics of green vegetation.

Based on the linear combination, four new indices corresponding to four bands of Landsat MSS are identified. Those are Soil Brightness Index (SBI), The Green Vegetation Index (GVI) and the Yellow Vegetation Index (YVI) and the Non-such Index (NSI). The coefficient of each combination index given below are taken from Kauth et al. (1979) and Thompson and Wehmanen (1980).

$$SBI = 0.322\ MSS_4 + 0.603\ MSS_5 + 0.675\ MSS_6 + 0.262\ MSS_7$$

$$GVI = -0.283\ MSS_4 - 0.660\ MSS_5 + 0.577\ MSS_6 + 0.388\ MSS_7$$

$$YVI = -0.899\ MSS_4 + 0.428\ MSS_5 + 0.076\ MSS_6 - 0.041\ MSS_7$$

$$NSI = -0.016 MSS_4 + 0.131\ MSS_5 - 0.452\ MSS_6 + 0.882\ MSS_7$$

Generally, the GVI and SBI indices contain most of the scene information (95% to 98%).

For the Thematic Mapper (TM), data in the six reflective bands were found to primarily occupy three dimensions, defining two perpendicular plane-like structures and also occupying a region between the two planes. One plane contains fully-vegetated samples, while the other contains bare soil samples. When both soil and vegetation are visible to the sensor, the data fall in the region between the two planes. Brightness and Greenness features, analogous to those defined for MSS data, were found to be appropriate for TM data as well, although the Brightness feature in TM data is not exactly equivalent to its MSS counterpart. In addition, a feature named Wetness was defined *to* correspond to the observed direction of soil moisture variation in the plane occupied by bare soils data. Here again, the three features typically capture 95 per cent or more of the total data variation.

In Landsat TM, the Tasseled Cap transformation is being used to produce six new layers from the six TM bands. The first two layers contain 95 to 98% of the total information (Jensen, 1996). The other layers, while useful for some applications such as moisture study, explain much less of the data. Soil brightness index (SBI) one of the two more important layers shows bare areas such as agricultural fields, beaches, and parking lots as the lightest features. Green vegetation index (GVI), the other one is an indicator of vegetation status since it displays areas with healthy, green vegetation as the lightest feature. The name "tasseled cap" comes from the fact that when the greenness and brightness of a typical scene are plotted perpendicular to one another on a graph, the resulting plot usually looks like a cap.

Originally constructed for understanding important phenomena of crop development in spectral space (Kauth and Thomas, 1976), the transformation has potential applications in revealing key forest attributes including species, age and structure (e.g., Cohen et al., 1995). The Tasseled Cap transformations for Landsat images were derived by looking at a small number of images and determining a new orthonormal basis for the bands that highlights differences in vegetation and soil (Horne, 2003).

Essentially two tasseled cap transformations have been developed based on Landsat Thematic Mapper (TM)—one based on digital number (DN) (Crist and Cicone, 1984) and the other on reflectance factor (Crist, 1985), which for simplicity will be referred to as DN based and reflectance factor based transformations, respectively. While the similar spectral characteristics of TM and the Landsat 7 Enhanced Thematic Mapper Plus (ETM+) might suggest direct applicability of both transformations to ETM+ images (Crist and Kauth, 1986), an at-satellite reflectance based transformation is needed for large area applications for the following reasons. First, the reflectance factor based transformation was developed based on ground measurements with little atmospheric effects (Crist, 1985). Applying this transformation to satellite images requires the images to be corrected for atmospheric effects. Although several atmospheric correction algorithms have been developed (e.g. Liang et al., 1997), for large area applications, many users are still concerned with possible unknown errors that may arise due to lack of or uncertainties in the ground and atmospheric data necessary for running these algorithms, and therefore, often choose not to perform atmospheric correction (e.g. Cohen et al., 2001). Second, use of the DN based transformation in multi-scene applications can be problematic, because changing sun illumination geometry strongly affects DN, and thus affects the derived tasseled cap value. As discussed in section 7.7.2, a large part of the impact of illumination geometry can be normalized by converting DN to at-satellite reflectance. Therefore, a transformation based on at-satellite reflectance is more appropriate for regional applications where atmospheric correction is not feasible.

Conclusions

An at-satellite reflectance based tasseled cap transformation was developed based on 10 Landsat 7 ETM+ scenes representing a variety of landscapes of the United States in both leaf-on and leaf-off seasons (http://landcover.usgs.gov/pdf/tasseled.pdf). This transformation is most appropriate for regional remote sensing applications where atmospheric correction is not feasible. The DN based transformation is often not appropriate for this type of application, because DN value is strongly affected by changing illumination geometry. A large part of such impact of illumination geometry can be normalized by converting DN to at-satellite reflectance. Applying the ground reflectance factor based transformation directly to at-satellite reflectance images is not appropriate because it may result in unreasonable tasseled cap values. The at-satellite reflectance based transformation generally does not have this problem when applied to clear and near cloud-free at-satellite reflectance images. The brightness, greenness and wetness of the derived transformation collectively explained over 97% of the spectral variance of individual scenes used in this study (Huang et al., 2002).

Tasseled Cap transformation is being used for enhancing spectral information content of Landsat TM data. Three of the six Tasseled Cap transformations namely, brightness, greenness and wetness that define the vegetation information content (Crist et al., 1986; Crist and Kauth, 1986) are often used.

Band 1: Brightnesss (measure of soil)—a weighted sum of all bands, defined in the direction of the principal variation in soil reflectance.

Band 2: Greenness (measure of vegetation)—orthogonal to brightness, a contrast between the near-infrared and visible bands, strongly related to the amount of green vegetation in the scene.

Band 3: Wetness—relates canopy and soil moisture (Lillesand and Kiefer, 1987).

Four separate images can be created by the linear combination of TM bands and associated weighting coefficients. Those are scene brightness, vegetation greenness, surface wetness, and atmospheric haze.

$$\text{Brightness} = 0.3037\ TM1 + 0.2793\ TM2 + 0.4743\ TM3 + 0.5585\ TM4 + 0.5082\ TM5 + 0.1863\ TM7$$

$$\text{Greenness} = -0.2848\ TM1 - 0.2435\ TM2 - 0.5436\ TM3 + 0.7243\ TM4 + 0.0840\ TM5 - 0.1800\ TM7$$

$$\text{Wetness} = 0.1509\ TM1 + 0.1973\ TM2 + 0.3279\ TM3 + 0.3406\ TM4 - 0.7112\ TM5 - 0.4572\ TM7$$

Hazeness = 0.8832 TM1 – 0.0819 TM2 – 0.4580 TM3 – 0.0032 TM4 – 0.0563 TM5 + 0.0130 TM7

Sharma et al. (1990) generated the coefficients for Green Vegetation Index and Soil Brightness Index for Indian Remote Sensing Satellite. They are given by:

SBI = 0.4233 B1 – 0.4328 B2 + 0.6490 B3 + 0.4607 B4

GVI = 0.1072 B1 – 0.1219 B2 – 0.4341 B3 + 0.8861 B4

where B1, B2, B3, B4 are digital numbers in four bands of IRS-IA LISS-II data respectively.

The Tasseled Cap transformation has been popularly used to a variety of applications, estimating moisture stress detection (Thompson and Wehmanen, 1980), crop condition assessment (Ajai et al., 1988) yield prediction (Barnett and Thompson, 1982), identification of wastelands (Kudrat et al., 1989; Saha et al., 1988). The linear combination of bands is very useful in feature selection during pattern recognition (Swain and Davis, 1978).

7.12 BAND RATIOING

Image division or spectral band ratioing is one of the most common mathematical operations applied to multi-spectral image data. It is a simple mathematical calculation combining the different DN values of the different bands.

The most widely used correction technique is band ratioing. It is defined as an enhancement procedure by which ratio images are generated dividing DN values of pixels in one spectral band(s) by the corresponding DN values in another band(s). Lillesand and Kiefer (1987); Holben and Justice (1981) have documented that a new variable is created with this technique by dividing the radiance value in one channel by the corresponding radiance value in a second channel. The rationale behind this technique is that the spectral radiance received by the sensor is equal to a multiplicative term plus an additive term. The multiplicative term is direct irradiance which is the product of direct spectral irradiance impinging the target at time t, target reflectance at time t and atmospheric transmittance at time t, and as such it is quantitative information. The additive term is basically haze which is added to the multiplicative term (Holben and Justice, 1981).

These ratios suppress brightness variation due to certain attributes such as topography, relief, shadows, illumination angle and intensity, etc. Thus, they can be used to reduce effect of these variables. Holben and Justice (1981) showed that band ratioing reduces but does not entirely eliminate the topographic effect. They specifically showed that ratioing is not effective for reducing the topographic effect on shaded surfaces which are illuminated solely by scattered light. They concluded that the unexplained residual effects are related to scattering from surrounding terrain, path radiance, and the inappropriateness of the implicit assumption that the undesirable multiplicative effects in the data are independent of wavelength. In addition, they sometimes, provide unique information not available in any single band (e.g. soil vegetation discrimination using MSS 7 image/MSS 5 image). Generally, lower the correlation between the bands, the greater is the information content of band ratioed image. However, sometimes ratioing bands may also enhance noise.

Algorithm

$$Y_{ij} = \frac{X_{ijk}}{X_{ij1}}$$

where, Y_{ij} = Output DN value of a pixel of the location (i, j)

X_{ijk} = Input DN of the pixel at the same location for band k

X_{ijl} = Input DN value for band 1 at the pixel location (i, j)

Integer Scaling Consideration

In order to preserve all numerical precision, the ratioed images are created in floating point. The domain of ratio X_{ijk}/X_{ijl} is from 1/255 to 255. Since computation with 0 DN value is not possible, the algorithms are modified as:

(i) Convert DN = 0 to DN = 1 or add 0.1 if DN = 0.

(ii) To encode the ratio values in 8-bit format, Normalising function is applied.

If ratio = 1, gray value of 128 is output

Ratio values 1/255 to 255 are re-assigned as

$$DN_{out} = INT\left(\frac{X_{ijk} \times 127}{X_{ijl}}\right) + 1$$

Ratio values 2 to 255 are assigned values as

$$DN_{out} = INT\left(\frac{128 + X_{ijk}}{2} \times X_{ijl}\right)$$

where INT stands for integer value of the expression.

(iii) Arctan Function.

Sometimes simple ratios are not effective when denominator becomes zero and the output pixel gets unbounded. In such cases $\tan^{-1}$ function of the ratio is taken to accommodate the output values within 0-255 range. $\tan^{-1}$ makes the scene independent of illumination on different parts of a scene.

$$DN_{out} = K\left[\arctan\left(\frac{X_{ijk}}{X_{ijl}}\right)\right]$$

when K = 162.34 is used, the output DN ranges from 0 to 255.

If there are n bands, then possible number of ratio images generated is n (n – 1).

Optimum Index Factor

Selection of the best band combination is usually the first step to compress remote sensing data while securing class separability. The Optimum Index Factor (OIF) is applied to remotely sensed data to evaluate the best band combination which can extract the maximum possible information content. Many color composite images were created using various band combinations, based on the statistical method of "Optimum Index Factor" (OIF) developed by (Chavez, 1982) which has been calculated using the following statistical formula:

$$OIF = \frac{\sum_{k=1}^{N} S_k}{\sum_{k=1}^{N} Abs\, C_{ck}}$$

where:

S_k is the standard deviation of digital numbers for band k and Abs C_{ck} is the absolute value of the correlation coefficient between any two bands out of N bands.

The high values for OIF indicate bands that contain much information with little duplication (ITC, 1997). The computation of OIF enables to select three best band combinations with the highest variance and the least correlated bands. This can be used as an input to fusion techniques which use limited number of bands. This combination is assumed to have the highest amount of information.

In general, mathematical operations addition and multiplication tend to enhance correlated information between bands, while band ratioing tends to enhance uncorrelated information between bands. Normally, for ratioing two uncorrelated bands are preferable (Jenson, 1986).

Band ratioing of the images communicate the spectral characteristics of the different features in the image regardless of varying scene illumination conditions. Variation in image brightness is caused due to certain attributes like terrain conditions and atmospheric conditions. By ratioing, this effect can be minimized and the spectral characteristics of the features in the image will be emphasized. Like normal FCC, three monochromatic ratio data sets can generate a false color composite. This kind of manipulation can help the analyst in interpreting the subtle reflectance differences.

The images before subjected to ratioing they must be freed from noise effect as the ratio may further intensify the noise patterns. The fundamental assumption of band ratioing is that the undesirable effects in the data are multiplicative and that the additive factors influencing the data are small in magnitude so they can be ignored. Generally, in a ratio, the common factors in both the numerator and denominator are cancelled while any additive factor can never be filtered out. Since atmospheric haze is an additive factor, it cannot be removed by ratioing. This should be removed prior to ratioing the images. Ratioing of two bands cancels out the common multiplicative terms, removing the undesirable effects (Woodcock, 1982).

Different materials having unique reflectance spectra but similar slopes in their reflectance curve may appear to be the same in the ratio image. This can create troublesome for the interpreter if these materials are contiguous and of similar image texture.

Landsat MSS Band Ratios

Different multispectral band combinations can aid in landscape evaluation and classification.

Landsat MSS contains 4 spectral bands. This implies there are 12 possible band ratio combinations. The ratios are 2/1, 3/1, 3/2, 4/1, 4/2, 4/3, and their six reciprocals:

(i) Characterizing arid region soils and vegetation conditions from remotely sensed imagery is often limited by the low brightness contrast between soil and vegetation and between different plant communities. This can be improved by band ratioing:

    The band ratios useful for discriminating soil and rock, and vegetation are 1/2, 1/4, 2/4, and 3/4. Vegetation will appear in dark tones with such an order.

(ii) Iron oxide or ferric iron can be detected using the ratio 2/1. If we take the reciprocal of it, then the same unit would appear in dark tones. The ratios 1/2 or 2/1 can also be used to study suspended sediment in water.

(iii) Spectral response curve for vegetation illustrates that there is a large difference in reflectance (%) between the visible (bands 1 and 2) and infrared regions (bands 3 and 4). In order to highlight the vegetation patterns, the band ratios 3/1, 3/2, 4/1, and 4/2 are useful. They can also be used to discriminate vegetation from the water bodies.

(iv) The relative greenness of vegetation, e.g., stressed plants versus unstressed and for estimating biomass is assessed using the band ratio 4/2.

(v) The 2/3 or 3/2 ratio is most often used for distinguishing general material types of soil and rock, vegetation, and water. Therefore, it is very useful for producing thematic maps of urban, soil and rock, vegetation, and water.

Landsat TM Band Ratios

As stated earlier, ratio images are derived by simple pixel-by-pixel division of digital values from two bands of data. There are many possible combinations of ratios that can be calculated. With the 6 bands of TM data alone there are dozens. One can combine any three ratios to form an RGB color image or even combine ratioed data with standard data.

Below are the common ratios and their properties.

TM1/TM2

Especially for deciduous forest, they are highly correlated, so the ratio image has low contrast.

TM3/TM4

Features such as water and roads depict in higher tones. This is because they reflect high in the red band (TM3) and little in the near infrared band (TM4). Vegetation appears in darker tones because of the relatively low reflectance in the red band and high reflectance in the near-infrared.

TM5/TM2

Vegetation appears in light tones because of its relatively high reflectance in the mid-infrared band (TM5) and its comparatively lower reflectance in the green band (TM2). However, certain vegetation types do not follow this trend due to their particular reflectance characteristics and they appear in very dark tones. So this ratio is useful for vegetation discrimination.

TM3/TM7

Roads and other cultural features appear in lighter tone due to their relatively high reflectance in the red band (TM3) and low reflectance in the mid-infrared band (TM7). Also differences in water turbidity are readily observable.

TM5/TM4

Coniferous forest damage shows clearly, with high ratios characterizing high damage sites and low ratios showing low damage sites. This means the dryer a leaf becomes, the higher the ratio. This is due to the fact that when a leaf becomes dryer its reflectance increases in the mid-infrared band (TM5) in contrast to the near-infrared band (TM4) that is relatively unaffected by changes in moisture content.

TM7/TM4

It is very sensitive to total leaf-water content of the vegetative canopy. In general mid-infrared band (TM7) is low when moisture is high, due to absorption and near-infrared band (TM4) is unaffected by changes in moisture content. Therefore, it is effectively used for forest damage estimation (works exactly as TM5/TM4) and vegetative biomass estimation. High TM7 values during a period of active growth imply a low productivity site because of moisture stress or less vegetation biomass at that site (*source*: www.forestry.umt.edu/academics/courses/for503/rs2.htm).

TM5/TM7 and TM3/TM1, TM4/TM5

Landsat Thematic Mapper data are extremely useful for mapping of Ultramafics and Associated Lithological Units (Chellaiah, 2003). He used Landsat TM band ratios 5/7 and 3/1 because clay and iron minerals have reflectance and absorptions features in these bands. Landsat TM band ratios 4/5 is used to emphasize silicate minerals compared to FeO-rich minerals (Abrams et al., 1983).

TM4*TM5*TM7/constant = 0-255

This is a very good discriminant of coniferous vegetation, deciduous vegetation, rock outcroppings and water.

Vegetation Indices

Research on spectral reflectance of vegetation started as early as 1913 when it was established that large amount of light transmitted into the spongy mesophyll gets reflected/critically internally reflected from the leaf cell walls where the refractive index changes from that of water (1.33) to that of air (1.00). This leads to highly efficient scattering of all wavelengths of light. Plant pigments such as chlorophylls strongly absorb light in the red part of the visible spectrum. Therefore, the reflectance of the green vegetated plants is low in the red region. However, on the other hand, mostly as a result of multiple scattering in the mesophyll, the reflectance of the vegetation in the near infrared region of the EM spectrum is very strong. This exhibits contrast between the low reflectance in the red and strong reflectance in the near infrared regions. Reflectance measurements in these regions are thus strongly correlated with the amount of photosynthetically active radiation (PAR) absorbed by the plant material. Thus, the finding that the strong reflectance contrast between the visible and the near infrared (NIR) is exploited for designing vegetation indices in the remote sensing of green vegetation.

One of the simplest types of band transformation is calculation of a vegetation index. A vegetative index is a value that is calculated (or derived) from remotely-sensed reflectance in the red band (typically 0.6-0.7 μm) and the reflectance in the near infrared band typically (0.8-1.1 μm). Various vegetation indices have been developed for vegetation monitoring (e.g., assessing biomass or leaf area index and discriminating between stressed and nonstressed vegetation) with different remote sensors and vegetation conditions (Tucker, 1979; Jackson, 1983; Philipson and Teng, 1988; Teng, 1990). Jensen (2000) defined vegetation indices as—dimensionless, radiometric measures that function as indicators of relative abundance and activity of green vegetation, often including leaf area index (LAI). Percentage green cover, chlorophyll content, green biomass and absorbed photosynthetically active radiation (PAR).

Though many vegetative indices exist, the most widely used index is the Normalized Difference Vegetative Index (NDVI). The NDVI, like most other vegetative indices, is calculated as a ratio between measured reflectivity in the red and near infrared portions of the electromagnetic spectrum. These two spectral bands are chosen because they are most affected by the absorption of chlorophyll in leafy green vegetation and by the density of green vegetation on the surface. Also, in red and near-infrared bands, the contrast between vegetation and soil is at a maximum.

A vegetation index should (Running et al., 1994; Huete and Justice, 1999):

(i) maximise sensitivity to plant biophysical parameter, preferably with a linear response in order that sensitivity be available for a wide range of vegetation conditions, and to facilitate validation and calibration of the index;

(ii) normalise or model external effects such as Sun angle, viewing angle, and the atmosphere for consistent spatial and temporal comparisons;

(iii) normalise internal effects such as canopy background variations, including topography (slope and aspect), soil variations, and differences in senesced or woody vegetation (nonphotosynthetic canopy components); and

(iv) be coupled to some specific measurable biophysical parameters such as biomass, LAI, or PAR as part of the validation effort and quality control.

Jackson and Huete (1991) categorized vegetation indices (VIs) into two groups: (i) slope-based and (ii) distance-based VIs.

Slope-based Vegetation Index

The slope-based vegetation indices are simple arithmetic combinations that focus on the contrast between the spectral response patterns of vegetation in the red and near-infrared portions of the electromagnetic spectrum. They are so named because any particular value of the index can be produced by a set of red/infrared reflectance values that form a line emanating from the origin of a bi-spectral plot.

The slope-based VIs are estimated to get information on the status and abundance of green vegetation cover and biomass. Different indices under this category include the RATIO, NDVI, RVI, NRVI, TVI, CTVI, and TTVI. Perhaps the best known of the classic VIs are the ratio vegetation index (RVI; Pearson and

Miller, 1972) and the normalized difference vegetation index (NDVI; Rouse et al., 1974) that are based on the reflectance in the Red and NIR part of the spectrum.

The Ratio Vegetation Index (RVI)

Spectral vegetation indices (VIs) calculated as linear combinations of near infrared (NIR) and VIS red reflectance have been found to be well correlated with canopy cover, leaf area index (LAI), and absorbed photosynthetically active radiation (APAR) (Elvidge & Chen, 1995; Myneni & Williams, 1994). However, it has been shown that most traditional VIs are sensitive to soil background, especially at low LAIs (Huete, 1989; Huete, Jackson and Post, 1985).

RVI is the ratio vegetation index which was first described by Jordan (1969). Rouse, et al. (1974) proposed RVI to separate green vegetation from soil background using Landsat MSS imagery. RVI is the slope of the line that joins the origin and the vegetation point in Red-NIR space. It is a simple ratio based index as given by:

$$RVI = \frac{NIR}{RED}$$

This is the most widely calculated vegetation index. The index ratio from the above conveys a clear capture of the contrast between the red and infrared bands for vegetated pixels. High index values are produced by combining high NIR and low red reflectance values. The RVI ranges from 0 to infinity. Since it is a ratio, it can minimize the variation of scene illumination conditions due to varying topography. This band ratio is widely being used to eliminate various albedo effects.

The Normalized Difference Vegetation Index (NDVI)

The most familiar index based ratio is the NDVI i.e. Normalised Differential Vegetation Index. The ratio is ascribed to Rouse et al. (1973), but the concept was first presented by Kriegler et al. (1969). Rouse et al. (1973) using Landsat MSS digital data found that the ratio NDVI separates green vegetation from its background soil brightness.

It is expressed as the difference between the near infrared and red bands normalized by the sum of those bands. The calculation of the NDVI is as follows:

$$NDVI = \frac{NIR - Red}{NIR + Red}$$

NDVI has the following value ranges:

Vegetation: $0 < NDVI \leq 1$

Soil: $0 \leq NDVI \leq 1$

Water: $-1 \leq NDVI \leq 1$

Pinty and Verstraete (1992) showed that a better minimizing of the relative influence of atmospheric effects is achieved by nonlinearly combining single-band reflectances into the vegetation index. They proposed the Global Environment Vegetation Index, GEMI, calculated as follows:

$$GEMI = \eta(1 - 0.25\eta) - \frac{red - 0.125}{1 - red}$$

$$\eta = \frac{2(nir^2 - red^2) + 1.5nir + 0.5red}{nir + red + 0.5}$$

The RVI and NDVI are functionally equivalent and related to each other by the following equation:

$$NDVI = \frac{RVI - 1}{RVI + 1}$$

In general, these indices (NDVI and RVI) tend to enhance the contrast between soil and vegetation while minimizing the effects of illumination conditions (Baret and Guyot, 1991). However, they have been shown to be sensitive to soil brightness effects (Baret et al., 1989; Huete, 1989; Roujean and Breon, 1995), especially at low vegetation cover. This index has the ability to minimize the topographic effects while producing a linear measurement scale. Furthermore, the measurement scale has the desirable property of ranging from –1 to 1 with 0 representing the approximate value of no vegetation. Thus negative values represent non-vegetated surfaces. In addition, division by zero errors is significantly reduced.

Analysis of Vegetation Vigour Using NDVI (Kumar, 2004)

The vegetation analysis of an area shows the land cover characteristics in terms of conservation measure, land use practices and soil moistures of the landscape. The vegetation has its own entity and is influenced by topography, water source and anthropogenic activities. It becomes the actual-time defined totality of the plant cover of a certain landscape and geographic region and therefore a concrete system through which one can walk and that one can measure, act and weigh (Naveh, 1984).

The vegetation vigor assessment is best illustrated by employing a digita image analysis method called Normalized Difference Vegetation Index (NDVI) a ratio output where the reflectance values of NIR region are maximized by separating the unwanted reflectance values in red wavelength of EMR. The NDVI is calculated as:

$$\text{NDVI} + \frac{(\text{Ch}_2 - \text{Ch}_1)}{(\text{Ch}_2 + \text{Ch}_1)}$$

where Ch_1 represents Red band of the visible channel (0.62 to 0.68 μm.) and Ch_2 represents near Infra Red Channel (0.77 to 0.86 μm.).

The vegetation status mapping using the above method requires limited ground truth and used for consistent vegetation discrimination. Higher NDVI values are associated with greater density, larger leaf area and large green biomass of the canopy and lower the value represents less vegetation area. Gaussian post normalisation is used in bringing the NDVI in the range of 0 to 255. Look up Table (LUT) is prepared for output gray values. The NDVI image is shown in **Figure 7.27.** The classification of gray values for vegetation vigor assessment is carried out as follows:

Gray Values	Land use category
0 - 90	Water bodies
91 - 117	Dry lands without vegetation
118 - 153	Agricultural land
154 - 165	Degraded forest/scrub lands
166 - 200	Medium vegetation areas
201 - 255	Dense vegetation

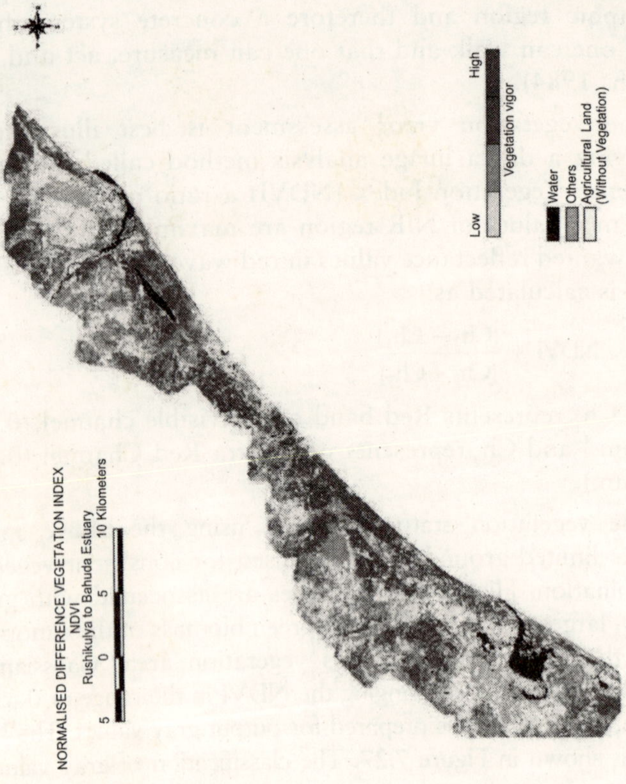
NORMALISED DIFFERENCE VEGETATION INDEX
NDVI
Rushikulya to Bahuda Estuary
5 0 5 10 Kilometers
Low
Vegetation vigor
High
Water
Others
Agricultural Land
(Without Vegetation)

Figure 7.27

The NDVI image clearly shows area of vegetation vigor based on the spectral data while the forested and plantation area show different patterns of vegetation growth, the scanty vegetation along the shore show the areas needing afforestation.

The Transformed Vegetation Index (TVI)

Deering et al. (1975) proposed the TVI by modifying the NDVI. They modified it by adding a constant of 0.50 to all its values and taking the square root of the results. In order to avoid the negative values of NDVI, they added the constant 0.50 the calculation of the square root is intended to correct NDVI values that approximate a Poisson distribution and introduce a normal distribution. With these two elements, the TVI takes the form:

$$\text{TVI} = \sqrt{\left(\frac{\text{NIR}-\text{RED}}{\text{NIR}+\text{RED}}\right) + 0.5}$$

The Corrected Transformed Vegetation Index (CTVI)

Perry and Lautenschlager (1984) proposed the CTVI in order to correct the TVI. Addition of 0.50, a constant to all NDVI values, does not necessarily eliminate all negative values as NDVI values may have the range -1 to +1. In order to resolve this situation the CTVI is being proposed, which is written as:

$$\text{CTVI} = \frac{\text{NDVI} + 0.5}{\text{ABS}(\text{NDVI} + 0.5)} \times \sqrt{\text{ABS}(\text{NDVI} + 0.5)}$$

Since the absolute value is taken, there will be no negative NDVI.

Thiam's Transformed Vegetation Index (TTVI)

Thiam (1997) proposed the TTVI after realising the fact that due to an overestimation of the greenness the resulting image of the CTVI was very "noisy". He suggested ignoring the first term of the CTVI equation in order to obtain better results. This is done by simply taking the square root of the absolute values of the NDVI in the original TVI expression to have a new VI called the TTVI.

$$\text{TTVI} = \sqrt{\text{ABS}(\text{NDVI} + 0.5)}$$

The Normalized Ratio Vegetation Index (NRVI)

Baret and Guyot (1991) proposed the NRVI, which is a modified equation of the RVI. In this equation, the result of RVI – 1 is normalized over RVI + 1.

$$NRVI = \frac{RVI - 1}{RVI + 1}$$

Similar to NDVI, it reduces topographic, illumination and atmospheric effects and it creates a statistically desirable normal distribution.

The Infrared Percentage Vegetation Index (IPVI)

The IPVI was first described by Crippen (1990). Its functioning is equivalent to that of NDVI and related to each other by the following equation:

$$IPVI = \frac{NDVI + 1}{2}$$

The IPVI was implemented because of the fact that in the numerator of the NDVI, the red which was subtracted thought to be irrelevant. Crippen therefore proposed this index as a way of improving calculation speed. Its value is also restricted between 0 and 1. IPVI can be calculated from the Red and NIR using the following equation:

$$IPVI = \frac{NIR}{NIR + Red}.$$

Distance-based Vegetation Index

The second broad category of classic vegetation indices is derived based on the perpendicular vegetation index (PVI). Distance-based vegetation indices are designed to eliminate the effect of the background soil brightness and only detect the features of the vegetation cover. The soil optical effect on the canopy reflectances is not minimized by using the RVI and NDVI. Richardson and Wiegand (1977) proposed the Perpendicular Vegetation Index (PVI) which is defined as the orthogonal (perpendicular) distance between a point representing the plant canopy (red_{veg}, nir_{veg}) and the soil line (**Figure 7.28**). These indices are distinct from the ratio based indices in that the greenness isolines in the Red-NIR space do not converge in the origin, but instead remain parallel to the principal axis of soil spectral variation.

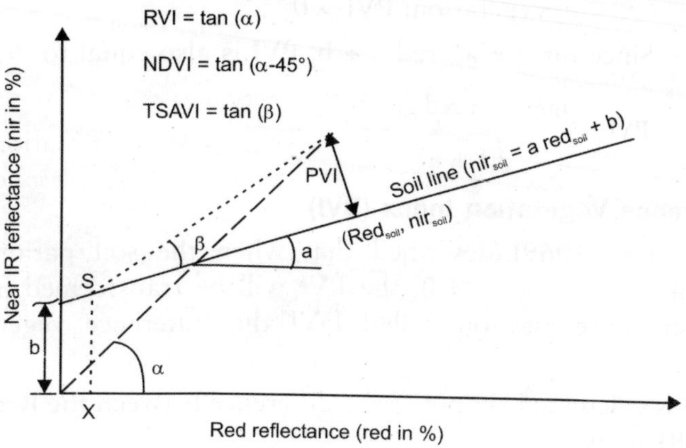

Figure 7.28

In the figure, soil line can be defined as a linear relationship between the near-infrared (NIR) and red (R) reflectance of bare soil as characterized by slope and intercept parameters. Vegetation indices use soil line parameters extensively in crop growth analyses.

In the Figure 7.28 the point of intersection of the soil line and the perpendicular is (red_{soil}, nir_{soil}) and RVI is the slope of the line joining the origin (0, 0) to the vegetation point (red_{veg}, nir_{veg}). It is, therefore, equal to tan (α) and NDVI is equal to tan (α–45°).

PVI is computed by:

$$PVI = \sqrt{\{(red_{soil} - red_{veg})^2 + (nir_{soil} - nir_{veg})^2\}}$$

where:

red_{soil} = soil reflectance in red band

red_{veg} = vegetation reflectance in red band

nir_{soil} = soil reflectance in nir band = a* red_{soil} + b, (a, b are parameters of the soil line)

nir_{veg} = vegetation reflectance in nir band

PVI has the following value ranges:

Bare soil: PVI = 0 (whatever the soil surface status)

Water: PVI < 0

Vegetation: PVI > 0

Since $nir_{soil} = a^* \; red_{soil} + b$, PVI is also equal to

$$PVI = \frac{nir_{veg} - a^* red_{soil} - b}{\sqrt{(1 + a^2)}}$$

Difference Vegetation Index (DVI)

Jordan (1969) described that when the soil parameters become a = 1 and b = 0, the PVI will be transformed into a simpler representation called DVI the difference vegetation index.

It is calculated simply as the difference between the Red and the NIR band.

$$DVI = NIR - Red$$

Richardson and Wiegand (1977) also suggested the equation of DVI in terms of Landsat MSS. The particularity of the DVI is that it weights the near-infrared band by the slope of the soil line. It is written:

$$DVI = g \; MSS\,7 - MSS\,5$$

where, g is the slope of the soil line.

With DVI, a value of zero indicates bare soil, values less than zero indicate water, and those greater than zero indicate vegetation.

The Ashburn Vegetation Index (AVI)

This vegetation index is a measure of green growing vegetation. Ashburn (1978) has suggested the algorithm in which the values in MSS 7 are multiplied by 2 in order to scale the 6-bit data values of this channel to match with the 8-bit values of MSS 5. The equation is written:

$$AVI = 2.0 \times MSS\,7 - MSS\,5$$

This scaling factor would not apply wherever both bands are 7-bit or 8-bit and the equation is rewritten as a simple subtraction.

The Soil-Adjusted Vegetation Index (SAVI)

Huete (1988) has introduced the vegetation index SAVI. This index attempts to be a hybrid between the ratio-based indices and the perpendicular indices. The intention was to minimize the effects of soil background on the vegetation signal by incorporating a constant soil adjustment factor *L* into the denominator of the NDVI equation.

The reasoning behind this index is that:

(i) the isovegetation lines are not parallel, and

(ii) they do not all converge at a single point.

$$\text{SAVI} = \frac{\text{NIR} - \text{Red}}{\text{NIR} + \text{Red} + \text{L}}(1+\text{L})$$

where L is a correction factor which ranges from 0 for very high vegetation cover to 1 for very low vegetation cover. The most typically used value is 0.5 which is for intermediate vegetation cover.

(1 + L) factor is used in SAVI and MSAVI (Modified Soil-Adjusted Vegetation Index) to cause the range of the vegetation index to be from –1 to +1. This is done so that both vegetation indices reduce to NDVI when the adjustment factor L goes to zero.

Transformed Soil-Adjusted Vegetation Index (TSAVI)

Baret et al. (1989) and Baret and Guyot (1991) have described the Transformed Soil Adjusted Vegetation Index (TSAVI) which is an improved version of PVI. It is computed as follows:

$$\text{TSAVI} = \frac{a(nir_{veg} - a \times red_{veg} - b)}{red_{veg} + a \times nir_{veg} - a \times b + X(1 + a^2)}$$

where:

a, b are parameters of the soil line

X = 0.08 (to minimize soil effects).

The Modified Soil-Adjusted Vegetation Indices ($MSAVI_1$ and $MSAVI_2$)

Based on a modification of the L factor of the SAVI, Qi et al. (1994) suggested the two MSAVIs (1 and 2). Both are intended to better correct the soil background brightness in different vegeta*ion cover conditions.

With $MSAVI_1$, L is selected as an empirical function due to the fact that L decreases with decreasing vegetation cover as is the case in semi-arid lands (Qi et al., 1994).

$$MSAVI_1 = \frac{NIR - RED}{NIR + RED + L}.(1+L)$$

where:

(L = 1 – 2g NDVI * WDVI)

Where γ is the slope of the background soil, and 2 is used to increase the L dynamic range (0 to 1)

$$MSAVI_2 = \frac{2pNIR + \sqrt{\{(2pNIR+1)^2 - 8(pIR - pRED)\}}}{2}$$

where:

pNIR = reflectance of the near infrared band.

pRED = reflectance of the red band.

Weighted Difference Vegetation Index (WDVI)

Clevers (1988) has introduced the WDVI, which has a relationship to PVI. It is a simpler version of PVI having unrestricted range. WDVI is very sensitive to atmospheric variations (Qi et al., 1994).

It is expressed as:

WDVI = nir–a*red

where a is the slope of the soil line. The WDVI although seems to be simple, it is as efficient as most of the slope-based VIs. Vegetation signal is maximized and soil brightness is minimised in the NIR region by multiplying the red band with the slope of the soil line (i.e. a*red).

Conclusion

Vegetation index is regarded as a simple remote sensing tool that is being used for more than 25 years. Plenty of literatures are available on numerous vegetation indices and those indices are still largely in operation today.

Various vegetation index models were introduced by various scientists to provide a quantitative assessment of green vegetation biomass. These models are applicable to both low and high spatial resolution satellite imageries and similar other such imageries those have been recorded in visible and NIR regions. The indices have been used in various applications to assess green vegetation biomass and also have been used in environmental change detection, especially in the context of drought (Kogan, 1990; Tripathy et al., 1996; Liu and Kogan, 1996) and land degradation risk assessment.

A vegetation index may be defined as a simple mathematical calculation used in remote sensing to estimate the likelihood that vegetation was actively growing at a particular location whenever it was observed. Most of the vegetation indices were empirically developed in such a way that larger values correspond to higher probabilities of actually finding live green plants at the selected location and time of observation.

In the areas with low vegetation cover, the vegetation indices work very badly. This is due to the fact that in those areas having low vegetation cover, the soil reflectance spectra dominate over the reflectance spectra of the vegetation. Huete et al. (1985) and Elvidge and Lyon (1985) demonstrated that the soil background can have a profound impact and vegetation index values with bright backgrounds are producing lower vegetation index values than dark backgrounds. Elvidge and Lyon (1985) documented that many background materials such as soil, rock, plant litter, etc. vary in their red-NIR slope, and these variations seriously impact measurements of vegetation indices. Then there is problem of non-linear mixing.

The advantages in implementing these vegetation indices are:

(i) simple to implement due to simple algorithms involved in computation.

(ii) low computational cost, which is an achievement of management objectives.

(iii) reduction in data volume for easy handling and interpretation of satellite data.

(iv) quick estimate to reduce analyst time.

However, the selection of an appropriate index for a particular application depends on the specific requirements of the analyst and on the observational data available. For the best estimation of a particular geophysical parameter based on data from particular sensor, optimized vegetation indices have been developed. Users should thus select the tool that will provide the desired information on the basis of available data.

7.13 PRINCIPAL COMPONENTS ANALYSIS

7.13.1 Introduction

Principal Components Analysis is a method of transforming a given set of observed variables into another set of variables. It is based on the fundamental assumption that some underlying factors, which are smaller in number than the number of observed variables, are responsible for the covariation among the observed variables. In other words, it is a statistical data reduction technique that linearly transforms an original set of variables into a substantially smaller set of uncorrelated variables that represent most of the information contained in the original set of variables.

Its goal is to reduce the dimensionality of the original set of variables and identify new meaningful underlying variables. A small set of uncorrelated variables is much easier to understand, interpret and use in further analysis than a large set of correlated variables.

For clear understanding, let us say, there are 'p' variables in a data set which are highly correlated. The 'p' correlated variables are then linearly transformed to 'k' uncorrelated variables. If 'k' variables are treated as independent, then they will maximize the prediction of the original 'p' variables. If 'k' derived variables which maximize the variance accounted for in the original variables, are called "principal components". Thus to recapitulate the above discussion:

(i) Principal components maximize the variance accounted for in the orginal variables;

(ii) Principal components represent the variation in the original variables;

(iii) PCA searches for a few uncorrelated linear combinations of the original variables that capture most of the information contained in the original variables.

Gauch (1993) described that the Principal Components (PCs) have the ability to simplify multivariate data by reducing the dimensionality of the data set. Loughlin (1991) pointed out that the features that are hidden in the data are brought out by PC analysis. Sabins (1997) while realizing its potential use in the analysis of multispectral digital data stated that Principal Components Analysis (PCA) is a spectral enhancement which can be used to compress the information content of a multispectral data set. Both Morrison (1976) and Preisendorfer (1988) described the PC Analysis as a translation and rotation of the original coordinate system into a new coordinate system that better reflects the principal modes of variability in the data set being analyzed. The first few PCs (mostly two or three) explain most of the variance in the original data set and the remaining components account only for other information and noise. Jensen (1996) in this regard stated that, the first principal component (PC-1) describes most of the variation of the brightness values for the pixels of the original bands. Subsequent components explain less and less of the data, with the final PC usually corresponding to atmospheric noise in the data rather than any ground features (Sabins, 1997). The main benefit of principal components analysis is that it can reduce the amount of data (bands) without losing much of the information and typically reducing redundancy (Jensen, 1996). For example, the three visible bands (1, 2 and 3) of TM images are usually highly correlated, meaning they look roughly the same and thus provide redundant information for classification purposes (Jensen, 1996). After running a principal components analysis on these three bands, we would find that the majority of the information contained within the three bands would be explained by PC-1. Thus, this one layer of data could replace the three original bands without much loss of information.

Although this technique has been frequently used in high resolution satellite imageries of both land and ocean, the technique can also be used to analyze the information content of lower-spatial-resolution weather satellite imagery and has been termed Principal Component Images (PCIs) by Hillger (1996). The principal component transformed images are often more interpretable than the original images (Kaneko, 1978; Byrne et al., 1980; Canas and Barnett, 1985).

The principal components analysis known as Karhunen-Loeve (K-L) transformation in remote sensing and pattern recognition literatures, is often used for determining the underlying statistical dimensionality of the image data set (Ready and Wintz, 1973) for image enhancement (Gillespie, 1980), for digital change detection (Byrne et al., 1980) and for characterizing seasonal changes in cover types (Townshend and Tucker, 1984).

7.13.2 Concept of PCA in Multispectral Satellite Data

Before appreciating the method of Principal Components Analysis and its application to multispectral as well as hyperspectral satellite data, the geometric concept of PCA is felt necessary. It is a simple statistical multivariate analysis. The essence of the method is to make linear combination of data from the multiple bands. This results in a small number (ideally 3) of components of transformed data, which contain most of the variation (independent data) of the original data set. Any three principal components can be displayed as RGB image and interpreted.

To extract maximum possible information that a multispectral image contains is the goal of the PCA. Any multispectral image contains certain bands that are highly correlated. If we consider two bands which are highly correlated, then both the data sets will convey essentially the same information. This fact is exploited in the principle of application of the PCA.

Figure 7.29a illustrates the scatter plot of the DN values from two bands (band 1 and band 2) of a multispectral imagery. Say, μ_1 and μ_2 are the mean values of band 1 and band 2 data sets. The original axes are then translated and rotated into a new coordinate system with (μ_1, μ_2) as the origin about which the new DN values are redistributed. In this case the relevant

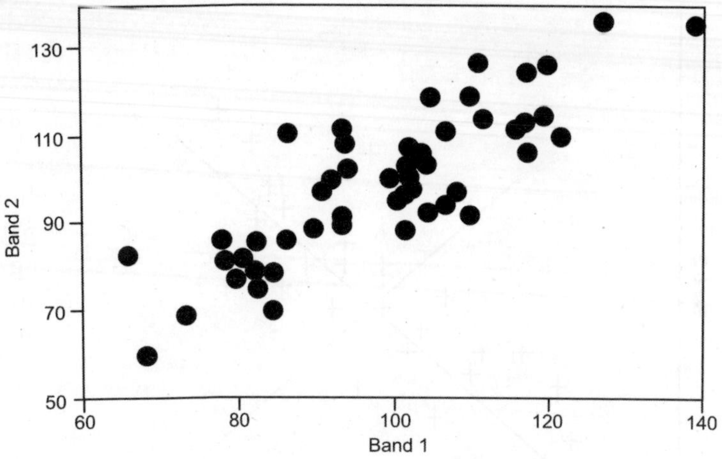

Figure 7.29a

means were subtracted from each value. **Figure 7.29b** depicts the shifting of origin to a new position (μ_1, μ_2). It is important to realise that it is the relationships between points that matters. Moving the axes does not change the underlying pattern in the data. In **Figure 7.29c** the axes have been rotated about their new origin. This rotation has been carried out in such a way that the position of one axis accounts for the largest amount of the data variance. Scattered points when rotated form an ellipse, the major axis of which represents the largest proportion of the total variance. This is the position of the new axis, termed as PC-1. The second axis perpendicular to PC-1 is the minor axis of the ellipse that represents PC-2. It must also be uncorrelated with the first one.

Lengths and orientations of these axes are given by the Eigen values and Eigen vectors of the correlation matrix.

Steps Involved in Implementing PCA

The following steps are involved while extracting the principal components:

(i) The mean value for each data set is first calculated.

(ii) The variance-covariance matrix for all the bands is computed.

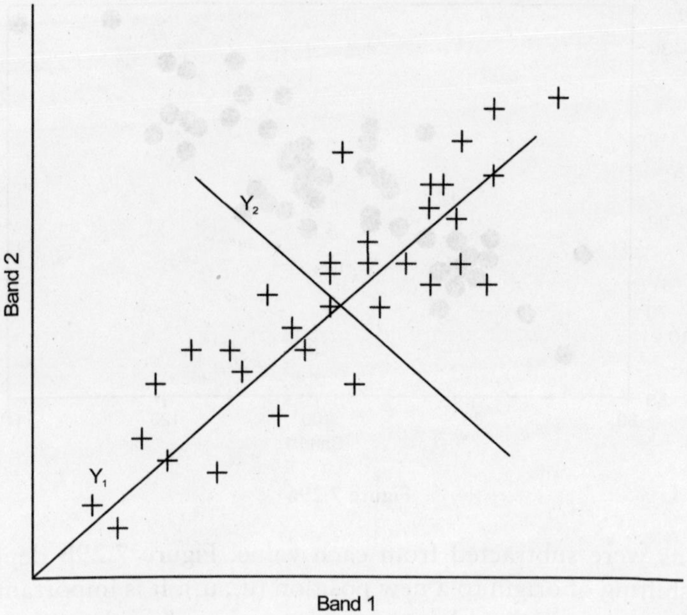

Figure 7.29b

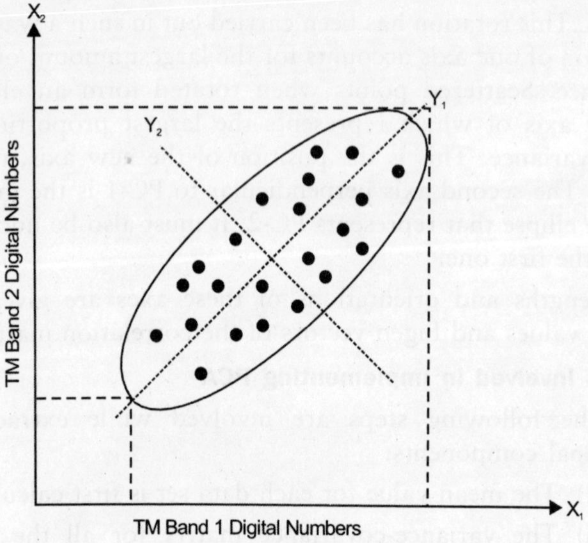

Figure 7.29c

For two band data sets, size of the matrix is 2 × 2 and the variance-covariance matrix is given by:

$$\begin{bmatrix} \sigma_{11} & \sigma_{12} \\ \sigma_{21} & \sigma_{22} \end{bmatrix}$$

Diagonal elements σ_{11} and σ_{22}, are known as the variance and the formula for calculating variance is given by:

$$\text{Variance} = \frac{\sum (X-X)^2}{N}$$

The matrix elements σ_{21} and σ_{12} are said to be covariance and the formula for computing covariance is given by:

$$\text{Covar}(X, Y) = \frac{\sum \left(X_i - \bar{X}\right)\left(Y_t - \bar{Y}\right)}{N}$$

where Covar (X, Y) = covariance between (X, Y)

X_i = gray values of any one band

$\bar{X}$ = mean gray value (X) of the same band

Y_i = gray value of other band

$\bar{Y}$ = mean gray value (Y) of the same band

N = total number of pixels.

(iii) Transformation equation is established, given by

$$[[A] - \lambda [I]]X = 0$$

where:

A: Variance-Covariance matrix

λ: Eigen value, magnitude of the PC (or length of the axes)

X: Eigen vectors (represent the direction of the axes)

I: Identity matrix.

(iv) After solving the above equation for λ, the eigen vectors for each λ is calculated.

(v) New DN values for PC-1 and PC-2 are evaluated

(vi) Percentage variance for PC-1 using the following formula is calculated:

$$\%\ \text{Var (PC-1)} = \text{Eigen value } (\lambda_1) \div \Sigma \lambda$$

Similarly % Var (PC-2) can also be calculated.

(vii) Finally, Factor Loading calculation is done adopting the following formula:

$$R_{kp} = \left(a_{kp}\sqrt{\lambda_p}\right) \div \sqrt{Var_k}$$

R_{kp} = Loading factor of band k and principal component PC-1

a_{kp} = Eigen vector

λp = Eigen value

Application of PCA in IRS multispectral data is shown in Figure 7.30.

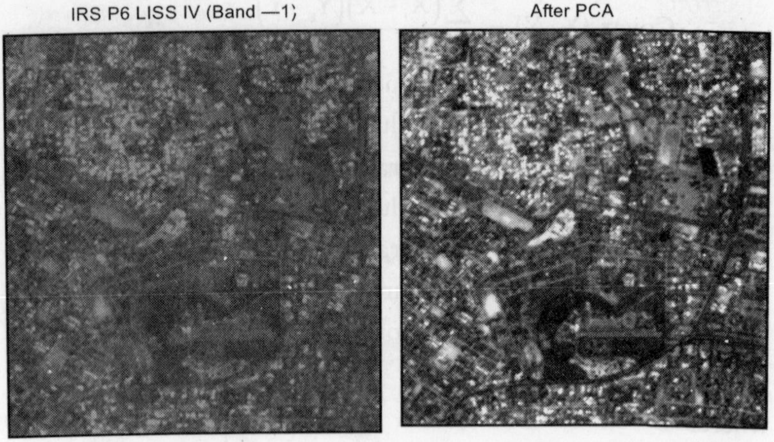

Figure 7.30

7.14. IMAGE CLASSIFICATION

7.14.1 Introduction

When we want to gather spatial information about the different features on the ground for environmental monitoring and decision-making purposes we need to examine color multispectral satellite imagery to identify and separate feature classes. Therefore, an important aspect of the digital image processing is to identify groups of pixels that have similar spectral characteristics and to determine the various features or land cover classes represented by these groups. This kind of analysis discriminating features on the image on the basis of their spectral character we call it as image classification. For example, since the reflectance of sand is higher than the reflectance of trees in one band, they could be classified based on their midpoint. Analyst's visual inspection of a multispectral image to classify reveals information on the visual elements (tone, contrast, shape, etc.) of different classes or categories which are in some way related to the underlying spatial objects of the region. Any attempt to use this information for identification, discrimination and classification of Earth surface features is based on the premise that different objects have unique reflectance/emittance properties in different parts of the electromagnetic spectrum. That is spectral signature of any object and or its conditions comprise a set of values for its reflectance and/or its emittance in different spectral bands. Unfortunatel; his is quite often not the case and provides the main physical imitation to remotely sensed habitat mapping.

A multispectral satellite image captures the signatures of various Earth s irface features. Classification of those signatures is simply a method of converting the spectral image into a finite number of individual classes or categories of data that represent the surface types seen in the imagery. These may be used to identify vegetation types, cultural details, natural resources, or transient changes in any of these properties.

Multispectral data are mostly used for classification, though any satellite data can be classified. The spectral pattern present within the data for each pixel is used as the numerical basis for categorization. A single band data is rather difficult to classify. This is because of more than one surface features exhibit the same digital number. Normally multibands are used for classification, and their combined pixel values are used to identify the spectral signatures of the scene classes present in the image. The more bands used to create a classification, the more likely the analyst will get a set of unique land cover classes.

Classification in earlier days was performed by visual inspection and interpretation of features on the paper print satellite imageries, and manual delineation of their boundaries. However, in today's digital world, human could able to use computer assisted classification and interpretation using various advanced digital image processing softwares and digital satellite data instead of paper prints.

7.14.2 Definition

Multispectral classification may be defined as the process of sorting pixels into a finite number of individual classes, or categories of data, based on their DN values. If a pixel satisfies a certain set of criteria, the pixel is assigned to the class that corresponds to those criteria. This process is also referred to as image segmentation (ERDAS Inc., 1999). Extraction of distinct classes or themes, land use and land cover classification categories, from satellite imagery is the image classification. Rees (2001) described image classification as the process of making quantitative decisions from image data, grouping pixels or regions of the image into classes intended to represent different physical objects or types. The output of the classification process may be regarded as a thematic map rather than as an image. The classification procedure is based on the assumptions that (i) Patterns of their DN, usually in multiband data (Spectral Classification), (ii) Spatial relationship with neighboring pixels, and (iii) Relationships between the data accquired on different dates. The result of a classification is that all pixels in an image are assigned to particular classes or themes (e.g. water, coniferous forest, deciduous forest, corn, wheat, urban, etc.), resulting in a classified image that is essentially a thematic map of the original image. Classification follows on feature extraction applied to the data.

The classification can be performed to monitor land use/ land cover patterns, geophysical processes, oceanic features, weather forecast, etc.

7.14.3 Spectral Classes and Information Classes

While classifying an image, analyst must discriminate between spectral classes and information classes. Spectral classes are nothing but grouping of pixels with similar characteristics i.e. the original multispectral image is imagined to be segmented into individual number of groups or classes. Each group contains pixels of nearly uniform spectral characteristics. In other words, spectral classes are classes that are uniform with respect to the spectral values that compose each pixel.

Informational classes are composed of numerous spectral classes—spectrally distinct groups of pixels that together may be assembled to form a unique descriptive class. Information classes are the various themes or groups the analyst is attempting to identify in an image. A classification process assigns to each pixel of a remotely sensed image a label denoting an information class, thus producing a thematic map. Information classes may include such classes as deciduous and coniferous forests, various agricultural crop types, urban or inland bodies of water. The objective of image classification is to match the spectral classes in the data to the information classes of interest.

7.14.4 Spectral Pattern Recognition

Spectral pattern recognition can be defined as identification and analysis of remotely sensed features using different portions of the electromagnetic spectrum.

Pattern recognition is a computer based methodology adopted to extract the information on features from a given satellite imagery. The term spectral pattern refers to pattern so constructed due to pixel's response to different portions of the EM spectrum. In other words, spectral pattern refers to the set of radiance measurements obtained in the various spectral bands for each pixel. Lillesand and Kiefer (1987) defined spectral pattern recognition as the family of classification procedures that utilizes this pixel by pixel spectral information as the basis for automated land cover classification.

A complete pattern recognition system consists of:

(i) a sensor

(ii) a feature extraction mechanism

(iii) a classification or description scheme.

A sensor records the relative intensities of an area over particular regions of the electromagnetic spectrum. Thus, gathers the observations to be classified or described. Numeric or symbolic information are computed from the observations using a feature extraction mechanism and finally, pixels are sorted into individual classes or categories based on a classification scheme.

Similar to spectral pattern recognition, there are spatial and temporal pattern recognitions. Spatial pattern recognition refers to sorting of image pixels into individual groups or classes based on their spatial relationship with the neighboring pixels. The classification scheme, in this case, is based on image texture, pixel proximity, feature size, shape, directionality, repetition and context. Temporal pattern recognition as the name indicates uses time as a basis while identifying the features. The variation in reflectance value as well as spatial changes during the growing cycle of an agricultural crop helps to identify it on multidate satellite imageries, which is not possible on attemping single date imagery.

Human eye can perform the process of pattern recognition on a spatially and spectrally enhanced image. Brain can affix information to the individual patterns by sorting pixels based on texture and colors. Thus, completes the process of classification of a multispectral satellite image.

Computer records the intensity values of the pixels (ground radiance value) in the form of data files. When displays it provides brightness values and statistics of the pixels. Computer sorts the pixels to form clusters based on certain mathematical criteria.

In the following section emphasis is given only to classification based on spectral pattern recognition, which form the backbone of the land cover mapping. Two scene features can never be distinguished on a remotely sensed imagery unless they have different spectra. Finding distinctive spectral response patterns is the key to most procedures for computer-assisted interpretation of remotely sensed imagery.

7.14.5 Image Classification Methods

Majority of conventional methods of image classification is based solely on delineating feature classes using only the spectral information of the individual pixels but not the image characteristics as texture and pattern. These techniques range from visual interpretation and direct mapping from an image (photo-interpretation) through multispectral image classification techniques viz. unsupervised and supervised classifications. The success with which this can be done will depend on two things: (i) the presence of distinctive signatures for the land cover classes of interest in the band set being used; and (ii) the ability to reliably distinguish these signatures from other spectral response patterns that may be present. Alternatively, new techniques such as neural networks based classification are being developed which include classification of pixels according to texture and pattern. However, the people rely on the standard multispectral classification, which is widely preferred to neural networks.

7.14.5.1 Photo-interpretation (Visual Interpretation)

Photo-interpretation is a method by which the analyst visually inspects the satellite image to detect and categorise the features with reference to ancillary information, local experience, common knowledge, field visits for ground truth collection and sound knowledge on theoretical aspects of radiation properties.

Remote sensing image may be considered as message from the space about happenings on the globe. The message is coded in the form of wavelength and intensity. This needs to be decoded before the message can be fully read (understood). The process of decoding is similar to image interpretation, which basically relies on the interpreter's knowledge on the electromagnetic radiation properties. Interpreter extracts meaningful information about the object/feature under investigation exercising his scientific skill, judgement as well as his local experience.

The image may be a photographic print or color hard copy of a satellite image or digital data viewed on a computer. The visual interpretation of satellite imagery involves three stages: (i) detection stage, (ii) recognition and identification stage, and (iii) feature boundary delineation stage.

Features are recognized and identified by the interpreter based on his day-to-day experience. Then they are delineated and labeled manually. Detection is the first stage followed by the recognition and identification stage. In the act of recognition, the interpreter follows strategically the general, local as well as specific levels of reference to assign objects into known categories. The general level of reference refers to the interpreter's knowledge of the phenomena and processes to be interpreted. Local level of reference includes local knowledge of the interpreter on the objects and their surrounding. Finally, the specific level is the interpreter's indepth knowledge of the processes and phenomena that he wants to interpret. In recognition and identification stage, the different objects are identified examining image elements (tone, texture, etc.) with the aid of analyst's visual interpretation key. The decisions reached at, are solely of the interpreter without any computational effort or statistical processing of the data. Delineation of boundaries having homogeneous patterns and characteristics is achieved using a tracing overlay and pencil or a mouse driven cursor on a computer screen. Objects in each polygon are classified following the process of induction (general inference from particular cases) and deduction (particular inference from general observations). For accuracy check, ground verification is carried out and final maps are prepared to display as spatial information for the use in Geographic Information System.

Conclusion

Manual preparation of a thematic map followed by visual interpretation of satellite imagery is outlined below:

(i) The habitats or information classes on the satellite image are first detected by the interpreter.

(ii) With the help of visual interpretation aid and interpreter's scientific skill, the habitat boundaries are delineated. In GIS, those habitat boundaries (areas) are termed as polygons.

(iii) Finally thematic map is prepared. In this stage, boundaries between the habitat polygons are digitized manually. The map so prepared is said to be in the vector format. The type of habitat within each polygon is interpreted from the original image and labeled.

(iv) The vector format in stage (iii) may be converted into

raster image with identified and labeled habitats in selective colors.

7.14.5.2 Multispectral classification of image data

Classification process in remote sensing attempts to sort all pixels in a digital image into various land cover classes, or "*themes*". This categorized data may then be used to produce thematic maps of the land cover present in an image. Normally, multispectral data are used to perform the classification and, indeed, the spectral pattern present within the data for each pixel is used as the numerical basis for categorization (Lillesand and Kiefer, 1994). Different land cover types (habitats) in a satellite image can be discriminated using image classification algorithms on spectral features i.e. the brightness and color information contained in each pixel. Each class of land cover is referred to as a "theme" and the product of classification is known as a "thematic map". The accuracy of the thematic map derived from remote sensing images should be verified by field observation, and/or other ancillary data such as aerial photographs.

Classification of multispectral satellite data generally comprises of four stages (**Figure 7.31**). Those are (i) training, (ii) signature evaluation, (iii) decision-making, and (iv) accuracy assessment.

Training

The first step of the classification of a multispectral imagery is to select training sites. It is the process of selecting particular features (e.g. agricultural field, forest types, etc.) which best describes the pattern. Hord (1982) described training as the process of defining the criteria by which these patterns are recognized. For this purpose, the computer system must be trained first to recognize the patterns in the data set. Training can be performed with either supervised or unsupervised classification.

Signature Evaluation

Each feature class is defined by spectral envelope i.e. by the minimum and maximum pixel values. For three band data the

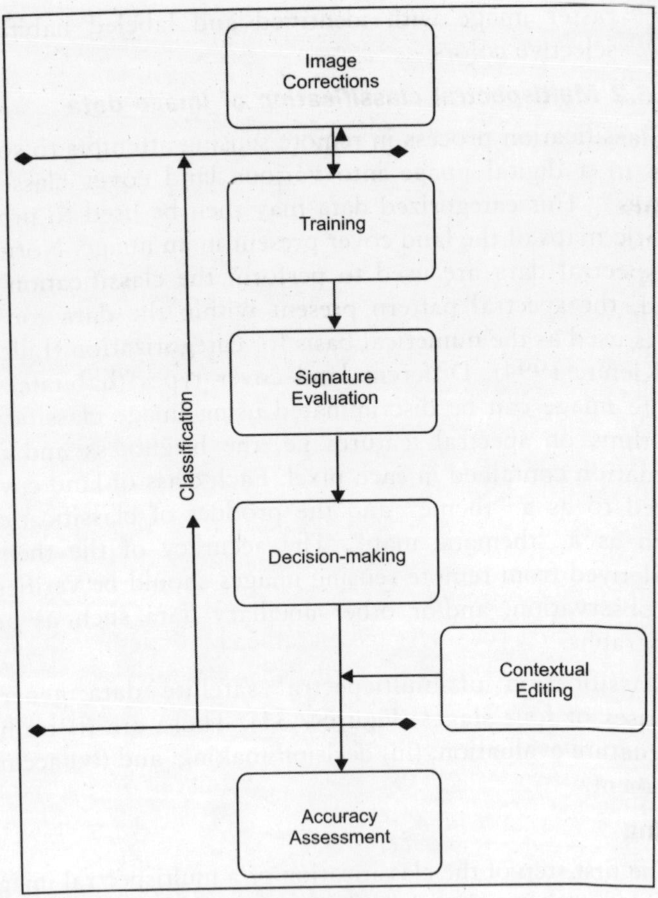

Figure 7.31

spectral envelope will represent the three dimensional cube. This simple statistical characterization of each information class is known as signature. In a multispectral digital image, signature is the area of visually homogeneous spectral response. A signature is a set of data that statistically defines the training site for each land category of interest. Each signature corresponds to a class and is used with a decision rule to assign the pixels in the image to a class.

Generating quality signature is a matter of concern as the entire classification scheme relies on it. The difficulty sometimes may be reflected due to heterogeneity within the study area and thus internal spectral variety within the training sites. The rural landscape is a mosaic of natural and human managed patches that vary in size, shape, and arrangement (Turner, 1990). In turn, this variety in the landscape is reflected by an array of different spectral characteristics. If training sites are characterized by inherent spectral variation, classification performance or accuracy will decrease as a result of its large spectral variability (Arai, 1992).

Signature data is the backbone of the image classification scheme. The importance of using signature data lies in the fact that it defines habitat classes in the feature space, which will subsequently be used to classify the data. The classification decision rule (classification algorithm) requires the signature definitions as input (ERDAS, 1999).

Each parametric signature is defined by statistical parameters of the pixels that are included in the training sites. A parametric signature is characterized by statistical attributes such as the number of bands in the input image, the minimum and maximum data value for each band for each training site, the mean data value in each band for each training site, the covariance matrix for training site, and the number of pixels in each training site (ERDAS, 1999).

Analysts determine signatures based upon the feature's intensity value and position in the electromagnetic spectrum. A signature is a set of statistics that defines the spectral characteristic of a target phenomenon or training-sites. Image analysts determine the measurement of signature separability by determining quantitatively the relation between class signatures. Signatures are refined by improved ground-truth and accuracy assessment analysis. By utilizing the developed signatures in multispectral classification and thematic mapping, the analyst generates new data for analysis (ERDAS, 1999).

Decision-making

Decision rule follows the signature evaluation of the training sites.

Decision rules are nothing but, certain mathematical algorithms used to determine qualifying of an image pixel into a particular class based on its signature characteristics. It performs the actual classification process of sorting image pixels into distinct classes using data contained in the signature. Although training and signature evaluation form the integral part of the process, decision-making is sometimes referred to as classification.

Classification Accuracy

It is the final stage of the classification scheme. In thematic mapping from remotely sensed data, the term accuracy is used typically to express the degree of 'correctness' of a map or classification. A thematic map derived with a classification may be considered accurate if it provides an unbiased representation of the land cover of the region it portrays. In essence, therefore, classification accuracy is typically taken to mean the degree to which the derived image classification agrees with reality or conforms to the 'truth' (Campbell, 1996; Janssen and Van der Wel, 1994; Maling, 1989; Smits et al., 1999). A classification error is, thus, some discrepancy between the situation depicted on the thematic map and reality.

To determine the classification accuracy, traditional methods employ random selection of sampling locations to visit on the ground for field verification of the true land cover type. Test site visit is compulsory to attain confident on the accuracy of the information classes extracted from the satellite image. The principal aim of the visit is to observe and collect information about the actual condition on the ground and determine the relationship between remotely sensed data and the object to be observed. Ground truth collection is carried out concurrent with the satellite overpass or at least within the time that the environmental condition does not change. For accuracy determination, the present day researchers apart from ground verification use techniques based on confusion or error matrix (Kappa and Tau coefficients).

Another approach for obtaining higher classification accuracy is adopting hierarchical structure in neural networks. In many pattern recognition applications, classification accuracy is often lowered due to patterns that are likely to be wrongly classified. Data which are close to boundaries between classes usually cause errors in classification. To classify these data more effectively, hierarchical structure is applied in classification using neural networks. The hierarchical approach means classification is done sequentially by multistage neural networks. When the input vectors which are difficult to classify are detected, their classification is done in the succeeding stage neural networks. In the current stage, input vectors having higher class separability are classified. This is effective to reduce misclassification and improve classification performance since successive classifiers are tuned to reduce remaining errors (Lee and Ersoy, 2006).

(A) Supervised Classification

Supervised image classification is a method in which the analyst defines small areas, called training sites on the image which are representative of each desired land cover category. Selection of training sites forms a basis for the effective implementation of the classification. The delineation of training areas as representative of a cover type is most effective when an image analyst has knowledge of the geography of a region and experience with the spectral properties of the cover classes (Skidmore, 1989). The image analyst then trains the software to recognize spectral values or signatures associated with the training sites. After the signatures for each land cover category have been defined, the software then uses those signatures to classify the remaining pixels (ERDAS, 1999). Various comparison methods are used to determine if a specific pixel qualifies as a class member. A broad range of different classification methods available are: Parallelepiped, Maximum Likelihood, Minimum Distance, Mahalanobis Distance, Binary Encoding, and Spectral Angle Mapper.

Various stages involved in supervised classification are shown in **Figure 7.32**. However, brief outline are as follows:

(i) Representative pixels from each of the information classes are first selected. The selected pixels are referred to as training data. Ground reference data are used

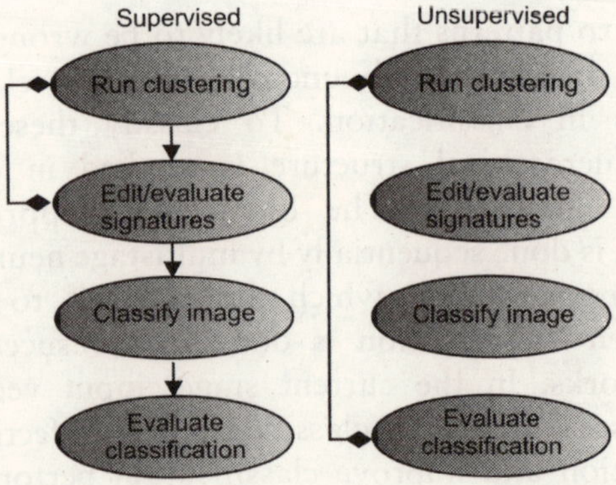

Figure 7.32

while selecting the training pixels. Usually the pixels in the training field for a given class are in a common region enclosed by a border.

(ii) Secondly, parameters which will be used for the particular classification algorithms are estimated using training data. With this procedure, the classifier is constructed and generalization is performed.

(iii) Finally, the images which are to be classified are inputted to the trained classifier, and final classification results are generated.

Supervised Training

The basic concept of supervised classification is to classify unknown pixels using samples of known identity. In supervised training, it is important to have a set of desired classes in mind and then create the appropriate signatures from the data. The analyst must also have some way of recognizing pixels that represent the classes that he wants to extract.

Supervised classification is advisable under the following situations:

(i) If the analyst searches for few classes.

(ii) If selected training sites are known to analyst that can be verified with the ground truth data.

(iii) If the analyst is able to identify distinct, homogeneous regions that represent each class.

It is mostly analyst controlled method. In this process, the analyst first detects and identifies the pixels that represent known land cover features. Then he made up his mind to search for the desired classes. After identifying the known patterns or features, the analyst instructs the computer to make groups or clusters so that each group contains all pixels in the image with similar spectral characteristics. If the classification is accurate, the resulting classes represent the categories within the data that analyst originally identified.

In supervised training it is necessary to have prior knowledge about the image data and land features to be mapped. This is known as a priori knowledge. A priori includes first-hand knowledge of the area (inputs from ground surveys), or possibly knowledge gained from interpreting aerial photographs or existing topographical surveyed maps. Supervised training solely depends upon the specialist's pattern recognition skills and a priori knowledge of the data to help the system determine the statistical criteria (signatures) for data classification.

Information about known features such as location of land cover types and their spatial and spectral characteristics etc. is gathered from the ground verification concurrent with the satellite overpass, which the remote sensing specialists call ground truthing. It refers to acquisition of information about the ground features and their conditions from field work, topographical maps, aerial photos, personal experience and other such reliable ancillary information.

Representative samples of known identity based on the above such criteria are selected as the training samples. The group of pixels which represent known features or habitats on the image are known as training samples. It is important that all pixels within the training samples must represent the feature class that is to be mapped.

Selection of Training Samples

There are various methods of identifying training samples based on the analyst's knowledge of the data, study region, and the desired classes fixed in his mind. Those are as follows:

(i) *Vector layer*

Vector layers are generated digitizing the features in the image that represent the training samples. They are later used to extract statistical information about the pixels within it.

(ii) *A seed pixel*

Identification of a single pixel in the image from which a region grows by way of a statistical algorithm is called as a seed pixel. It acts as the representative of the training sample and treated as model pixel in the image. Contiguous (i.e. similar values) pixels are compared to it and accepted if they are similar to it or rejected if they are dissimilar to it. These criteria including geographical constraints are fixed by the analyst. The accepted pixels having similar spectral characteristics form a homogeneous layer representing a feature class (training sample). When digitized it converts into a vector layer of information for further analysis to extract statistical information.

(iii) *Feature space*

It is a scatterplot of DN values of one band against the DN values of another band. It creates an image having same data structure as the original input image. Both the original and the feature space images are linked by a cursor if we use a satellite data processing software called ERDAS IMAGINE. This helps to determine location of a training sample in the feature space image. The signature of the training sample collected using the feature space image is nonparametric. The decisions made in the classification process have no dependency on the statistics of the pixels. This helps improve classification accuracies for specific non-normal classes, such as urban and exposed rock (Faust et al., 1991).

(iv) *Training samples from unsupervised classification*

If a priori information be adequate to accept some of the classes produced from the unsupervised classification, then those signatures may also be used along with the signatures derived from the above methods in the supervised classification. Thus, in supervised classification, signatures from both the classifications may also be used in combination.

Evaluating Signatures

Brief fundamental on signature evaluation has already been explained in the stages of classification. However, in this section attempt has been made to explain the different methods of evaluating signatures.

A signature that we described previously is a set of data that defines a training sample. There are two types of signatures: (i) parametric, and (ii) non-parametric.

A parametric signature is based on statistical parameters (e.g., mean and covariance matrix) of the pixels in the training sample. It is used to train a staticsally based classifier to define the classes. However, a non-parametric signature is not based on statistics, but on discrete objects in a feature space image.

Once signatures are created, it is important to identify and delete the aberrant training samples. Training samples of similar spectral characteristic are merged together and renamed before subject to decision-making stage of the classification. When the signatures are created, several statistical measures, associated with the pixel values in training samples were calculated. This information can be used to compare signatures.

For evaluating signatures, statistics can be done by viewing two dimensional ellipses of signature statistics, calculating the Euclidean distance between training samples, or calculating the statistical separability of two signatures.

Evaluation of signature involves three steps:

(i) *Contingency matrix*

A contingency matrix is created to test the viability of the selected classes. It classifies only the pixels in the image training sample, based on the signatures. This algorithm reviews all of the selected signatures to determine the degree of overlap found between each of the signatures using the minimum distance, maximum likelihood, or Mahalanobis distance decision rule. The decision rule evaluates the variance (range of DN values) and the covariance (correlation or overlap between signature classes) of each of the established classes in order to derive the probability of each pixel falling into a particular class. When the probability of each pixel's potential has been calculated, the pixel is then assigned to the class with the highest probability (Lillesand et al., 2004). If the signatures are distinct from each other, then a high percentage of each sample's pixels will be classified. The output of the contingency matrix is a matrix of percentages that allows showing how many pixels in each training sample were assigned to each class.

An example of a contingency matrix prepared for the four habitats Reef, Algae, Seagrass and Sand is given in **Table 7.4**. It is found that eighty-five per cent of the pixels in the training sample for class Reef classify as Reef, 10% as Algae and so on. In this example, the signatures from the training sample Sand need to be evaluated further because most of these pixels actually classify as class Algae.

TABLE 7.4: Contingency matrix—an Example

	Reef	Algae	Seagrass	Sand
Reef	85	10	0	5
Algae	7	79	5	9
Seagrass	0	7	91	2
Sand	2	58	2	36

Source: Green et al., 2000.

(ii) *Viewing ellipses and scatterplots*

This is one of the signature evaluation methods representing the training sample signatures in two dimensional feature space. Two dimensional scatterplots, in the form of feature class ellipses, were calculated from the means and standard deviations derived from the range of pixel values in each training site in two different bands. Scatterplot of the training data in two bands takes the shape of an ellipse. Viewing of ellipses in two dimensional feature space simplifies to take a decision based on how the different training sample signatures that we selected are distinct or separable from each other. In the evaluation process, the mean and standard deviation of each signature are used to plot ellipses in two dimensional feature space (**Figure 7.33**). The figure depicts ellipses of joint normal distributions due to plot of data from two bands.

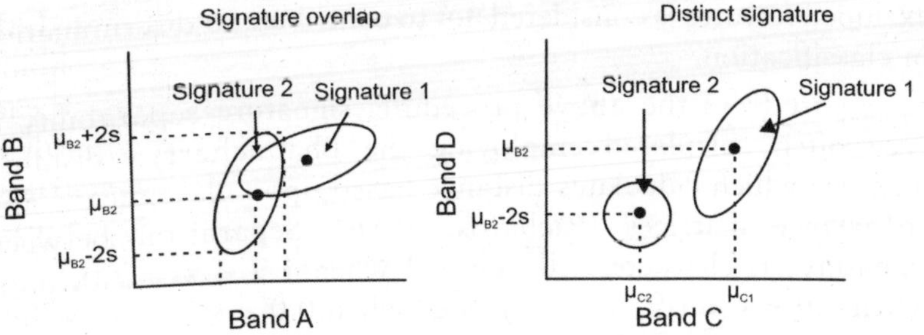

Figure 7.33

By comparing the ellipses for all signatures in two bands, we can be able to determine which signatures and which bands provide accurate classification results. If there is a large amount of overlap, then they are spectrally similar in these particular wavelength bands. However, they may be separable in some other combination of bands. If there is minimal ellipse overlap, in any two band combination, the signatures represent distinct and separable pixel groups.

(iii) *Separability*

Separability is normally used for feature selection. Classification performance depends on the class separability of input data as well as how classification system is designed well. Class separability implies the ease with which patterns can be correctly associated with their classes using statistical pattern classification. If the separability is higher, the classification becomes easier and more correct.

The procedure of signature separability is based on the Euclidean spectral distances between their means. The larger the Euclidean distance between signatures, the more distinct the signatures are from each other, and the more success in accurate classification. The analyst has to determine the minimum value of signature separation to use for the evaluation. Features which are not helpful for discrimination, by contributing little to the separability of spectral classes, are to be discarded. This is referred to as feature selection. By feature selection, the dimensionality of a data set is reduced. If the separability is not lowered even though some features are excluded, then the set of excluded features is considered not to contribute to discrimination in classification.

Apart from the above procedure, signature separability is very often calculated employing the Bhattacharrya distance measure which calculates distance based upon class means and covariance matrices (Richards, 1986). Separation between signatures is characterized as good when 1.9 < x < 2.0, poor when 1.0 < x < 1.9, and very poor when 0.0 < x < 1.0.

(iv) *Transformed Divergence*

This procedure computes the statistical distance between pairs of signatures. The equations used to find the statistical distances are related to the formula used in the maximum likelihood classification algorithm, making it best suited to predict the results of a maximum likelihood classification.

The formula for computing Transformed Divergence (TD) is as follows:

$$D_{ij} = \frac{1}{2} \operatorname{tr}\left[(C_i - C_j)(C_i^{-1} - C_j^{-1})\right] + \frac{1}{2} \operatorname{tr}\left[(C_i^{-1} - C_j^{-1})(\mu_i - \mu_j)(\mu_i - \mu_j)^T\right]$$

$$TD_{ij} = 2000\left(1 - e^{\frac{D_{ij}}{8}}\right)$$

where:

i and j = the two signatures (classes) being compared

C_i = the covariance matrix of signature i

μ_i = the mean vector of signature i

tr = the trace function (matrix algebra)

T = the transposition function

(*Source*: Swain and Davis, 1978)

Note that to find the best band combination, the number of bands used can be changed. The transformed divergence has an upper bound of 2000 and a lower bound of 0. If the calculated transformed divergence value is equal to 2000 then the signatures are completely separable in the bands listed. A value of zero means the signatures are not separable. Usually transformed divergence values of 1700-1900 indicate good separability. Below 1700, the separation is poor (Jensen, 1996).

Decision-making

This is the final step of thc multispectral classification scheme. **Figure 7.34** shows the flow of an image pixel through the classification decision-making process (Kloer, 1994). The term decision-making is explained in the introductory section. In this step, the spectral characteristic of each pixel in the image is compared with the evaluated cover type signatures. If a pixel resembles mostly with a particular cover type i.e. if it satisfies a certain set of criteria fixed for a particular cover type, then the pixel is assigned to the cover type that corresponds to that criteria. This is achieved using a decision rule that frames a certain set of criteria. The classifier using the decision rules, it allows a candidate pixel to a class if it qualifies the fixed criteria for that class or else it remains unclassified. In this manner, all the pixels in the multispectral image are categorized into the evaluated signature classes, thus portraying at the end a thematic map. This process is known as spectral pattern recognition.

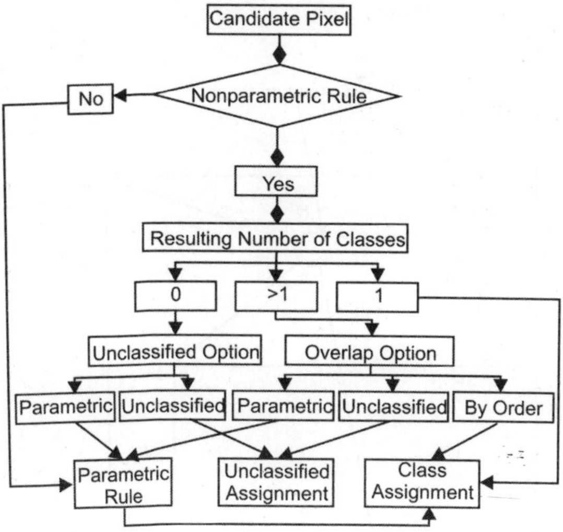

Figure 7.34

The most common classification algorithms used to classify the spectral data include:

1. Minimum-distance-to-Means
2. Parallelepiped
3. Gaussian maximum likelihood.

Each of these algorithms has various advantages and limitations. Regardless of the algorithm used, the result will be a "thematic" map, based upon the training statistics defined by the analyst.

1. Minimum-Distance-to-Means Classifier

This is one of the parametric decision rules, a simplest classification algorithm. The spectral distance between a candidate pixel and the mean signature of each class is calculated (Swain and Davis, 1978). The pixel is then assigned to the spectrally closest class. Every pixel is thus classified.

In **Figure 7.35**, μ_1, μ_2 and μ_3 are the respective means of the signature classes in two dimensional feature space image defined by Band "A" and Band "B". The lines joining the candidate pixel and the means of the signature classes are termed as the spectral distances. As per the decision rule, the candidate pixel is assigned to the closest signature class i.e. shortest distanced class.

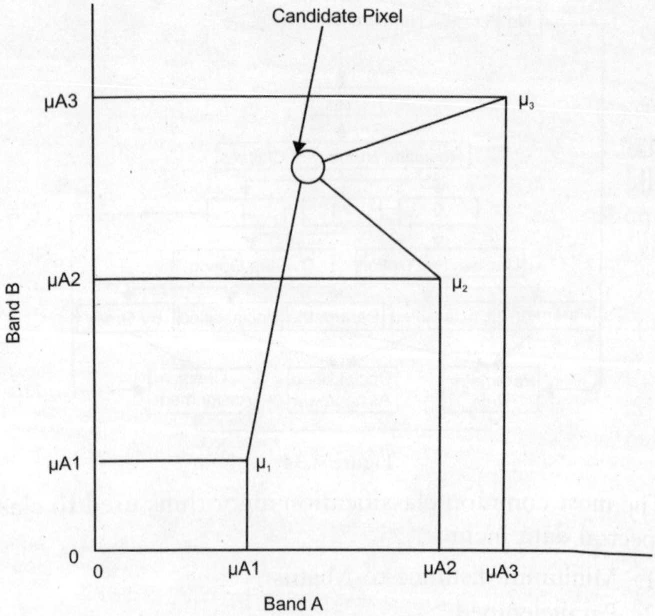

Figure 7.35

In Figure, each axis indicates reflectance values of each band. The position of the means are placed in the feature space using the mean reflectance on these bands as X, Y coordinates. Similarly, the position of any unclassified pixel can also be placed in this space by using its reflectance on the two bands as its coordinates.

The algorithm for classifying by spectral distance is based on the equation for Euclidean distance:

$$SD_{xyc} = \sqrt{\sum_{i=1}^{n}\left(\mu_{ci} - X_{xyi}\right)^2}$$

where:

- n = number of bands (dimensions)
- i = a particular band
- c = a particular class
- X_{xyi} = data file value of pixel x, y in band i
- μ_{ci} = mean of data file values in band i for the sample for class c
- SD_{xyc} = spectral distance from pixel x, y to the mean of class c

(*Source*: Swain and Davis, 1978)

Using the above equation, the spectral distance of the candidate pixel from the means of all possible classes is computed. The candidate pixel is assigned to the class for which SD is the lowest.

Advantages:

1. No pixel in this decision rule remains unclassified since every pixel is nonetheless assigned to the closest.
2. This decision rule is computationally not intensive and so can be performed rapidly, except the parallelepiped classification rule.

Disadvantages:

1. The pixel which should be unclassified is classified by this method. However, this effect can be reduced by using thresholding, a method of setting limits for which

pixels assigning outside the limits form a separate class i.e. unclassified.

2. The main disadvantage of the minimum-distance decision rule is that it does not consider class variance and hence insensitive to different degrees of variance in the spectral response data. Therefore, this classifier is not widely used in applications where spectral classes are closed to one another in the measurement space and have high variance (Lillesand and Kiefer, 1994).

2. Parallelepiped Classifier

The parallelepiped classifier is a non-parametric decision rule applicable to both the supervised and unsupervised classifications.

The parallelepiped classifier uses the range of values in each category training set to determine if a given pixel falls within the class or not. The range is determined by the analyst. The range may be maximum and minimum values, mean ± various standard deviations (μ ± 1s, μ ± 2s, μ ± 3s) or any other limit defined by the analyst based on the knowledge of the data acquired during signature evaluation. With multispectral image data, these ranges form an enclosed box-like polygon of expected values known as a parallelepiped. **Figure 7.36** illustrates a two-dimensional example of a parallelepiped classification, where the upper and lower limits of the training sets are fixed at μ ± 3s. The DN values of the candidate pixel are compared to the upper and lower limits of the training set fixed by the analyst. After the limits have been fixed, unknown pixels are assigned to a class if they fall within the limits of the class or decision region. If the candidate pixel does not fall within any limit of the class or within the decision region, then it is left unassigned. Figure 7.35 illustrates this effect. The unclassfied pixels (symboled as ⊕) are further classified according to the decision rules set up by the analyst. There are three options. The first option is: the signatures are ordered and the unassigned pixel is qualified to the signature class with the highest rank. Analyst determines the rank based upon knowledge of which is the most likely class. The second option is: the pixel can be

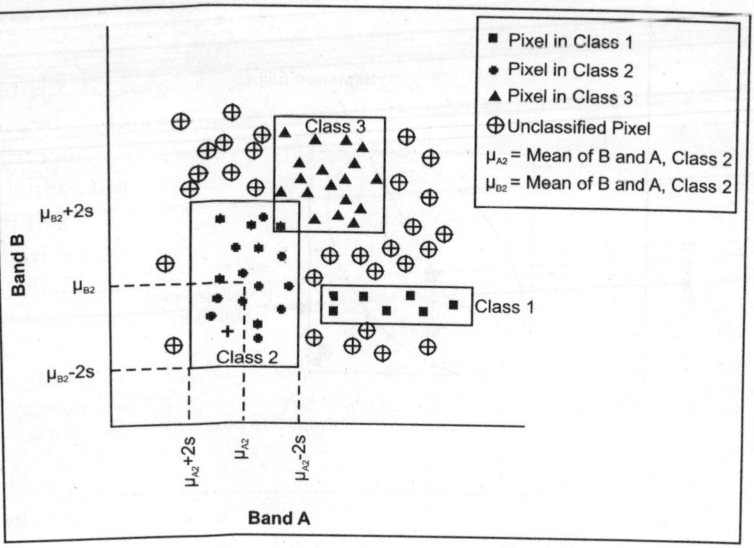

Figure 7.36

defined according to a selected parametric decision rule. The last option is: the pixel may be left unclassified.

The parallelepiped classifier claims the advantage of speed and the ability to take into account the differing variability of classes. This classifier takes care of variance but not of covariance. In the presence of covariance the rectangular decision regions fit the category training data very poorly. However, the classifier generally performs rather poorly because of the potential for overlap of the parallelepipeds. Also, unclassified or unidentified pixels sometimes may be undesirable for some applications.

There are also situations (**Figure 7.37**), where a candidate pixel falls at the corner of the parallelepiped, which is quite distant from the mean of the signature class. The said pixel nonetheless is classified in it. This may lead to misclassification of the pixel.

3. Gaussian Maximum Likelihood Classifier

As the name indicates, the Gaussian maximum likelihood classifier is based on the assumption that the distribution of the cloud of pixels forming the category training data is Gaussian

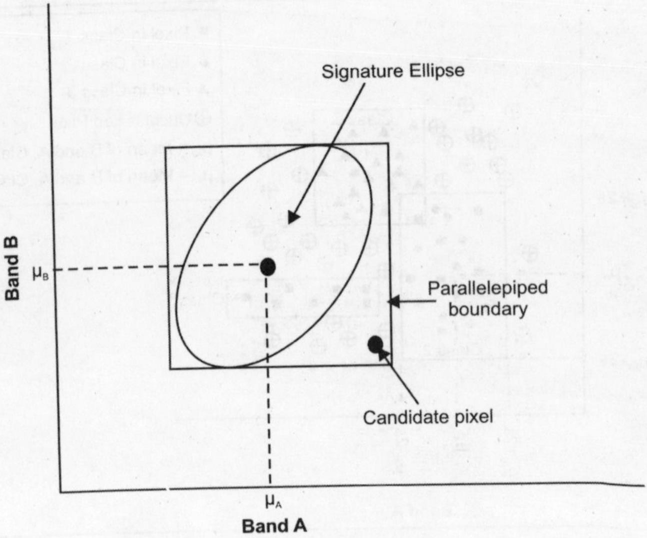

Figure 7.37

or normal. Based on the normal distribution of the data sets, the classification algorithm computes both the variance and covariance of the training class pixels. Since the training pixels follow the Gaussian distribution, their variance about the central value can be explained by the mean vector and covariance matrix. With these two parameters, the probability density function can be computed. **Figure 7.38** depicts the probability density function computed by the classifier for different signature classes. The statistical probability of a candidate pixel belonging to each signature class is then computed and compared i.e. the probability of occurrence of the pixel is compared with the Maize class, the urban class, sand class and so on (Figure 7.38). The candidate pixel is then assigned to the class to which it belong most likely or else labeled unknown if its value downs below the threshold value of the analyst.

In this decision-making rule, the classifier delineates ellipsoidal "equiprobability contours" in the scatter diagram (Lillesand and Kiefer, 1994). These decision regions are shown

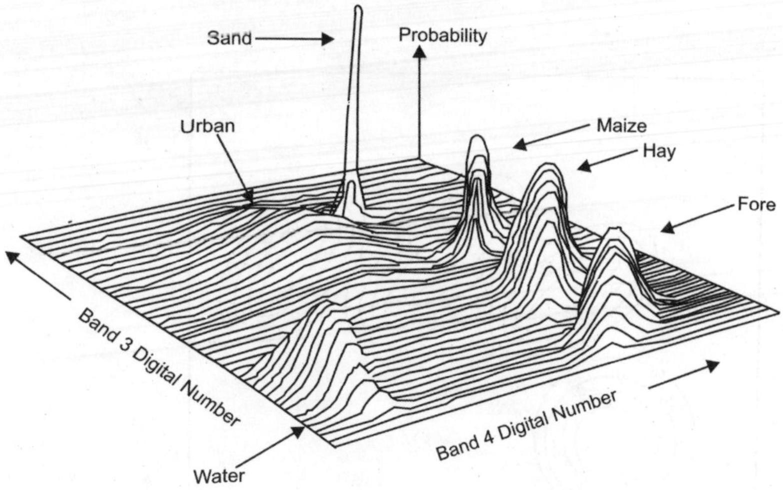

Figure 7.38

in the **Figure 7.39** and take care of the sensitivity of the classifier to covariance. As a result of which the pixel which was not appropriately classified in parallelepiped classification would be correctly assigned to the category it actually belongs.

This is regarded as the most sophisticated parametric decision-making classifier as it takes into account the most variables. On the other hand, the major disadvantage of the maximum likelihood classifier is the large number of computations involved to classify each pixel. Although it is an expensive, modern computers allow it to be widely used.

Equation of Maximum Likelihood Classifier

The Maximum likelihood decision rule is based on the probability that a pixel belongs to a particular class. The basic equation assumed that these probabilities are equal for all classes, and that the input bands have normal distribution.

The equation for the maximum likelihood classifier is as follows:

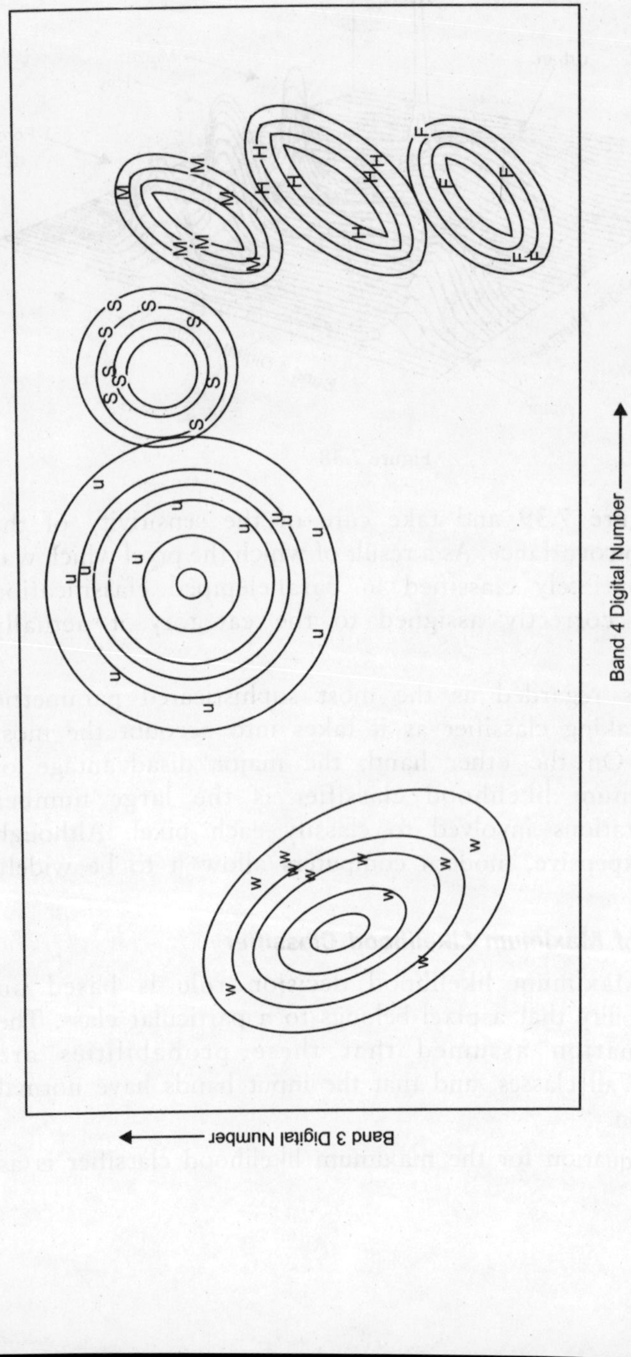

Figure 7.39

$$D = \log(a_c) - [0.5 \log(|Cov_c|)] - [0.5 (X-M_c) T (Cov_c^{-1}) (X-M_c)]$$

where:

D = weighted distance (likelihood)

c = a particular class

X = the measurement vector of the candidate pixel

M_c = the mean vector of the sample of class c

a_c = per cent probability that any candidate pixel is a member of class c

(Default to 1.0, or is entered from a priori knowledge)

$|Cov_c|$ = determinant of Cov_c (matrix algebra)

Cov_c = the covariance matrix of the pixels in the sample of class c

Cov_c^{-1} = inverse of Cov_c (matrix algebra)

T = transposition function (matrix algebra)

The pixel is assigned to the class c, for which D is the lowest.

Unsupervised Classification

Unsupervised classification is more computer-automated. The major stages of implementing unsupervised classification are shown in Figure 7.32. In this classification, training data are not used as the basis for classification. The analyst specifies certain parameters such as (i) number of classes (ii) number of iterations (iii) convergence threshold that the computer uses as guidelines to reveal statistical patterns that are inherent in the data. The result is simply a number of classes based on natural groupings or clusters of pixels with similar spectral characteristics. The basic premise on which the classifer functions is that the spectral values within a class should be close together in the feature space and spectral values in different classes should be comparatively well separated.

Normally this method is in use when adequate information about the data is not available before classification. It is then the analyst's responsibility, after classification, to attach meaning to the resulting classes (Jensen, 1996). Unsupervised classification is useful only if the classes can be appropriately interpreted. The classes can later be merged, disregarded, otherwise manipulated, or used as the basis of a signature.

The unsupervised classification scheme is particularly implemented in an inaccessible area, where collection of sufficient training samples is not possible, which is required for decision-making or interpretation key to train the classifier. In agricultural scene, the widely used maximum likelihood classifier cannot be reliably implemented. The impediment is that the normal distribution assumption upon which the likelihood classifier functions is dishonoured. The reasons behind the deviation of pixels from unimodal to multimodal probability distribution are (i) different dates of sowing, (ii) variation of crop management practices, and (iii) heterogeneous topographic soil conditions (Swain and Davis, 1978). Similar problem is also common in case of water classes or wetland environments due to pollution, silt contents and the presence of various types and amount of aquatic vegetation. In order to overcome such difficulties, in stead of maximum likelihood classifier, the unsupervised classification based on the clustering techniques is chosen as the alternative and particularly it is being used in studying vegetation in a bigger way (Sokal and Sneath, 1963; Duda and Hart, 1973). Numerous clustering algorithms that can be used for unsupervised classification are available in remote sensing and pattern recognition literature (Duda and Hart, 1973; Fu, 1976; Haralick and Fu, 1983). But the clustering algorithm used for unsupervised classification for remote sensing data vary according to the efficiency and objective of the analysis (Jenson, 1986). Different criteria of efficiency lead to different approaches.

Unsupervised Training

Natural grouping of pixels in the feature space using the specified parameters of the analyst is otherwise called clusters or unsupervised training. Clusters are formed by clustering algorithm which often uses all or many of the pixels in the image for its analysis.

Unsupervised training uses statistical clustering techniques to derive the dominant spectral clusters within an image based on the analyst's specified parameters. The classes produced are meaningfully interpretated with the aid of field survey data.

Tou and Gonzalez (1974) described the Iterative Self Organizing Data Analysis Technique (ISODATA) clustering method to perform an unsupervised classification. It is a more sophisticated version of the K-Means classifier which allows classes to be created and destroyed.

The ISODATA clustering technique uses the minimum spectral distance formula to form clusters. This is one of the widely used methods of the unsupervised classification. The abstract of the clustering process is as follows:

(i) The algorithm generates number of arbitrary clusters in the multi-dimensional feature space and calculates their means accepting some initial parameters including the number of clusters defined by the analyst.

(ii) The algorithm computes the spectral distance between each pixel in the image and each of these arbitrary cluster means. Each pixel in the feature space image is then assigned to the cluster whose mean vector is closest.

(iii) The process of clustering begins for the second iteration. Once all pixels in this manner have been assigned to the clusters of their respective closest, revised mean vectors for each cluster are then recalculated. A new mean thus produced, shifts from its original position to new position and determines the inclusion of a pixel.

(iv) Computer calculates the spectral distance between each pixel in the image and each new mean vector of the clusters formed. Each pixel is then allocated to the cluster whose mean vector is closest to it. It may be observed that some pixels in the subsequent iterative process will be closer to a different mean and assigned to a different cluster accordingly.

(v) The clustering process by ISODATA repeats until there is no significant change in the location of the cluster mean vectors between successive iterations of the algorithm or it reaches a level set by the analyst, called the convergence threshold. The convergence threshold is the maximum percentage of pixels that has cluster assignments that can go unchanged between iterations. A convergence threshold of 95% means that 95% of the image pixels were not assigned to new classes in the last iteration of the algorithm. If the convergence threshold is not reached then the program terminates once the maximum number of iterations specified by the analyst is reached.

Once the classification has been completed, it is very important to evaluate the ground reality of the classification as the process is fully computer automated. This can be evaluated comparing certain pixels in thematic image to reference pixels for which the land cover type is known. The image analyst may also compare the classes with ground truth information, field reports, published documents, aerial photography, etc., and create an error matrix to get a quantified accuracy evaluation.

B1. K-Means Clustering Algorithm

K-means (MacQueen, 1967) is one of the simplest unsupervised training algorithms that solve the well-known clustering problem. The procedure follows a simple and easy way to classify a given data set through a certain number of clusters (assume k clusters) fixed a priori.

The algorithm assumes that the pixels form a vector space. The main idea is to define k-means, one for each cluster such that some metric relative to the means of the clusters is minimized. That is the objective is to minimize total intra-cluster variance, or, the squared error function

$$V = \sum_{i=1}^{k} \sum_{x_j \in S_j} \left| X_j - \mu_i \right|^2$$

Where there are k clusters S_i, $i = 1,2,...,k$ and m_i is the mean point of all the points $x_j \in S_i$.

The algorithm starts with grouping the image pixels into k number of clusters either at random or using some heuristic data. For k such clusters it calculates k-means. The k-means should be placed as much as possible far away from each other. The next step is that it takes each image pixel and associates it to the nearest mean. Then the means are recalculated for the new clusters. The algorithm repeats the same steps until the means no longer move. This produces a partitioning of the pixels into groups from which the metric to be minimized can be calculated.

The algorithm has remained extremely popular because it converges extremely quickly in practice. The result of application of this algorithm depends largely upon the initial number of clusters. Since it is extremely fast, it is advisable to run the algorithm several times and return the best clustering found.

7.14.6 The Classification Accuracy

The term accuracy is described previously as an introductory concept. An overall review of the literatures on accuracy, concept of the error matrix (contingency matrix) and metrics (Kappa and Tau coefficients), user's and producer's accuracy, etc. are highlighted in this section.

The accuracy of spatial data has been defined by the United States Geological Survey as: "The closeness of results of observations, computations, or estimates to the true values or the values accepted as being true" (USGS, 1990). The non-positional characteristic of spatial data is termed as thematic accuracy. If data have been subjected to multispectral classification, then thematic accuracy is also known as classification accuracy (Stehmen, 1997). The classification accuracy refers to correspondence between the class label and the true class, which is generally defined as what is observed on the ground during field survey. In other words, it is a process of assessing how much of the feature class belonging to a particular group or category is in actual in the ground while making field verification.

Calculation of Classification Accuracy

Accuracy assessment or validation is an important step in the processing of remote sensing data. It determines the information value of the resulting data to a user. Meaningful utilization of spatial data is only possible if the quality of the data is known. Furthermore, integrated processing of different types of spatial data cannot be effective if the data quality is not known. During the last two decades, numerous studies have been published concerning accuracy assessment of land cover classifications (Congalton, 1996; Rosenfeld and Fitzpatrick-Lins, 1986; and Foody, 1992).

The quality of spatial data sets is a very broad issue that may relate to a variety of properties (Worboys, 1998) but frequently the property of interest is map or classification accuracy. As any map is simply a model or generalization, it will contain error (Brown, Loveland, Ohlen and Zhu, 1999; Dicks and Lo, 1990; Maling, 1989; Smits, Dellepiane, and Schowengerdt, 1999). Thus, although a thematic map provides a typically unquestioned simplification of reality, it has flaws and is only one model or representation of the depicted theme (Woodcock and Gopal, 2000). For example, as the mapping processes involve generalization there is some loss of information and so completeness (Maling, 1989). It is important, therefore, that the quality of thematic maps derived from remotely sensed data be assessed and expressed in a meaningful way. This is important not only in providing a guide to the quality of a map and its fitness for a particular purpose, but also in understanding error and its likely implications, especially if allowed to propagate through analyses linking the map to other data sets (Arbia, Griffith and Haining, 1998; Janssen and Van der Wel, 1994; Veregin, 1994). Although classification accuracy assessment is now widely accepted as a fundamental component of thematic mapping investigations (Cihlar, 2000; Cohen and Justice, 1999; Congalton, 1994; Justice et al., 2000; Merchant, Yang and Yang, 1994), it is not uncommon for map accuracy to be inadequately quantified and documented (Dicks and Lo, 1990).

Many methods of accuracy assessment have been discussed in the remote sensing literature (e.g., Aronoff, 1982, 1985;

Kalkhan, Reich and Czaplewski, 1995; Koukoulas and Blackburn, 2001; Piper, 1983; Rosenfeld and Fitzpatrick-Lins, 1986). The most widely promoted and used, however, may be derived from a confusion or error matrix.

Green et al. (2000) explained the calculation of classification accuracy from a confusion or error matrix in an image that has been classified into two categories: (i) seagrass, and (ii) coral reef which are having similar reflectance spectra. In such a situation, the image could not be accepted as 100% true representation of the ground reality as similar spectra of seagrass and reef would cause misclassification of some reef as seagrass and *vice versa*. A method using error matrix therefore is followed to quantify classification errors by estimating how many reef pixels are in reality seagrass and how many seagrass pixels are reef.

As a simple cross-tabulation of the mapped class label against that observed in the ground or reference data for a sample of cases at specified locations, the error matrix provides an obvious foundation for accuracy assessment (Campbell, 1996; Canters, 1997). An error matrix is a square array of numbers organized in rows and columns which express the number of sample units (i.e. pixels and clusters of pixels) assigned to a particular category relative to the actual category as indicated by reference data (Congalton, 1996). Each cell contains the number of sampling sites (pixels or groups of pixels) assigned to a particular category.

The accuracy is essentially a measure of how many ground truth pixels were classified correctly. Confusion matrix in this regard provides various measures of classification accuracy. One such measure is the most popular percentage of cases correctly allocated. This is an easily interpretable guide to the overall accuracy of the classification. If attention focuses on the accuracy of individual classes, then the percentage of cases correctly allocated may be derived from the confusion matrix by relating the number of cases correctly allocated to the class to the total number of cases of that class. This may be achieved from two standpoints, giving rise to the somewhat awkwardly termed (Janssen and Van der Wel, 1994) user's and producer's accuracy, depending on whether the calculations are based upon the matrix's row or column marginals (Campbell, 1996; Story and Congalton, 1986).

The calculation of percentage correct, user's and producer's accuracy and metrics such as Kappa coefficient and Tau coefficient is illustrated below in the light of **Table 7.5**. In the derivation of these indices, a number of fundamental assumptions are typically made. For example, it is generally assumed implicitly that each case (e.g., pixel) to be classified belongs fully to one of the classes in an exhaustively defined set of discrete and mutually exclusive classes (Congalton et al., 1998; Congalton and Green, 1999; Lewis and Brown, 2001; Townsend, 2000).

TABLE 7.5: The confusion matrix

Predicted Class	Actual Class				
	A	B	C	D	Σ
A	n_{AA}	n_{AB}	n_{AC}	n_{AD}	n_{A+}
B	n_{BA}	n_{BB}	n_{BC}	n_{BD}	n_{B+}
C	n_{CA}	n_{CB}	n_{CC}	n_{CD}	n_{C+}
D	n_{DA}	n_{DB}	n_{DC}	n_{DD}	n_{D+}
Σ	n_{+A}	n_{+B}	n_{+C}	n_{+D}	n

Table 7.5 represents a generalized form of an error matrix representing the reference data in X-axis and predicted data (field data set) in Y-axis. The diagonal elements of the matrix represent the cases where the predicted classes from the imagery and reference data set agree. For example in **Table 7.6**, the 85 represents the number of reefs correctly predited from the multispectral image. Whereas the off-diagonal elements contain those cases where there is a disagreement in the labels.

TABLE 7.6: The hypothetical error matrix for Reef and Seagrass

	Reference Data		
Classification Data	Reef	Seagrass	**Row Total**
Reef	85	25	110
Seagrass	15	75	90
Column	100	100	200
Total	85%	75%	
Producer Accuracy			

User Accuracy for Reef = 85/110 = 77%

User Accuracy for Seagrass = 75/90 = 83%

Source: Green et al., 2000.

The general formula for calculating per cent correct, Use's accuracy, and Producer's accuracy are given below:

$$\text{Per cent correct} = \frac{\sum_{k=1}^{q} n_{kk}}{n} \times 100$$

$$\text{User's Accuracy} = \frac{n_{ii}}{n_{i+}}$$

$$\text{Producer's Accuracy} = \frac{n_{ii}}{n_{+i}}$$

Table 7.6 shows user's accuracy and producer's accuracy calculated using the formulae given above. The probability that a pixel classified on the image actually represents that category *in situ* is termed the user's accuracy for that category. In the example, the omission errors for the reef and seagrass classes are 15% and 25% respectively. The producer's accuracy is the probability that any pixel in that category has been correctly classified and in this example is 85% for reef and 75% for seagrass.

7.14.6.1 Kappa Statistics

It is an index which compares the agreement against that which might be expected by chance. Kappa can be thought of as the chance-corrected proportional agreement. It is a discrete multivariate technique used to assess classification accuracy from the error matrix. Kappa analysis generates Kappa coefficient that has a possible range from +1 (perfect agreement) through 0 (no agreement above that expected by chance) to -1 (complete disagreement). The formula for calculating Kappa coefficient is given by:

$$\text{Kappa Coefficient} = \frac{N\sum_{i=1}^{r} X_{ii} - \sum_{i=1}^{r}(X_{i+}.X_{+i})}{N^2 - \sum_{i=1}^{r}(X_{i+}.X_{+i})}$$

where r = number of rows in a matrix, x_{ii} is the number of observations in row i and column i, x_{i+} and x_{+i} are the marginal totals of row i and column i respectively, and N is the total number of observations (accuracy sites) (Bishop et al., 1975). K expresses the proportionate reduction in error generated by a classification process compared with the error of a completely random classification. For example, a Kappa coefficient of 0.89% implies that the classification process was avoiding 89% of errors that a completely random classification would generate (Congalton, 1991).

The importance and power of the Kappa analysis is that it is possible to test if a land use and land cover map is significantly better than if the map had been generated by randomly assigning labels to areas (Congalton, 1996). It is widely used because all elements in the classification error matrix, and not just the main diagonal, contribute to its calculation and because it compensates for change agreement (Rosenfeld and Fitzpatrick-Lins, 1986). The Kappa coefficient represents the proportion of agreement obtained after removing the proportion of agreement that could be expected to occur by chance (Foody, 1992).

Although informative, measures such as the percentage of cases correctly classified have often been criticized. A major problem for some users is that some cases may have been allocated to the correct class purely by chance (Congalton, 1991; Pontius, 2000; Rosenfeld and Fitzpatrick-Lins, 1986; Turk, 1979). To accommodate for the effects of chance agreement, Cohen's kappa coefficient has often been used and some commentators argue that it should, in some circumstances, be adopted as a standard measure of classification accuracy (Smits et al., 1999). The kappa coefficient has many attractive features as an index of classification accuracy. In particular, it makes some compensation for chance agreement and a variance term may be calculated for it enabling the statistical testing of the significance of the difference between two coefficients (Rosenfeld and Fitzpatrick-Lins, 1986). This is often important, as frequently, there is a desire to compare different classifications and so matrices. To further aid this comparison, some have called for the normalization of the confusion matrix such that each row and column sums to unity (Congalton, 1991; Smits et al., 1999).

Overall Accuracy

In order for understanding the degree of accuracy for the whole image across all classes, the overall accuracy is calculated. It is calculated by dividing the sum of diagonal classes (diagonal elements representing classes only) by the total number of accuracy sites (i.e. column total or row total). In Table 7.6, overall accuracy is (85 + 75)/200 = 80%. Off-diagonal elements are used for calculating the Kappa coefficient. K will be less than the overall accuracy unless the classification is exceptionally good i.e. the numerator of the overall accuracy must be low. Thomlinson et al. (1999) set a target of an overall accuracy of 85% with no class less than 70% accurate. The overall accuracy is a more accurate estimate of accuracy.

7.14.6.2 *Tau coefficient*

The concept behind the implementation of Tau coefficient in preference to Kappa coefficient is originated by Ma and Redmond (1995) for ease interpretation of the index value. The computation formula for Tau coefficient of Ma and Redmond (1995) is given below as:

$$T = \frac{P_0 - P_T}{1 - P_r}, \text{ where } P_r = \frac{1}{N^2}\sum_{i=1}^{M} n_i X_i$$

where P_0 is the overall accuracy; M is the number of habitats; i is the ith habitat; N is the total number of sites; n_i is the row total for habitat i and X_i is the diagonal value for habitat i (i.e. number of correct assignments for habitat i).

If the Tau coefficient is 0.80, say for example, then it can be interpreted as 80% more pixels were classified correctly than would be expected by chance only. The distribution of Tau coefficient approximates to normality and Z-test.

7.14.6.3 *Comparison between Error matrix and Tau coefficient*

The Tau coefficient should be used to test for significant differences between error matrices.

Z-tests between Tau coefficients 1 and 2 (T_1 and T_2) can be conducted using the following equations:

$$Z + \frac{T_1 - T_2}{\sqrt{\sigma_1{}^2 + \sigma_2{}^2}}$$

Where σ^2 is the variance of the Tau coefficient calculated from:

$$\sigma^2 = \frac{P_0\,(1 - P_0)}{N(1 - P_r)}$$

7.14.7 Conclusion

The overall discussion on multispectral classification and accuracy assessment leads to conclude that the thematic maps are not the true representation of the ground reality. However, they are widely and comfortably used in many application fields successfully. In case of land use/land cover classification of a remotely sensed image, the end product is a thematic map with each pixel assigned to a unique category or class. This result is compared with the reference data (ground data) for accuracy assessment.

The value of the map is clearly a function of the accuracy of the classification (Foody, 2002). The researchers mostly rely on confusion matrix to derive many measures of classification accuracy. Determination of classification accuracy is, however, not straightforward. Divergence of opinion exists with respect to the method of accuracy assessment or reporting. Review of literatures indicates that reporting on classification accuracy by various authors is often flawed. Accuracy assessment has now become a fundamental part of any thematic map preparation and so also a challenging area of research. Although there have been many recent advances, the current status of accuracy assessment indicates that numerous problems remain to be solved.

8

Geographical Information Systems

8.1 INTRODUCTION

Before going into the details of Geographic Information Systems (GIS), it is imperative to think on what 'geography' means. In the classical sense, the word geography is derived from the Greek words 'geo' meaning 'Earth' and 'graphy' meaning "to describe" or "to write" or "to map". Thus geography literally means writing about the Earth. Geographers call it as the study of Earth and its features and of the distribution of life on the Earth. Another definition of geography focuses on man's relationship with the land. Domain of the geography is, therefore, thought to be a set of Earth's features, objects, phenomenon or any other information which are spatial in character. In order to process, analyse and present the spatial data we basically need a tool. A geographical map is referred to a key tool to represent the spatial objects of the Earth. It is a two dimensional geometrically accurate representation of a three dimensional space containing a graphic abstract of the real world features. Since it is a two dimensional graph in a flat surface, it is easily handled. The most important feature of a map is the scale, since the transferred features are scaled down from three dimensional Earth to a two dimensional flat surface. Since a map is drawn to scale, and reproduces the correct position, true proportions of the length, breadth and area of the different features shown, it is a source of much more definite information than any other type of geographic representation. Thus the map is a source of spatial information about the features on the Earth's surface.

The information contained in a map are: (i) Spatial data of one or more themes, (ii) points, lines, polygons and symbols describing the features, (iii) Captions: Title, Legends or Keys,

(iv) Textual information or attributes of the Geographic features and (v) Map projection, Map scale and Tics showing the geographic co-ordinates along the border of the map. All information in the real world is represented by such maps. A collection, storage, retrieval, analysis and display of spatial information content of such maps in a computer system is said to be Geographic Information System.

Over the last two decades, the rapid growth in information technology has provided scientific communities and other related professionals with new tools to process analyse and present spatial data. One set of such tools is the Geographic information systems, a particular type of information system applied to geographical data. The use of Geographic Information Systems (GIS) was first introduced in the 1960s when the technology was limited to a handful of researchers. Today, with the advancement of computer technology information system, the users of GIS are now many. It is one of the fastest growing areas of information technology. In the last decade, Geographic Information System has got its recognition as a key tool for rural development and urban planning, natural resources mapping and management. Its extraordinary propagation of applications in the diversified fields lies in the fact that it has the capability to store, retrieve, analyse, model and map large areas with huge volumes of spatial data. Geographic information systems are now used for land use planning, utilities management, ecosystems modeling, landscape assessment and planning, transportation and infrastructure planning, market analysis, visual impact analysis, facilities management, tax assessment, real estate analysis and many other applications.

Geographical information management (GIM) is an exciting and rapidly growing branch of Information Technology (IT) incorporating satellite remote sensing, aerial photography and other spatial data, such as land use/land cover information, to derive information which is essential for the management of the Earth's resources. The development of GIS has relied on innovations made in many different disciplines: Geography, Cartography, Photogrammetry, Remote Sensing, Surveying, Geodesy, Civil Engineering, Statistics, Computer Science,

Operations Research, Artificial Intelligence, Demography, and many other branches of the social sciences, natural sciences, and engineering have all contributed.

Today people are talking about the digital world which is described as a set of largely independent layers and spatial elements, each representing some component of the environment and some set of environmental concerns as well. These layers might include landforms, soil types, vegetation patterns, geomorphic features, ground water, and other resources. GIS helps to explore relationships between different layers of information about the features that are distributed unevenly over space. This is possible using advanced techniques of query, selection, analysis, and display. GIS technology is advancing day by day towards meeting the various problem-solving situations.

8.1.1 History of GIS

GIS stores information about the world as a collection of thematic layers that can be linked together by geography. The GIS is an evolution of the long traditional map making. Development of GIS was influenced by a specialised group of researchers, academics, government, corporations and individuals. A detailed history of GIS is not well understood because GIS technology evolved through multiple parallel but separate applications across numerous disciplines (Pickles, 1999; Antenucci et al., 1991). In the field of GIS development, Canada is recognized as a world leader. It is the first country, where the actual GIS development took place which was spread over the entire world in the later period. The GIS development in Canada began in 1963 and completed in 1971. The GIS history dates back to 1960 when the U.S. Census Bureau developed the large-scale adoption of digital mapping by the government. Other organizations took the initiative are the US Geological Survey, the Harvard Laboratory for computer graphics and Environmental Systems Research Institute (ESRI). Canadian Geographic Information Systems in Canada, Natural Experimental Research Center, and Department of Environment are such notable organizations in U.K. involved in early developments. Canada GIS initiated with developing land management plan for large areas of rural Canada analyzing data collected by Canada Land Inventory (CLI). Harvard Laboratory for computer Graphics and Spatial Analysis established in 1964 contributed a major share in producing experts in GIS during 1970s and 1980s. The GIS did a marvelous job in producing the Census TIGER files, one of the most important socio-economic spatial data sets in use today. Important geographic work was also being done at universities throughout the 1950s and 1960s. A grid-based mapping program called SYMAP, developed at the Laboratory for Computer Graphics and Spatial Analysis at the Harvard Graduate School of Design in 1966, was widely distributed and served as a model for later systems (Mark et al., 1997; Antenucci et al., 1991).

In the 1970s, private vendors began offering off-the-shelf GIS packages. M&S Computing (later Intergraph) and Environmental Systems Research Institute (ESRI) emerged as the leading vendors of GIS software (Antenucci et al., 1991). In 1981, ESRI released Arc/Info, a standard package which ran on mainframe computers (ESRI, 2001). In 1992, ESRI released Arc View, a desktop mapping system with a graphical user interface that marked a major improvement in usability over Arc/Info's command-line interface (ESRI, 2001). By the early 1990s, GIS initiatives existed in all fifty states (Harris and Elmes, 1993).

In India the major developments have happened for the last one decade with significant contribution coming from Department of Space emphasizing the GIS applications for natural resources management. Recently the commercial organizations in India have realized the importance of GIS for many applications like infrastructure development, facility management, and business/market applications.

8.1.2 Definitions of GIS

Geographic Information System is commonly used in the context of geo-base information system, natural resources information system, geo-data system, geographical system. They all stress the spatial analysis capacities of their system.

Geographic Information System (GIS) is defined as an

information system that is used to input, store, retrieve, manipulate, analyze and output geographically referenced data or geo-spatial data, in order to support decision-making for planning and management of land use, natural resources, environment, transportation, urban facilities, and other administrative records.

A geographic information system (GIS) is a system for capturing, storing, analyzing and managing data and associated attributes which are spatially referenced to the Earth. In the strictest sense, it is a computer system capable of integrating, storing, editing, analyzing, sharing, and displaying geographically-referenced information. In a more generic sense, GIS is a tool that allows users to create interactive queries (user created searches), analyze the spatial information, edit data, maps, and present the results of all these operations (en.wikipedia.org/wiki/GIS_data).

A Geographic Information System (GIS) can be defined as a computerized database system for the capture, storage, retrieval, analysis and display of tabular and spatial data.

Many authors define GIS in many ways. Therefore, it is confusive to select a particular definition that serves the overall significance of the term. Attempts have been made to define the term by various authors. They are given below.

Maguire (1991) defines GIS in the following manner:

- Process-/function-oriented—emphasises the information handling capabilities of GIS (e.g., storage, retrieval, manipulation, and display of geographic data);
- Application—divides information systems according to the problems they seek to address (e.g., soil, land, and planning information system);
- Toolbox—emphasises the generic aspects of GIS as a toolbox to manipulate spatial data;
- Database—regards GIS as a database system, reflecting the influence of database theory and practice on GIS.

Rhind (1989: 28) proposes that GIS is 'a computer system that can hold and use data describing places on the Earth's surface'.

Burrough (1986: 6) defines GIS as 'a set of tools for collecting, storing, retrieving at will, transforming, and displaying spatial data from the real world for a particular set of purposes'.

Dangermond (1988) describes GIS as it consists of five basic elements: "data, hardware, software, procedure and people".

Dickinson and Calkins (1988) GIS is made up of three components: "GIS technology, GIS data base, GIS infrastructure".

Carter (1989: 3) GIS is "an institutional entity, reflecting an organizational structure that integrates technology with a database, expertise and continuing financial support over time".

Burrough (1990) GIS has three components: "computer hardware, sets of application software modules, and a proper organizational context".

Maguire (1991: 15) GIS comprises "four basic elements which operate in an institutional context: hardware, software, data and liveware".

The US Department of the Environment (1987: 132) 'a system for capturing, storing, checking, integrating, manipulating, analyzing and displaying data which are spatially referenced to the Earth'.

In general, the definition of GIS covers three main components. These definitions reveal that GIS is a computer system which implies more than just a series of computer boxes sitting on a desk, but including hardware (the physical parts of the computer along with its associated peripherals like plotters and printers), software (the computer programs that run on the computer) and appropriate procedure (or techniques and orders for task implementation).

From these definitions we can also analyze that GIS uses spatially referenced or geographical data and that, GIS carries out various management and analysis task on this data, including their input and output.

The Department of the Environment (1987) lists the capabilities that a 'well-designed GIS' should be able to provide:

(i) Quick and easy accessibility to large volumes of data.

(ii) The ability to:
- Select detail by area or theme;
- Link or merge one data set with another;
- Analyse spatial characteristics of data;
- Search for particular characteristics or features in an area;
- Update data quickly and cheaply; and
- Model data and assess alternatives.

(iii) Output capabilities (maps, graphs, address lists and summary statistics) tailored to meet particular needs.

In a short way, it can be said that GIS can be used to add value to spatial data by allowing data to be organized and viewed efficiently, by integrating them with other data, by analysis and by the creation of new data that can be operated on in turn. GIS can create useful information to help decision-making.

8.2 COMPONENTS OF GIS

The key components of GIS are a computer system for analysis and display of spatial data, spatial data set and the user (**Figure 8.1**).

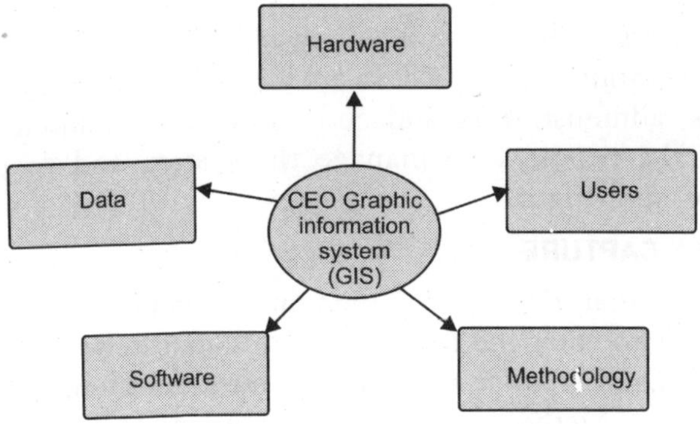

Figure 8.1

8.2.1 Computer System

GIS operates on a computer system, which is a key component to store, process, display and print spatial data. A wide range of hardwares are used ranging from centralized computer servers to desktop computers used in stand-alone or networked configurations.

8.2.2 Softwares

Modules for organizing the GIS database, integration operations are done and outputs are processed. This includes the GIS software, database and drawing software.

8.2.3 Geospatial Data

It includes data components viz. geographic data and attributal data. They can be collected from field, existing topographical maps, satellite images and aerial photos, available tabular data and from other such commercial data providers. Using the DBMS module in GIS softwares a database is created and maintained to organize and manage data.

8.2.4 Procedures

The defined methods used to analyse the data and produce accurate results. The procedures include access protocols, standards and guidelines.

8.2.5 Users

The users of the GIS technology are responsible for the day-to-day operations of the GIS. Users include technical specialists, planners, administrators and end users. It is of limited value without the people who manage the system and to develop plans for applying it.

8.3 DATA CAPTURE

Before analyzing spatial data in a computer system, the first step is to enter the topographical map, satellite image and other available spatial information into the computer's GIS environment. Method of data acquisition and storage in GIS is illustrated in **Figure** 8.2.

Geographic data are created from paper maps by scanning and digitising. Geospatial data are also created by taking physical measurements from field surveys. The present technology is so advanced that, most geospatial data are captured in digital form at the point of initial survey. In this regard the PDA (Personal Digital Assistant) with GIS ARC-Pad software facilitates to capture spatial data in real time (field GIS). By capturing the image using scanner and digitised using the GIS software, it can be stored on the computer and displayed on screen.

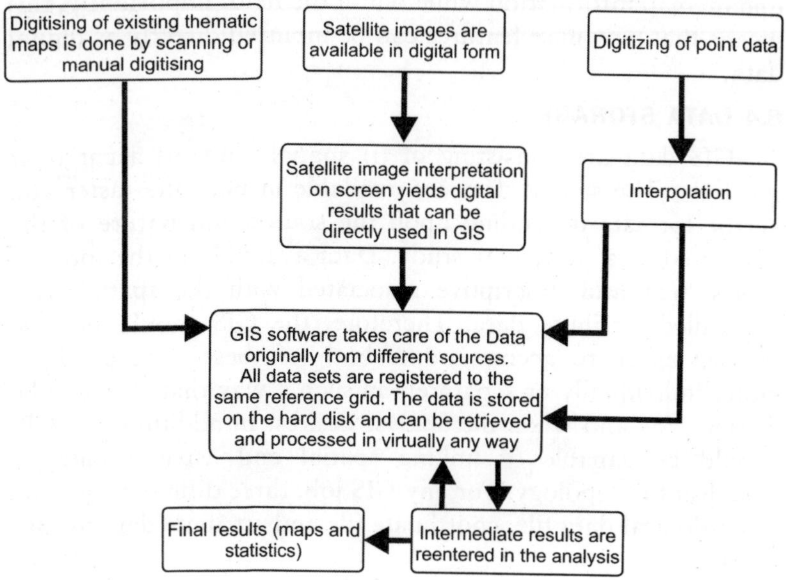

Figure 8.2

Some much specialised systems using GIS software are able to convert raster data to vector data by recognising patterns in the image.

Spatial data are captured from surveying and remote sensing. For ground measurements, the surveyors are now using the recent electronic devices instead of the conventional tools. Real time geographic information is collected using the aerial photos and satellite images. Positioning of objects on or above the Earth's surface can easily be collected using Global Positioning System (GPS). GPS can be used almost anywhere in the world. Along with mapping, GPS measurements are obtained in the

GPS coordinate system: World Geodetic System 1984 (WGS84). This position usually needs to be converted into the local coordinate system for the region. GIS data collectors can make use of the free GPS service provided to locate map objects and features directly in the field. The surveyors capture field data using the most advanced equipment like hand held pen computer equipped with flat, touch-sensitive screens. These computers allow the surveyor to draw and click directly onto the screen to update map information while out in the field. Importantly, this means that new map features can be inputted directly as digital data.

8.4 DATA STORAGE

GIS data are consisting of (i) spatial, and (ii) aspatial or attribute. The spatial data are available in both the raster and vector formats depending upon the source and nature of the data used in a particular study. Data available in the form of tables, text and descriptive associated with the spatial data are called attribute data. Therefore, the GIS should provide storage space to accommodate both of these data products either individually or separately in such a way that they can be retrieved as and when the analyst desires. In addition, the GIS should be capable of linking spatial and aspatial data to establish the topology. For any GIS job, three different files viz. a topological data file, coordinate file and attribute data file are created.

Computer system needs sufficient space to accommodate the GIS software and other image files and to process the geospatial data. GIS software on a microcomputer can occupy tens of megabytes even without data, and on a workstation perhaps hundreds of megabytes. Since more raster layers are used due to availability of high resolution satellite data, the GIS system is in need of more space in the gigabyte range size.

Spatial data are generally stored in two formats: (i) tabular format, and (ii) GIS format. Spatial data storage in tabular format needs less disk space and can easily integrate with other tabular data sets. Tabular data which have adequate georeferencing may then be converted to a GIS format when this is necessary for display and spatial analysis. Stream-based data are commonly stored in this method because linear geographic information is readily compatible with tabular formats. Point data can be stored in the tabular data format. Since polygon requires specific geographic referencing, it is less amenable to tabular storage.

8.5 DATABASE MANAGEMENT SYSTEM

Essentially, a GIS is a database management system for geographic data. Spatial Data in either vector and/or raster formats is the core of any GIS. Collection and organization of data referenced to locations on the Earth is known as geodatabase. Geodatabases are grouped into two different types: vector and raster. In addition, each spatial object has an associated attribute (e.g. a name, a number, a range of values, etc.). Documentation of GIS datasets is known as metadata (data about data).

Data management concerns with how the data collected about the location of spatial features (points, lines and polygons representing objects on the Earth surface), attributes and linkages are structured and organized in a computer system. Traditionally, the computer program used to input, manage and analyse attribute data is known as database management system (DBMS). Apart from managing attribute data, it is also required to have spatial database management. Thus, a GIS typically incorporates not only a traditional DBMS, but also a variety of utilities to manage the spatial and attribute components of the geographic data stored.

The function of DBMS in GIS is to enter attribute data such as tabular data and statistics, and in turn, extract specialized tabulations and statistical summaries to provide new tabular reports. In other words, the DBMS provides us with the ability to analyse attribute data. Many map analyses have no true spatial component, and for these, a DBMS will often function quite well.

8.6 DATA RETRIEVAL, ANALYSIS AND DISPLAY

One of the major capabilities of GIS is data retrieval. GIS supports the retrieval of features by both their attributes and their spatial characteristics. Data may be retrieved according to its location on a map layer or its spatial relationship to other mapped data. This process is known as spatial retrieval. Spatial retrieval produces a spatial representation (map). Spatial features and their attribute information are closely associated in a geographic analysis. Traditionally the retrieval and analysis of geographic data and information were not inherently associated with a particular format. The GIS softwares are now capable of building linkage between the two performing more sophisticated analysis. Queries which form an important component of geographic analysis and problem-solving can be graphics- or data-driven. Graphics-driven queries involve spatial based searches for objects and retrieval of the associated attribute data. Data-driven queries involve the use of data values selectively to display the matching spatial features.

GIS provides operations to retrieve, summarise, selectively display and analyse geographically referenced data. Examples of the types of spatial analysis include:

8.6.1 Choropleth Mapping

Choropleth map is a thematic map in which areas are shaded or patterned in proportion to the measurement of the statistical variable being displayed on the map. Choropleth mapping involves classifying area features according to some attribute or set of attributes. A legend acts as a look-up table with each range of attribute values being associated with a particular color or shading pattern. GIS retrieved choropleth map provides an easy way to visualize how a measurement varies across a geographic area.

8.6.2 Buffer Generation

Buffer Analysis is one of the most important functions of spatial analysis in GIS. A buffer is a region of memory used to temporarily hold output or input data. The units of buffering in GIS are points, lines, and polygons. Buffering refers to creating a zone of a specified width around a point or a line or a polygon area. The resulting buffer is a new polygon, which can be used in queries to determine which entities occur either within or outside the defined buffer zone. It is also referred to as a zone of specified distance around coverage features. Boundaries are created around points, lines or areas at an equal distance in all directions. There are two types of buffers: constant width buffers and variable width buffers. Both types can be generated for a set of coverage features based on each features attribute values. An example of buffering is to calculate zones of different noise intensity around an airport.

8.7 POLYGON OVERLAY

The most familiar and fundamental operation in GIS applications is Overlay operation. The functionality of this operation is to integrate separate feature classes together into new feature classes that represent the combination of two inputs. There are two types of overlay operators: (i) arithmetic operators, and (ii) logical operators.

Arithmetic operators are used to add, subtract divide or multiply values in one layer by values in the corresponding location in the second layer. Logical operators are used to identify a specific set of conditions in one layer with a feature or condition in another layer. Logical operators use Boolean functions to operate on the spatial and attribute properties of the features. Vector overlay can be performed on point, line or polygon features.

Point in polygon overlay: This method is used to find which points fall inside which polygons. The attributes from the polygon overlay are added to the point attribute table of the output layer. In this example the Boolean operator NOT is used to find the point features that are not in the polygon layer [**Figure 8.3 (a)**].

Line in polygon overlay: This method is used to find the common space between a line feature class and a polygon feature class. The arcs from the input polygon coverage are split at the overlay coverage polygon boundaries and new lengths are calculated for these arcs. Attributes from the polygon overlay are added to the attribute table for the new arcs. In this example the Boolean operator AND is used to find features that common to both layers [**Figure 8.3 (b)**].

Polygon on polygon overlay. This method is used to find the common area between two polygon feature classes. The polygons of the input coverage are split at the overlay polygon boundaries and new arcs are calculated for the polygons of the input coverage. Attributes from the polygon overlay are added to the polygon attribute table. In this example the Boolean operator OR is used to find features that are in both layers [**Figure 8.3 (c)**].

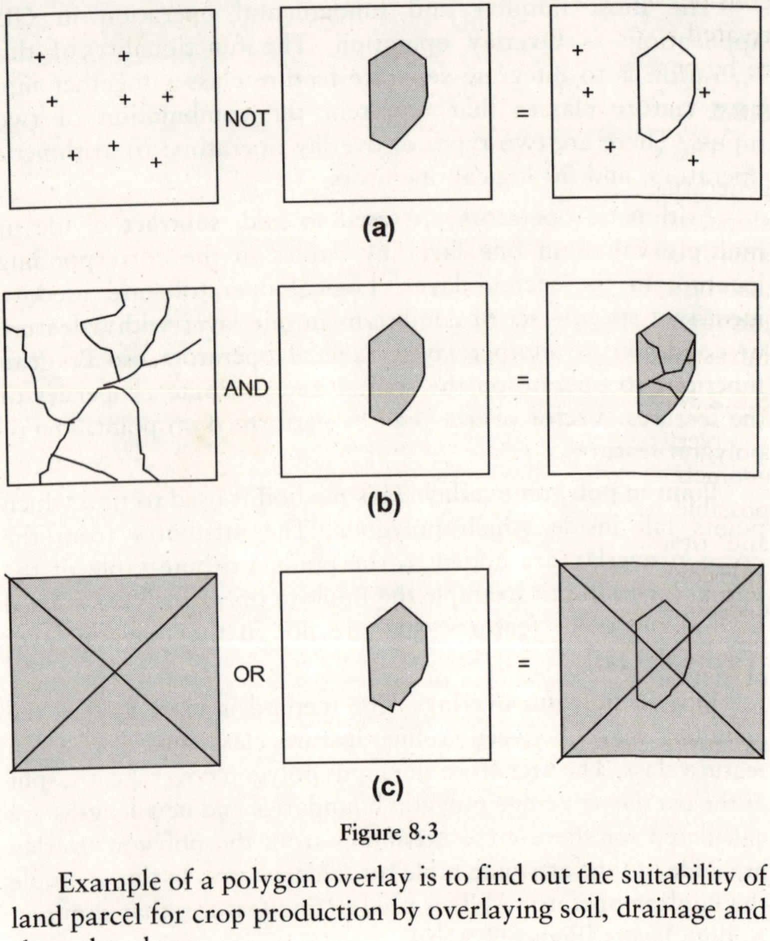

Figure 8.3

Example of a polygon overlay is to find out the suitability of land parcel for crop production by overlaying soil, drainage and slope data layers.

8.8 CONTOURING AND SURFACE GENERATION

Relief features and different kinds of slopes are represented on maps by contours. Contour lines are imaginary lines joining places which have the same height above mean sea level. Contouring and surface generation relates to interpolation of lines or surfaces from measurements made at surrounding sites (spot height data) to fill in the gaps at unsampled points. In order for generating such surfaces the variable under consideration must be continuous. The new surfaces can be treated as another layer in a GIS to support the spatial analysis to be performed.

8.8.1 Terrain Analysis

The terrain analysis module in GIS provides facility for generating DTM from contour map. It facilitates to calculate slope, aspect and generates relative relief map using DTM. The area visibility and profile plotting of generated DTM can also be carried out. A variety of applications include watershed management, groundwater studies, land use planning, infrastructure development, etc.

8.8.2 Network Analysis

Network analysis is the mathematical processing of the geometry of a link/node layer, enabling the identification of all possible routes around that network, along with the distances and times involved. Simply, it is the analysis of a set of interconnected lines. A GIS using networking can improve decision-making; but it cannot make decisions for us, thus in this way it helps to improve everyday life. Typical examples of network analysis are flows of traffic, streams, pipelines, etc. Network analysis can be used to evaluate in at least four situations such as:

(i) Address geocoding, in which the location is estimated in the GIS coordinate system taking address as an input. One such example is generating driving directions to a given address.

(ii) Optimal routing, the network to delineate the best (shortest) route to get from one location to one or more locations.

(iii) Finding closest facilities, a special type of optimal routing, where the analysis is performed to estimate the closest point to a given location. Here the specialists call points as facilities and the given location as event location. A typical example may be the best ambulance station to respond to an accident report event.

(iv) Resource allocation, a network analysis of solving allocation problem of finding the supply nodes for the demand nodes. It is the allocation of resources from supply centers to customers on a network. A suitable example is a hostel in a university allotting rooms to students, a well supplying water to an irrigation network, etc.

8.8.3 Area and Length Calculations

These are simple calculations used to highlight only those features with attributes satisfying certain selection criteria. Typical examples are: calculation of the area of parkland within a city which needs to be maintained, calculation of the length of pipeline within a sewage network which needs replacing, etc.

8.8.4 Data Output

The data output functionality allows the user to generate graphic displays, normally maps, and tabular reports and charts representing derived information products. Graphic output from GIS is produced using the automated cartographic techniques. Output devices commonly used are pen plotters, ink-jet printers, electrostatic plotters, laser printers and screen copiers. The derived information product is exported to multimedia for presentation, statistical packages, and Microsoft word by converting internal data into other specific standard formats.

8.9 SPATIAL DATA MODELING

The capability of GIS so far we understood, is the ability to digitize spatial data and attach attributes to the features stored to analyze these data based on those attributes, and to map out the result. Below described approaches to represent spatial data in GIS.

8.9.1 Geographic Data

Geographic data can be defined as data that describes any part of the Earth's surface or the features found on it. The geodata can, therefore, be used to describe our living planet in the manner such as city planning, flood prediction and relief, emergency service routing, environmental assessments, wind pattern monitoring and many other land and ocean applications. To manage and implement such a vast geographic data in a purposeful way to meet the societal need and health of the planet is the principal aim of the geographic information system. Some of the common areas of application of GIS are scientific investigations, resource management, asset management, Environmental Impact Assessment, Urban planning, cartography, criminology, history, sales, marketing, and route planning.

Geographic data includes not only cartographic and scientific data, but also business data, land records, photographs, customer databases, travel guides, real estate listings, legal documents, videos, etc. It has been found that over 80% of all the data used day-to-day in business and government can be related directly to geographic features like zip codes, street addresses, factory locations, census tracts, cities, states and countries.

8.9.1.1 Spatial data

Spatial data form the core of the GIS, without which GIS is meaningless. Every geographic data is spatial. A geographic feature in the Earth's surface is identified by (i) geometric location, and (ii) an attribute describing what the feature is and other characters of the feature. Spatial data stores the geometric locations and attributes of the geographic features. The locational information of the features is stored in a vector or raster data structure whereas the corresponding attribute information is stored in a set of tables related geographically to the features they describe. This is also known as a georelational data structure.

To sum up, spatial data contains two types of data structures (i) geo-locational information in vector or raster, and (ii) attributal information in geo-relational data structure. Any information about the location and shape of, and relationships among, geographic features can be explained by spatial data.

In land use/land cover classification, the geographic coordinate of a forest cover is spatial data, while attribute data are the forest characteristics e.g. cover group, dominant species, crown closure, height, etc.

Vector Data

Positional data in the form of (x, y) coordinates is called as vector data. Vector data records spatial information as (x, y) coordinates in a rectangular coordinate system. Point features are recorded as single x, y locations. Line features including the outlines of polygons are recorded as an ordered series of x, y coordinates.

Vector data records the geographic locations of discrete features such as streets, parcel boundaries, streams, telephone poles, etc. In ArcInfo GIS software, ArcInfo coverages are vector data. Drawing files in CAD are also vector data.

Actual ground locations of geographic features are accurately recorded in vector formats. However, accuracy is largely controlled by the (x, y) coordinate points that are chosen for the features. Natural features like streams and coastlines should have accurate coordinates so that they can be accurately recorded in vector formats.

In any GIS software (familiar GIS ArcView, ArcInfo, ArcGIS, etc.), when we are working with spatial data, both locational and attribute information are simultaneously accessible. For example, if we select particular features to display, the software will automatically highlight the records containing the attributes of these features.

Spatial data are always georeferenced as they are registered to known locations on the Earth's surface. Spatial data when displayed it provides a coordinate, unit of measurement, map projection and finally a scale just like scale in a topographical map. Spatial data is primarily feature based. GIS employs spatial data to manage, manipulate and analyse the Earth features using specific modules to meet a wide range of people's need. Other types of graphic data may be oriented solely towards presentation and display, and may store features such that they can only be accessed in a limited number of ways. Geo-referenced images are usually associated with features.

Spatial data is stored and organized to present in the form of different themes or layers called thematic layers. There is one theme for each set of geographic features or phenomena for which information will be recorded. For example, streams, land use, elevation, and buildings will each be stored as a separate spatial data sources, rather than trying to store them all together in one. This makes it easier to manage and manipulate the data.

Examples of data in vector format include site plans, survey maps and Computer-Aided Design (CAD) drawings.

Raster Data

Data expressed as a matrix or array of grid cells or pixels is called raster data. Raster data records spatial information in a regular grid or matrix organized as a set of rows and columns. Each cell within this grid contains a number representing a particular geographic feature, such as soil type, elevation, land use, slope, etc. Raster data stores information about geographic features that vary continuously over a surface such as reflectance, elevation, ground water depths, etc. ArcInfo grids are raster data. Satellite image data are in the form of raster as they are recorded in the form of a regular grid, each grid cell or pixel has a number that represents DN value. The resolution of a grid cell controls the volume of raster data. The size of the grid cell is fixed so that if you zoom it on a display device, you can see the shape of the cell.

Examples of data in raster format include scanned aerial photographs, satellite images and scanned documents or maps.

The Difference between Vector and Raster

The two basic data structures in GIS for storing and manipulating images and graphics data are raster and vector.

Raster maps store data in the form of a matrix or grid. Each grid cell is identified with a coordinate and a value called pixel value. The pixel value indicates the attribute, such as color, elevation, or an ID number. Vector-based maps are built from digitized points that may be joined to form lines, or vectors, and polygons, or closed shapes (**Figure 8.4**). Points are stored using the coordinates and lines are stored as a series of point pairs, where each pair represents a straight line segment, for example, (x_1, y_1) and (x_2, y_2) indicating a line from (x_1, y_1) to (x_2, y_2).

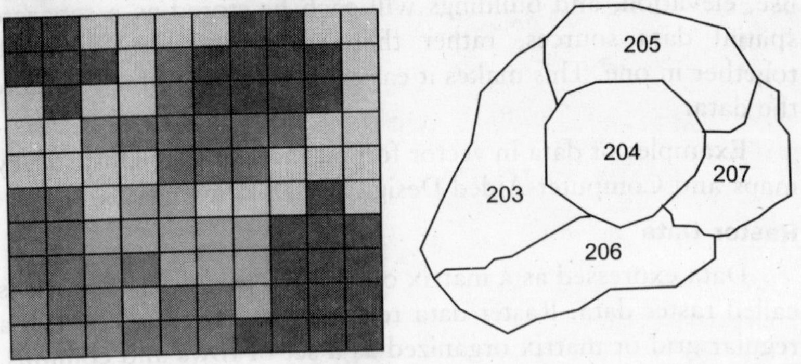

Figure 8.4

Raster method for spatial data structure needs huge space to store and process all pixels in an image to get the same level of spatial resolution obtained by vector data structures which stores only point coordinates. Vector data is easier than raster data to handle on a computer because it has fewer data items and it is more flexible to be adjusted for different scale. This makes vector data structure the apparent choice.

The vector model is extremely useful for describing discrete features, but less useful for describing continuously varying features such as soil type or accessibility costs for hospitals. The raster model has evolved to model such continuous features.

Lines in raster displays may appear jagged if resolution is not high enough. Rescanning an image at a finer resolution greatly increases file size. Vector files are good at showing lines but are labour intensive due to the need to clean and edit vector data (Faust, 1998). Most databases in urban areas use vector format; examples include street centerlines, precinct boundaries, and census geography. Vector files are not very good for managing continuous distributions, such as temperature or land elevation. Another way of differentiating is that "Vector systems produce pretty maps, while raster systems are more amenable to geostatistical analysis" (LeBeau, 1995).

8.9.2 Points, Lines and Polygons

The Geographic Information System is an attempt to model the real world in its entirety in a computer system so that with little cost, time and effort any planning, management and predictive models could be possible. However, modeling the real world in its entirety is rather more complex. Therefore, specific areas of interest should be selected for inclusion within a given GIS application. Once area of interest is selected, the desired geographic features are recorded in terms of their geographic co-ordinates and corresponding attributal information. All data captured should be in digital form to use in GIS. While modeling the real world, it is necessary to model each real world feature to the best possible. The geographic features on Earth are essentially represented by five different types of spatial objects, called as entity, feature or facility, within GIS. Those are point, line, polygon, surface, and network. The first three are regarded as the basic spatial objects.

A point or node is defined as a vertex that starts or ends an arc segment. Joining two nodes, a vector line is formed, which is referred to as arc. Point features are defined by (x, y) coordinate pair. A set of closed coordinate pairs define polygon features. In vector representation, the storage of the vertices for each feature is important, as well as the connectivity between features, e.g. the sharing of common vertices where features connect.

8.9.2.1 Points

The spatial positions of geographic features are stored as database in GIS for spatial date modeling. They are described by point, line and polygon.

A point can represent a geographic object of Earth which has only one reference coordinate or a point that has no area such as a mountain peak (**Figure 8.5**). Examples include trees, pylons, rainfall gauges, health clinics and hotels. In a large scale representation a hotel may be shown as area whereas it may

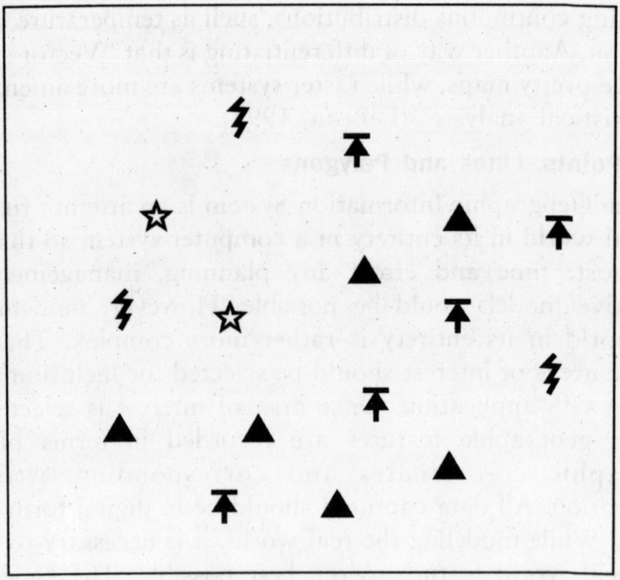

Figure 8.5

only a point (symbol) if the scale is reduced. A linear feature can be represented by vector data using a sequential points or vertices, called nodes. Each point or node is represented by a single x, y (lat., long.; or easting, northing) coordinate pair.

Each point on the map is created by latitude and longitude coordinates, and is stored as an individual record in the shapefile of GIS software.

8.9.2.2 Lines/Polylines

Line is formed by connecting two data points (**Figure 8.6**). A line (polyline) is a set of line segments and it represents a linear geographic feature such as a river, road, or utility line. Lines can also represent non-geographical boundaries, such as voting districts, school zones, contour lines, etc.

In GIS line can be said as a vector connecting two x-y coordinates (X = longitude, Y = latitude). The more points used to create the line, the greater the detail. System stores one end of the line as the starting point and the other as the end point, giving the line "direction".

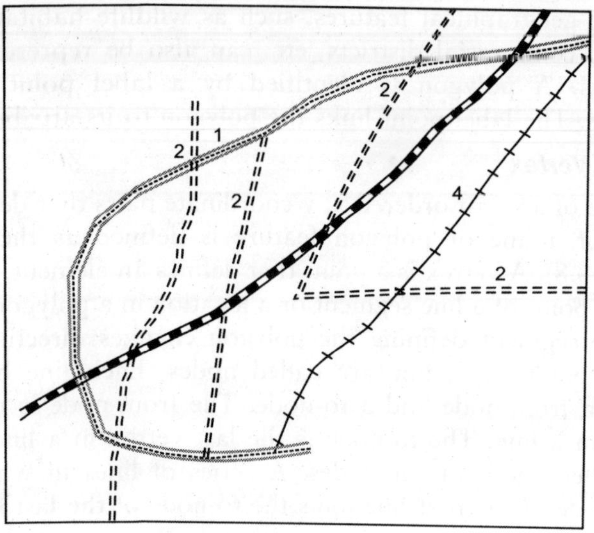

Figure 8.6

8.9.2.3 Polygons

A polygon is defined by the lines that make up its boundary and a point inside its boundary, a label point for identification. It is a closed shape defined by a connected sequence of x, y coordinate pairs (**Figure 8.7**), where the first and last coordinate pair are the same and all other pairs are unique. Because lines have direction, the system can determine the area of the polygon. Each polygon contains one type of data i.e. a polygon is defining a homogeneous area, for example soil type, land use, or water body. All of the data points that form the perimeter of the polygon must connect to form an unbroken line.

Figure 8.7

Non-geographical features, such as wildlife habitats, state borders, commercial districts, etc. can also be represented by polygons. A polygon is identified by a label point that it contains. The label point links the polygon to its attributes.

8.9.2.4 Vertex

One of a set of ordered x, y coordinate pairs that define the shape of a line or polygon feature is defined as the vertex (**Figure 8.8**). A vertex is a point that defines an element, such as the end point of a line segment or a location in a polygon where the line segment defining the polygon changes direction. The ending points of a line are called nodes. Each line has two nodes: a from-node and a to-node. The from-node is the first vertex in a line. The to-node is the last vertex in a line. Lines join other lines only at nodes. A series of lines in which the from-node of the first line joins the to-node of the last line is a polygon.

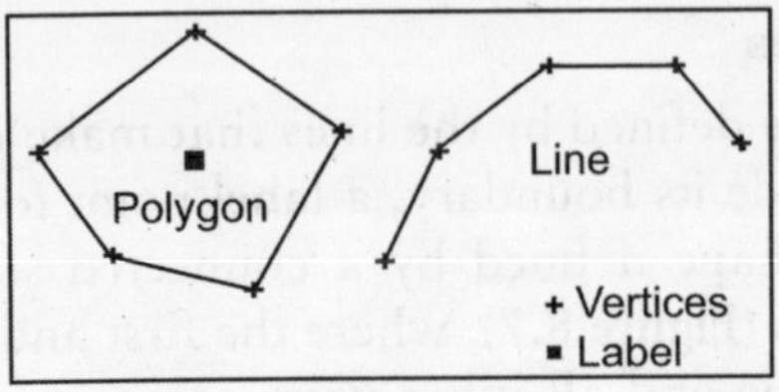

Figure 8.8

8.9.3 Spatial Data Models

From the discussions we made so far, we observe that the key aspects in GIS are:

(i) What is present in our surrounding?

It concerns with spatial data.

(ii) Where it is?

It concerns with corresponding attribute data.

Based on those, there are typically two spatial data models:

(i) Raster (the satellite remote sensing pixel view of the world)

(ii) Vector (the cartographic view of the world)

Two fundamental conceptual models of spatial data are:

(i) Objects

(ii) Fields

Objects are considered discrete entities having distinct boundaries. Examples are: buildings, roads, pipes, properties, etc. Vector GIS is suitable for the representation and analysis of objects.

Fields are continuous phenomena such as elevation, temperature and soil chemistry; they exist everywhere (every point has an elevation or temperature); they are not discrete entities. Since the fields are continuous phenomena, the raster GIS is suitable for the representation and analysis of field data. However, both objects and fields can work in vector as well as raster GIS taking into account the limitations involved.

Basic entities like points, lines and areas can be used to represent more complex entities like surfaces and networks. Raster and vector models represent surfaces differently in the form of Digital Elevation Models (DEMs) and Triangulated Irregular Networks (TINs) respectively. Vector models can represent network entities more effectively because of topological structure.

8.9.3.1 Topology

Topology involves the mathematical study of spatial relationships. It describes the characteristics of a geometric figure that do not change under continuous transformation. In a graph, the number of line segments, intersection points, and polygons, and their relationship to each other, are constant as the plane in which they exist is stretched or distorted. In GIS applications, topology is the means to describe, manage, and retrieve these relationships explicitly without resorting to time-consuming spatial comparisons (Ramage and Woodsford, 2002).

The principles of topology are utilized to implement a system that provides for:

- Rapid spatial data retrieval.
- Enhanced spatial data analysis.
- Spatial data editing and clean up.

- Improved data consistency.
- Management of shared geometry (Hoel et al., 2003).
- Definition and enforcement of data integrity rules (Hoel et al., 2003).

Some GIS applications use "persistent topology", that is, they structure the data according to topological principles so that the topological relationships are stored and available persistently in the database. In addition to the above benefits, this persistent topology approach also provides for:

(i) More efficient storage of spatial data.

(ii) Reduction or elimination of redundancy of spatial data.

(iii) Easier implementation of certain spatial business rules in the database.

(iv) Improved management of hierarchical geographic relationships (Ramage and Woodsford, 2002).

(v) A more straightforward way to address coordinate precision and tolerance issues that sometimes lead to gaps or slivers (Egenhofer et al., 1989).

The manipulation and analysis of large quantities of spatial data and their nature of relationships is the primary role of Geographic Information Systems. The nature of relationship between the geographical features is although important to understand, yet it is a complex problem in GIS. Spatial relationships include:

(i) Absolute and relative location

(ii) Distance between features

(iii) Proximity of features

(iv) Features in the "neighbourhood" of other features

(v) Direction and movement from place to place

(vi) Boolean relationships of "and", "or", "inside", "outside", "intersecting", "non-intersecting", etc.

The spatial relationships between features in a vector layer are defined using topology. In topological vector data, a mathematical procedure is used to define connections between features, identify adjacent polygons, and define a feature as a set of other features (e.g., a polygon is made of connecting lines) (Environmental Systems Research Institute, 1990). In GIS, topology is the term used to describe the geometric characteristics of objects which do not change under transformations such as stretching or bending and are independent of any coordinate system (Bernhardsen, 1992). The topological characteristics of an object are also independent of scale of measurement (Chrisman, 1997). Topology is a mathematical approach that allows us to structure spatial data based on the principles of feature adjacency and feature connectivity. Topology as it relates to spatial data consists of three elements adjacency, containment, and connectivity (Burrough, 1986):

(i) adjacency—shared boundary

(ii) connectivity—shared node in arc-node topology

(iii) containment—accounts for polygons within polygons "islands".

The geometric relationship which exists between area features is described by the adjacency and containment. When two area features share a common boundary, areas can be described as adjacent. Containment is the act of keeping an area feature fully within the boundary of another area feature such as an island within a lake. It is an extension of the adjacency theme. Connectivity is a geometric property used to describe the linkages between line features. Roads are connected together to form a road network through which traffic can flow.

An understanding of the topological relationship is important for analysis and integration in GIS. The topological data structure logically determines exactly how and where points and lines connect on a map by means of nodes (topological junctions). Without a topologic data structure in a vector based GIS, most data manipulation and analysis functions would not be practicable or feasible.

8.9.3.2 Raster GIS models

After collecting so much of geographic data the complexity arises how to encode and store the data in a computer system to reconstruct the spatial data model in digital form. There are many different data structures in use in GIS. To exchange spatial data between different GIS softwares is problematic due to such a diversity of data structures. Despite this diversity, data structures are classified into raster and vector.

In the raster world, individual cells are used as the building blocks for creating images of point, line, area, network and surface entities. The shape and character of an entity is created by the grouping of cells. The size of the grid cell is very important as it influences how an entity appears.

To model geographical variation in the real world in terms of discrete spatial objects is a cumbersome task, as it involves enormous quantities of data. A set of rules used to convert real geographical variation into discrete objects is the data model. Tsichritzis and Lochovsky (1977) define a data model as "a set of guidelines for the representation of the logical organization of the data in a database...(consisting) of named logical units of data and the relationships between them".

As discussed previously, raster data model represents the entire study region in terms of regular grids of cells in specific sequence. Raster data encodes from the top left corner in a sequence row by row. Each grid cell contains single value. In this model every location in the study area corresponds to a cell in the raster grid. There is no cell lying vacant as the raster data is continuous. A raster database contains layers of information e.g. land use, land cover, soil type, elevation, etc. Each layer corresponds to a set of cells and associated values.

How to Create a Raster

The form or format of the raster data stored and manipulated on the computer is the raster data structure. There are several methods for storage and representation of spatial entities in the computer.

A simplest method of coding raster data is shown in **Figure 8.9**. In this method, a grid is laid over a thematic map (satellite classified image or map). A raster file [Figure 8.9 (c)] is created by coding each cell (assigning a value) corresponds to a particular feature type. When finished, every cell will have a coded value. The first line of the file (8, 8, 1) indicates the image contains 8 rows and 8 columns with maximum cell value of 1.

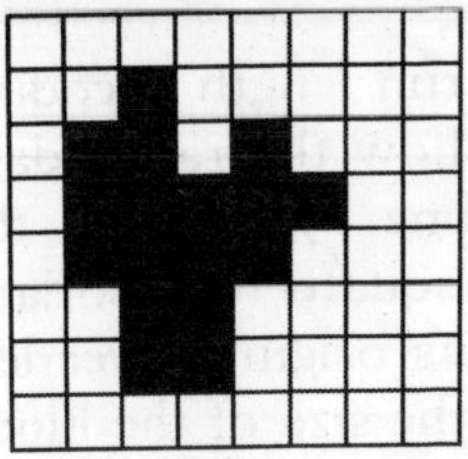

(a)

0	0	0	0	0	0	0	0
0	0	1	0	0	0	0	0
0	1	1	0	1	0	0	0
0	1	1	1	1	1	0	0
0	1	1	1	1	0	0	0
0	0	1	1	0	0	0	0
0	0	1	1	0	0	0	0
0	0	0	0	0	0	0	0

(b)

8, 8, 1

0	0	0	0	0	0	0	0
0	0	1	0	0	0	0	0
0	1	1	0	1	0	0	0
0	1	1	1	1	1	0	0
0	1	1	1	1	0	0	0
0	0	1	1	0	0	0	0
0	0	1	1	0	0	0	0
0	0	0	0	0	0	0	0

(c)

Figure 8.9

In this type of raster data structure, different spatial features must be stored as separate data layers. If there are 'N' different spatial features, data will be stored in 'N' different data layers of information. In the given figure, the value 0 has been recorded for cells where the entity is not present (no data), and a value of 1 for cells where the entity is present [Figure 8.9 (b)].

Run-length encoding

The first method as explained above needs large amount of space to store the data and is more confusing since it involves data redundancy. This problem is sorted out in run-length encoding method which is comparatively efficicent. This method is otherwise called raster image compression technique. If a raster contains groups of cells with identical values, run-length encoding can compress storage. It does not store each cell. It stores data as pairs, (i) a value, and (ii) a number of cells with that value (see **Figure 8.10**).

A	A	A	A	A	0	0	0	0
A	A	A	A	A	A	0	0	0
A	A	A	A	0	B	0	0	0
A	A	A	A	0	0	0	0	0
A	A	A	0	0	0	C	C	C
C	C	C	C	C	C	C	0	0
C	C	C	C	C	0	0	0	0
C	C	C	C	0	0	0	0	0
0	0	0	0	0	0	0	0	0

A, 5 0, 9

A, 6 0, 9

A, 4 0, 5 B, 6 0, 9

A, 4 0, 9

A, 3 0, 6 C, 9

0, 4 C, 7 0, 9

C, 5 0, 9

0, 9

Figure 8.10

Block coding

This approach is the extension of the run length encoding method to two dimension. Figure depicts how the raster data is compressed by block codes. This is done by dividing the mapped area into a series of hierarchical square blocks. Each block is described by the co-ordinates of its origin (lower left corner) or centre and by a number telling the size of the block expressed as square cells. **Figure 8.11** illustrates that the area is subdivided into 10 blocks. These are seven unit cells, two four-cell squares and one nine-cell square. Their respective coordinates are given in the figure 8.11.

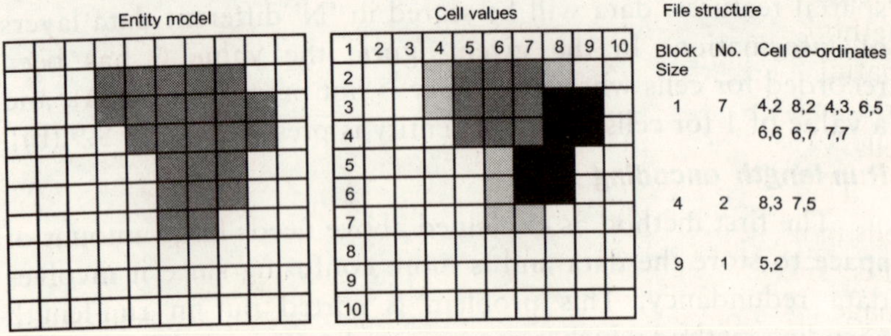

Figure 8.11

Chain coding

It is a compression method of data encoding. It describes an area as a series of unit vectors emanating from a defined start point. The boundary is defined as a sequence of unit cells starting from and returning to a given origin. **Figure 8.12**

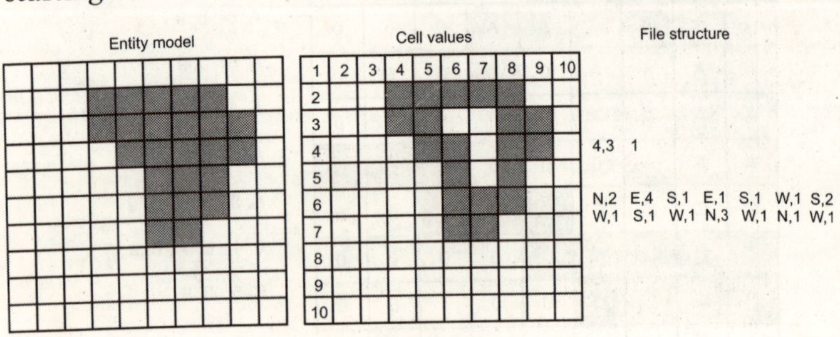

Figure 8.12

illustrates the method of chain coding, where the direction of shifting unit cells is given by:

0 = North, 1 = East, 2 = South, and 3 = West.

In the Figure 8.12 the first line in the file structure is represented by 4, 3 and 1. This indicates chain coding starts at 4, 3 and there is only one chain.

Quadtrees

This method is the most complex form of storing reduced format raster data. This method is also difficult to understand. One of the advantages of the raster data model is that each cell can be subdivided into smaller cells of the same shape and orientation (Peuquet, 1990). The advantage of using this method over the previously described ones is that for higher spatial resolution data, quadtree format supports smaller grid sizes while for large uniform areas it can support larger cells.

The principle of data encoding in this method is that the cells in a raster image are subdivided into four equal quadrants. This subdivision process continues until each cell in the image can be classed as having the spatial entity either present or absent within the bounds of its geographical domain (see **Figure 8.13**).

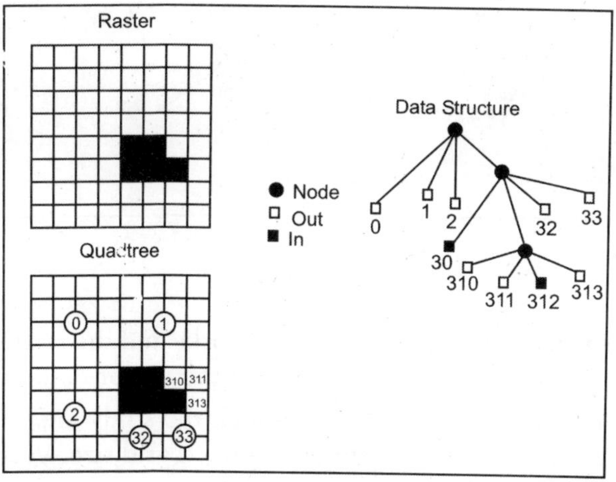

Figure 8.13

8.9.3.3 Vector GIS models

The word vector represents a straight line having a starting point and an end point with known geographic co-ordinates. These starting and ending points are called nodes. In vector models, spatial data are represented by the vectors. The fundamental element in vector data model is a dot or point. Dots can join together to construct the shape of an object. The dots are joined by straight lines. In the vector model dots are called vertices. Line segments or arcs join at vertex, which is called a node. A polygon is formed when a series of arcs, while joining return to the starting node. A vector that demarcates two areas is often called edge, arc or chain. **Figure 8.14** explains the geometry of point, line segments and polygon.

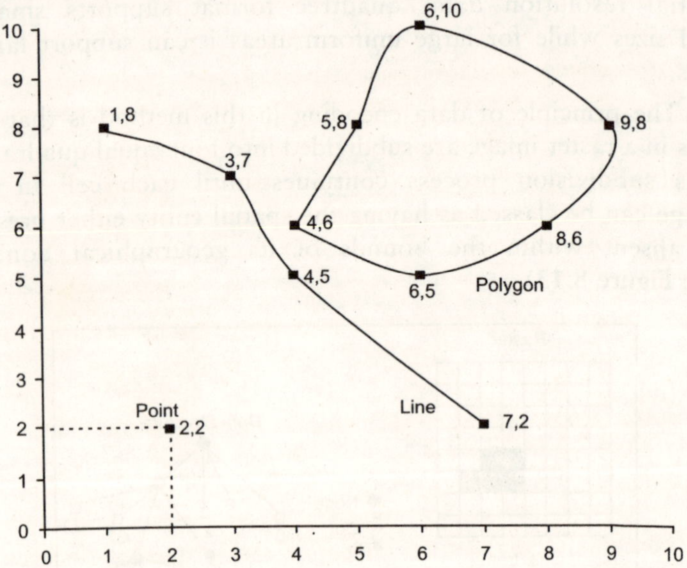

Figure 8.14

Linear features such as cable networks and the relationships between areas are well represented in the vector data model.

In comparison to the raster model, the study area does not require to be divided in a grid with a fixed grid size. Vector data also need not require scale. Each point (vertex) in the vector model is treated as its smallest component. Each vector is assigned with a unique Id-number, which can be used to link the geographical co-ordinates to one or more attributes in a database.

In the vector data model, spatial entities are stored as points, lines or areas. A brief note on storing points, lines and polygons in the computer is described below.

Storing points

Each point and line represents the geographic features (e.g. house, the roads and rivers) on the Earth's surface. The locational information for these features is stored using geographical coordinates.

Points

(i) These are the simplest features. Consider a hotel to be stored. The position of the hotel can be stored by a pair of x-coordinate and y-coordinate. We can create a simple table if we do the same for all such point objects.

(ii) The table can be stored on the computer in an Excel spreadsheet or as a table in a dBase database.

(iii) Using the coordinate information in the table we can draw a new version of our map with all such points. From the map, we can easily compute the distance between any two points.

(iv) By storing the location of the points as a pair of coordinates we can do three things:

(a) We are storing the locational data from the map on a computer.

(b) Using this information we are reproducing the map.

(c) We are able to make simple calculations from the data.

However, difficulties arise if our study area covers more than one map. In this situation, measuring the coordinates from the origin is not useful. Instead, the UTM (Universal Transverse Mercator) coordinates give a standard way of representing positions for any point in the country as mostly, every map has a grid laid over it.

(v) Along with the locational information, we also need to store the attributes of each point. This can be done by adding one extra column in the existing table. The new column is known as feature code, indicating what category a feature falls into (e.g. hotel for building 1, church for building 2 and so on). This information can be used later to assign symbol when redrawing our map. The other attribute information specific to each point (to describe each feature) can be added using extra columns.

Storing lines

(i) A linear feature (e.g. road) can be represented by a series of line segments. To store the location of the line segments, we simply need the coordinates of the points between each line segment. We can plot the position of each point and join successive points with a straight line. Since road is a straight linear feature, most correctly it can be stored. However, say in case of streams, which is a curved linear feature; this representation by a series of straight sections will only be an approximation of the reality.

(ii) Length of any of the straight sections can be calculated by calculating the distance between the points at each end; adding all these up gives us an estimate of the linear geographic feature.

(iii) For correct matching between the real line (on the survey map) and series of straight segments we have to add extra points along the line. To get the closer match, more points need to be added. This results in larger files on the computer.

(iv) The locational information i.e. the X, Y coordinates of the lines can be stored in a spreadsheet or a database package. However, the job will be difficult if we store attribute information in the same table.

(v) When we attempt to store a linear feature, we will come across a large number of X and Y coordinates rather than a single pair we had for a point feature. To overcome the messy table of data record, numerous methods have been adopted. Many of the database systems commonly used today will not allow tables where each row has a different number of columns, or where a single feature is represented by more than one row—this is certainly true of the relational database systems often used in GIS.

(vi) To avoid the messy tables of data record, the locational and attribute data are stored separately. The attribute data is stored in a standard database package, while the locational data are stored in specialised software. The GIS package Arc/Info has the facility to store the attribute data in a standard database—INFO—and locational data in specialised software—Arc.

(vii) To link both the location and attribute data, each feature has a unique identifier, which is stored with, both the locational and attribute data.

Storing areas

The third major type of spatial feature is the area. The areas are stored in two ways in vector model; those are (i) polygon storage, and (ii) arc storage.

Polygon storage

(i) Every polygon is stored as a sequence of X, Y coordinate values. Although we know how to store a line, in this case, the difference is that the end of the line joins the start to make a closed boundary.

(ii) The points would be stored in order with the coordinates of the last being the same as the first.

(iii) Since the attributes are associated with the interior of an area rather than the boundary, a centroid of the area is generally stored, which is a point within the polygon boundary. The centroid is commonly used to give a position for labels. Each centroid indicates storing of one polygon in GIS. In GIS a single area can be stored by two things—the line defining the boundary, and the point defining the centroid.

(iv) The centroid of an area is associated with an identifier that links the attribute table for the area.

(v) Polygon shares part of its boundary with adjacent polygons i.e. most boundaries (line segments) of a polygon are shared between two adjacent areas. The result is that we are storing nearly twice as much data as necessary.

(vi) While storing the same line as boundary two times, they may not coincide. It may result in small areas of overlap and small gap between the two areas. These mismatches are called silver polygons, as they are very small and thin.

(vii) When we perform analysis rather than map drawing, another type of problem we come across with this method of storing area boundaries. Suppose we wish to produce a new map with urban and non-urban areas, then we have to merge together all the polygons which are not urban areas. This can be done by dissolving the boundaries of the polygons retaining only two categories: (i) urban, and (ii) non-urban. This operation is called a polygon dissolve in GIS. The simple method of storing area boundaries, however, cannot overcome this difficulty.

Arc storage

(i) Every arc is stored as a sequence of coordinates.

(ii) Each area boundary is made up of a series of arcs which meet at junctions, called nodes. The line segments between two nodes are called arcs or chains or links. For each link we can store the location of the nodes and a series of positions along the line that build up the link. To know which link belongs to which areas, the identifier of the polygons on either side of the link is stored for each link.

(iii) By using this method, only one version of each internal shared boundary is input and stored.

Vector GIS is created in the following manner:

(i) Inputting or digitising spatial data.

(ii) Attributes are coded and entered into a database.

(iii) Geographic or locational data are linked to attribute data (topology building).

Attribute tables store topology in which different objects are linked to the attributes via their Id-numbers.

The simplest vector data structure that can be used to reproduce a geographical image in the computer is a file containing (x, y) coordinate pairs that represent the location of individual point features (**Figure 8.15**).

8.10 SURFACE ANALYSIS IN GIS

Recent advances in GIS technology help to provide a powerful data integration tool in many areas such as landscape science, ocean science, atmospheric science, environmental management, or oil and gas, dealing with complex three-dimensional spatial data. Many GIS users treat 3D GIS as a three-dimensional view of standard two-dimensional spatial data. This three-dimensional view is often seen as an aesthetic addition to GIS. Although the data may be visualized in three dimensions, the majority of the spatial analysis, editing, and other common GIS tasks are performed in a strictly two dimensional environment. Although the vast majority of geographic information systems (GIS) work only in two dimensions, certain applications require the addition of a third dimension due to the complex nature of the features to be modeled or to allow for spatial analysis in three dimensions.

The basic components of a GIS model are: (i) points (or nodes), (ii) lines (or edges), and (iii) polygons (or faces), and solids. It is quite evident, therefore, that the geometric primitives of a three-dimensional data model are essentially those in a standard two-dimensional data model.

Surface in GIS is defined as any raster layer or vector layer that describes all locations in a study area. Surface analysis in the GIS environment deals with the attributes which are quantitative and vary continuously over space. A raster Digital

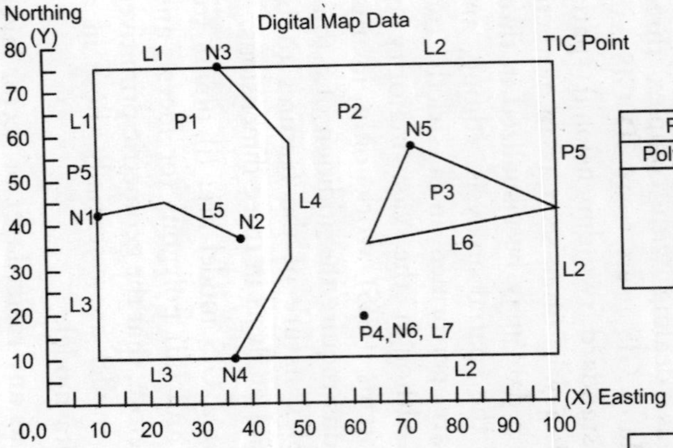

Polygon Topology	
Polygon-ID	Vectors
P1	L1, L4, L3
P2	L2, L4
P3	L6
P4	L7
P5	L1, L2, L3

Node Topology	
Node-ID	Vectors
N1	L1, L3, L5
N2	L5
N3	L1, L4, L2
N4	L2, L3, L4
N5	L6
N6	L7

Vector Topology				
Vector-ID	Start Node	End Node	Left Polygon	Right Polygon
L1	N1	N3	P5	P1
L2	N3	N4	P5	P2
L3	N4	N1	P5	P1
L4	N3	N4	P2	P1
L5	N1	N2	P1	P1
L6	N5	N5	P2	P3
L7	N6	N6	P2	P2

Vector Coordinate Data			
Vector-ID	Start X, Y	Intermediate X, Y	END X, Y
L1	10.0,42.5	10,75	35,75
L2	35,75	100,75; 100,42.5; 100,10	37,10
L3	37,10	10,10	10,42.5
L4	35,75	48,58; 48,31	37,10
L5	10,42.5	23,45	38,37
L6	71,57	63,35; 100,42.5	71,57
L7	62,19	—	62,19

Figure 8.15

Elevation Model (DEM), for instance, is such a surface, where elevation varies continuously over space. Other examples are surfaces might describe NDVI, population density, or temperature. In these types of surfaces, each pixel may have a different value than its neighbors. According to this definition, a land use/land cover map cannot be a surface as the values are not quantitative, and they do not vary continuously over the map. Use of the term 'surface model' in stead of using 'surface' alone signifies that at every scale there is mismatching of surface layer to the reality. Although different models exist for the same phenomenon, selecting a suitable model depends upon number of factors such as the application, accuracy requirements, and availability of data.

A surface, as per the definition, is a continuous field of values that may vary over an infinite number of points. For example, surface of the Earth is treated as a floor of continuous objects rather than assembly of discrete objects. Therefore, to model the Earth surface, one needs an infinite number of points which in turn, require massive space for data storage. If we take point observations on the Earth's surface, they certainly vary spatially, for example, elevation, proximity to a feature, or concentration of a pollutant. In a three-dimensional x, y, z coordinate system, such an observation is represented on the z-axis and is often called z-value. Surface analysis in GIS involves analyzing the distribution of a variable which can be represented as the third dimension of spatial data. We have X and Y coordinates of features plus a third quantity which represents the variation in the surface. The third variable, as discussed, is called Z variable. Elevation is the example of a third dimensional variable.

Any event on the surface of the Earth has continuous points on its spatial extent for which observations could be made. Those observations at the infinite number of points across the surface are derived from a limited set of sample values. These may be based on direct measurement, such as height values for an elevation surface, or temperature values for a temperature surface. The values for the points in between are assigned by interpolation. Surfaces can also be mathematically derived from other data, such as slope and aspect. Since observations are taken at discrete interval, a surface model to present the three dimensional shape; $z = f(x, y)$ should be built to allow the interpolation of value at arbitrary points of interest.

For handling extra dimension in 3D spatial data modeling, the development of specific data structures and algorithms is required. To generate a Digital Elevation Model, for example, the following four types of sampling point structure are used.

Grid at regular intervals

Bi-linear surface with four points or bi-cubic surface with sixteen points is commonly used.

Random points

Points can be regularly or irregularly distributed across a surface. They are usually used as input to interpolation to create raster or triangulated irregular network (TIN) surfaces.

Contour lines

These are sets of lines of equal value across a surface. They are frequently created to represent surfaces on a map. Interpolation based on proportional distance between adjacent contours is used. TIN is also used.

Profile

Profiles are observed perpendicular to an alignment or a curve such as high ways. In case the alignment is a straight line, grid points will be interpolated. In case the alignment is a curve, TIN will be generated.

Every object on the Earth surface more or less exists in three dimensional space. However, for many purposes it is possible to ignore the third dimension and produce a model which still represents the approximation of the reality. For example maps represent 3D objects in two dimensional planar surface. But topographical survey maps provide third dimension data i.e. the contours and spot heights from which slopes and direction of slopes can readily be estimated. Contours are not a true three dimension representation. They can only represent the upper surface of the landscape. They cannot represent 3D view of overhangs or certain geological phenomena such as geological strata.

8.10.1 Digital Elevation Model

The Earth's surface is a continually varying surface rather than a surface of discrete objects. Such surfaces are represented by isolines (contours). These contours can be effectively regarded as sets of closed, nested polygons. Set of isolines are very suitable for the display of a continually varying surface; they are not particularly suitable for numerical analysis or modeling. So other methods have been developed in order to be able to represent and to use effectively information about the continuous variation of an attribute (usually attribute) over space.

A digital elevation model (DEM) is a digital representation of ground surface topography or terrain. A DEM is an ordered array of numbers that represents the spatial distribution of elevations above some arbitrary datum on the landscape (Moore et al., 1991). In principle, a DEM describes the elevation at any point of a given area in digital format and contains information of the so-called 'skeleton lines'. Skeleton lines are lines of slope reversals (drainage, crests) and breaks of slope. In contrast, a DEM is used to generate a "digital terrain model (DTM)". A digital terrain model includes the spatial distribution of terrain attributes. A DTM is a topographic map in digital format, consisting not only of a DEM, but also slope, aspect, the types of land use, settlements, types of drainage lines and so on, (Meijerink et al., 1994).

The quality of DEMs and their derivatives generally depends more on the quality of the data input than on the topic of data structure and interpolation algorithms. The data generally used are contours and spot heights which are digitized from topographic maps. The quality of the final DEM will be influenced by the human errors that occurred when the digitizing was done; the map scale of the topographic maps; and the interpolation method used to extract the height information from the contours and spot heights for the creation of the DEM. Many attempts have been made to define the accuracy of DEMs quantitatively over the years (Li, 1988; Tempfli, 1980; Shmutter and Doytsher, 1990). Most of these studies were concerned with the accuracy of the interpolation of the DEM from a set of data points (Carter, 1989; Dias and Nery, 1992); and the change in accuracy with data density (Li, 1992).

Various considerations are taken into account when generating a DEM. Terrain characteristic is one of them. In areas where the terrain fluctuates to great extent, the accuracy of the derived DEM depends upon the contour interval, (effectively the number of contour lines), and the available spot height data. In some instances the requirements for uniform topographic map formats within a country adopts a single contour interval (or possibly two) and as a consequence, the representation of the high and low areas of relief is compromised and incorrectly represented in the DEM.

Another consideration is sampling methods used to collect height information data. It is the desire to collect the optimum amount of data such that there is a balance between the density of sampling and the volume of data, which would minimize the cost of data collection and at the same time be sufficient to enable an accurate re-construction of the terrain surface, (Robinson, 1994). A balance in the sampling process is necessary because if one looks at the extreme circumstances—"Over-sampling" will lead to the generation of redundant information while "Under-sampling", will result to unwanted effects of generalizing the landform.

A more common consideration used in the development of a DEM is the interpolation methods used in the generation of the final output. The reason for having interpolation in the creation of DEM is that, it provides a scientific method through which unknown data can be generated based upon the data that are in the neighborhood. There are various interpolation methods available in most DEM generation software. However, despite the individual method of approaches, the implicit assumption which is common in all of the methods is that, arbitrary points lying between sampled positions can have elevation values estimated to a certain degree of accuracy by some form of interpolation. According to Robinson (1994: 808), strictly speaking, "interpolation is only valid if the surrounding points are sufficiently correlated. If the surface has been grossly under-sampled, then large errors in the estimated values will occur. From a statistical point of view, 'interpolated' values can only be estimated from the local trend at the point in question with an error given by the variation of the known data points from the trend to that region."

As the mathematical model of the terrain surface, DEMs are the terrain expression according to some specific methods based on the set of measured data points. There are many data structures used for DEM creations, and as such, depending upon the intended use of the DEM, the choice of the data structure greatly influences the results, therefore, an evaluation into their pros and cons is necessary. At present there are two main methods for terrain modeling i.e. TIN based and regular grid based (Li, 1990; Zhu and Chen, 1998).

(i) Grid structure

Grid structure usually consists of a regularly spaced square grid with an elevation value specified at each grid square, called a cell. Most of the DEMs available now are in grid structure format (Garbrecht et al., 2001). The elemental area is the cell bounded by four adjacent grid points. In comparison, the raster based GISs uses the square grid networks. When the terrain is highly irregular, the grid density or cell size must be manipulated in order to get the best approximation of a DEM to represent reality. However, this causes redundancy in those areas with a flat or gently sloping topography. The quadtree database structure is one way of avoiding data redundancy, and its functionality is based on the regular decomposition of a feature into quadrants and sub-quadrants until a single pixel is obtained.

The advantage of the regular grid method is the simplicity of the data storage, usually done as sequential height information "z" coordinates along the "x" (or "y") planar direction with a specified starting point and grid spacing.

(ii) Triangulated Irregular Network (TIN)

Digital representation of part of the Earth's surface in the shape of a model of the reality is the important subject of GIS. The digital model of the terrain is stored in various ways. Those are: (i) a set of contour vectors [**Figure-8.16(a)**]; (ii) a regular grid of spot heights (Digital Elevation Model: DEM—Figure 8.16(b); and (iii) an irregularly spaced set of points connected as triangles [Triangular Irregular Network: TIN—Figure 8.16(c)].

A triangulated irregular network (TIN) is the most commonly used digital data structure used in a geographic information system (GIS) for the representation of a surface. Data points with a set of known values (e.g. elevation, spot height) are used as the vertices to construct sets of triangles. In this method surface is constructed from a set of irregularly spaced data points. The key to the use of the TIN is the fact that the height of the three corners is known. With this information, it is possible to work out which way the triangle is facing, how steeply it is sloping and the height at any point in the triangle. In other words, the triangles apart from representing the height of the sample points, they model the whole surface.

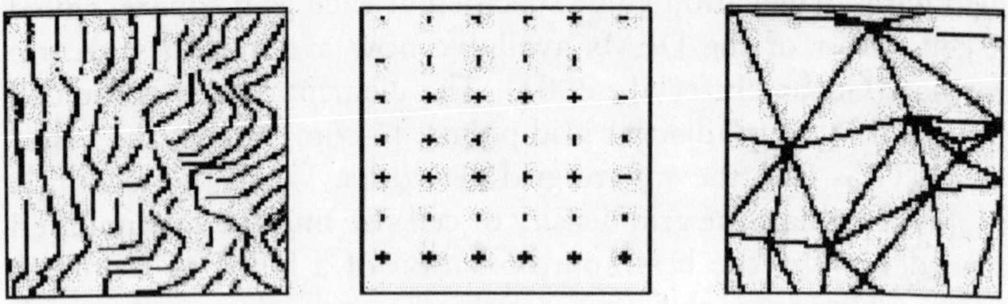

(a) **(b)** **(c)**

Figure 8.16

The TIN data structure is based on two basic elements: points with horizontal (x, y) and height (z) values. It takes the original data points and connects them to form a series of irregular triangles [Figure-8.16(c)]. The relationship between the points is stored using this triangular network. The TIN allows extra information to be gathered in areas of complex relief without the need of huge amount of redundant data to be gathered from areas of simple relief. While sampling the terrain, data must include skeleton information such as peaks, pits, lines of slope, drainage divides, etc. which helps to control the generated DEM. The skeleton lines appear as arcs in the TIN surface which are used as edges for the triangles. Linear features where there is an abrupt change in surface occurs such as ridge-lines, streams, or roads can be incorporated into TINs as breaklines. The values at the points between nodes can be estimated using the linear interpolation from the nearest nodes.

There are many different methods of triangulation. The Delaunay triangulation process is most commonly used in TIN modeling.

Delaunay Triangulation

Delaunay triangulation can divide a surface into regions that are particularly well-suited for image processing applications. Triangulation is a process of dividing a surface into regions with certain common characteristics. The type of these characteristics depends on the type of triangulation. This method divides a surface into a set of triangles. Each edge of a triangle is entirely shared by two triangles.

The Delaunay triangulation of a set of points in the plane is a set of triangles connecting the points satisfying an "empty circle" property: the circumcircle of each triangle does not contain any of the points. It is in some sense the most natural way to triangulate a set of points.

A Delaunay triangulation is defined by three criteria: (1) a circle passing through three points of any triangle (i.e., its circumcircle) does not contain any other data point in its interior, (2) no triangles overlap, and (3) there are no gaps in the triangulated surface. **Figure 8.17** shows examples of Delaunay and non-Delaunay triangulations. The figure on the right side is not satisfying the criteria (1), as a point belongs to its interior. Therefore, it is called non-delaunay triangulation.

Delaunay triangulation is suitable for image processing applications. Due to the inherent geometric qualities, an image when divided into a set of triangulated regions; it is possible to model a wide variety of geometries. Triangles are the simplest building block for such models. With this method, an image can be built up or segmented into triangulated regions. The advantage of using this method over other types is that it maximizes the minimum angles of the triangles, so that the triangles will tend toward equiangularity. Such a situation will be ideal for interpolation algorithms which attempt to avoid distortions unlike other methods of triangulation.

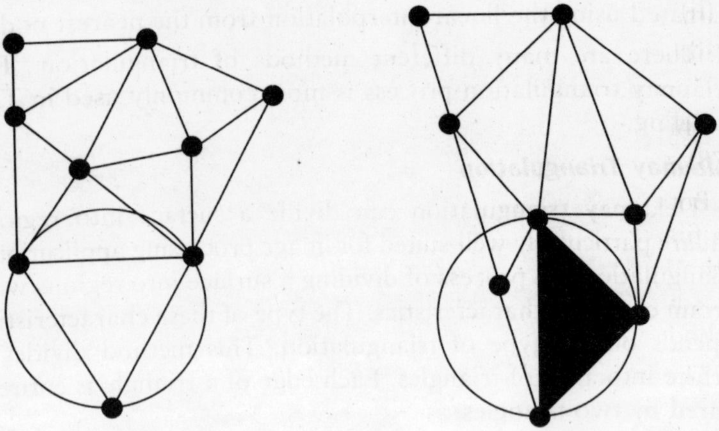

Figure 8.17

Structure of TIN

There are various ways in which TINs are stored in the database. Goodchild and Kemp (1990) propose two structures which are based on triangles and nodes. The triangle structure is efficient for slope analysis, while the node structure allows easy contouring and other traversing procedures requiring little storage space. Apart from the above two methods, other structures have been devised, such as the TIN side based structure, which is efficient for contouring.

(a) Storing TIN triangle by triangle:

In this method of storing, each record contains (i) a reference number for the triangle (ii) the x, y, z, coordinates of the three vertices, and (iii) the reference numbers of the three neighboring triangles.

Since a vertex participates in, on the average, six triangles, repetition of coordinates can be avoided by creating a separate vertex file and referencing them in the triangle files.

(b) Storing TIN by points and their neighbors.

The alternative method to store every vertex is:

(i) an identification number

(ii) the xyz coordinates

(iii) references (pointers) to the neighboring vertices in clockwise or counter-clockwise order.

This structure was the original TIN structure (Peucker et al., 1978).

Both structures are necessary depending on the requirements. The first method is used for slope analysis, whereas the second method is preferred for contouring and other traversing procedures. In comparison to the first method, the second method generally needs less storage space.

The most widely used structure is based upon the triangle. Lists of coordinates, nodes and of the tri-neighbors are related through pointers. The use of pointers creates a dynamic internal character. McKenna (1997) proposed a flexible data structure, which has tables of nodes, triangles and connected nodes, without the neighbouring triangles.

8.10.2 Terrain Analysis

To create a terrain, we need surface measurement data in the form of features stored in a geodatabase. Terrain analysis involves the processing and graphic simulation of elevation data. A map providing topographic information is not new. However, in recent days, with the Differential Global Positioning System point elevations have been collected in precise digital form to simulate the terrain surface. The purposes of terrain analysis include visualization, preparation of data for e.g., hydrological, geomorphological, and biological applications.

Topographic data are essential for studies of trafficability, route design, non-point source pollution, intervisibility, siting of recreation areas, etc. (Welch, 1990). Slope, aspect and shaded relief are considered as these are the most commonly used applications of terrain models used in GIS.

- slope images—illustrates changes in elevation over

distance. Slope images are usually color-coded according to the steepness of the terrain at each pixel.

- aspect images—illustrates the prevailing direction that the slope faces at each pixel.
- shaded relief images—illustrates variations in terrain by differentiating areas that would be illuminated or shadowed by a light source simulating the sun.

Calculating slope and aspect

By the term gradient, we mean slope. Gradient is expressed or measured by stating how far one must travel to reach a given height. If we rise 1 meter in covering a distance of 20 meters, then the gradient is 1 in 20. Slope is the steepness or gradient of unit of terrain, usually measured as an angle in degrees or as a percentage. In raster DTMs slope is calculated using a 3 × 3 window that is passed over a database to determine a best fit tilted plane for the cell at the center of the window. It is the maximum rate of change from a cell to its neighbors. The slope is typically used to indicate the steepness of terrain.

Below given 3 × 3 window, where each pixel has a ground resolution of 30 × 30 meters. Let 'e' is the elevation of the pixel X, Y, and a, b, c, d, f, g, h, i are the elevations of the pixels surrounding 'e'.

a	b	c
d	e	f
g	h	i

First, the average elevation changes per unit distance in the X- and Y-direction Δx and Δy are calculated as:

$\Delta x_1 = c - a$ $\quad\quad$ $\Delta y_1 = a - g$

$\Delta x_2 = f - d$ $\quad\quad$ $\Delta y_2 = b - h$

$\Delta x_3 = i - g$ $\quad\quad$ $\Delta y_3 = c - i$

$$\Delta x = \frac{(\Delta x_1 + \Delta x_2 + \Delta x_3)}{3 \times x_s}$$

$$\Delta y = \frac{(\Delta y_1 + \Delta y_2 + \Delta y_3}{3 \times y_s}$$

where:

a...i = elevation values of pixels in a 3 × 3 window

x_s = x pixel size = 30 meters

y_s = y pixel size = 30 meters

The slope at pixel x, y is calculated as:

$$S = \sqrt{\{(\Delta x)^2 + (\Delta y)^2\}/2}$$

$$\text{Slope in degrees} = \tan^{-1}(s)\left(\frac{180}{\Pi}\right)$$

Aspect is derived from a raster surface and is considered as slope direction. It is the direction in which a unit of terrain faces, usually expressed in degrees from north, clockwise, from 0 to 360°. It starts 0 degree at north, 90 degrees at East, 180 degrees at South and 270 degrees is due west. However, 361 degrees identify flat surfaces such as water bodies. It identifies the down slope direction of the maximum rate of change in value from each cell to its neighbors. The values of the output raster will be the compass direction of the aspect. The aspect of a surface typically affects the amount of sunlight it receives.

If Δx = 0 and Δy = 0, then the aspect is flat. Otherwise, θ is calculated as:

$$\theta = \tan^{-1}\left(\frac{\Delta r}{\Delta y}\right)$$

Visibility analysis

Another aspect of the terrain analysis is the visibility analysis; the identification of areas of terrain that can be seen from a particular point of observation on a terrain surface. Visibility analysis is used in a number of applications ranging from the locating radio transmitters for maximum coverage to minimizing the impact of commercial forestry in protected areas (DeMers, 1997; Fishwick and Clayson, 1995). This analysis uses an elevation model that does not consider trees or buildings that may obscure the view. To avoid the obscure, one has to add a raster layer that represents the locations of towers, buildings and trees (and their heights) to the DEM so that a more realistic surface can be viewed.

In a GIS, viewshed map is created using a DEM and a map containing the point location, or locations from which the viewshed will be calculated. DeMers (1997) describe the functioning of these analyses. The location of the observer is connected to each possible target location in the terrain. The line, or ray, is followed for each target back to the observer, looking for locations that are higher. Higher points will obscure what is behind them. Thus through repeated 'ray tracing' a viewshed map can be built (Clarke, 1990).

The result of the visibility analysis is a viewshed map. Both raster and vector methods calculate a visibility surface and yield similar type of results presenting in the form of a viewshed map. Heywood et al. (2002) explain the viewshed map as a Boolean polygon map that indicates which areas of terrain are visible from the location of interest. Some image processing and GIS packages such as ERDAS Imagine and ArcGIS provide draping facility through which the viewshed map may be more easily interpreted and understood by draping over the top of a terrain surface for visualization.

Using the DEM, the GIS draws a straight line from the viewpoint cell/s to each cell in the raster. It then checks the elevation of each cell that this line passes through in order to determine if any of the elevations are great enough to block the line of site between the viewpoint cell and the target cell (ESRI, 1995). If the target cell is visible the GIS places a "1" in that cell in the viewshed map, if it is not visible the GIS assigns a "0" to the cell. The result is a map of zeros and ones, with the ones designating the viewshed of the viewpoint/s. In addition, factors such as vegetation or the introduction of towers can be accounted for in order to more accurately model the real world.

Conclusion

Geographic Information System is defined as a collection of computer hardware, software, and geographic data for capturing, managing, analyzing, and displaying all forms of geographically referenced information. GIS provides a framework and various tools for integration and handling of spatial and attribute information together. Spatial data describe the location of objects with respect to a known coordinate system, while attribute data describe the nature or characteristics of spatial phenomena.

GIS ends with aesthetic cartographic maps. A map helps to work with geographical data of our surrounding. GIS can provide a great deal more problem-solving capabilities than using a simple mapping program or adding data to an online mapping tool. GIS, therefore, is regarded as an application-led technology.

A GIS can be considered in three ways: (i) the database view, (ii) the map view, and (iii) the model view.

The database view: A GIS is a set of intelligent maps and other views that show features and feature relationships on the Earth's surface. Other views are: interactive maps, 3D scenes, summary charts and tables, time-based views, and schematic views of network relationships.

The map view: Maps are used in many ways. They convey geographic information about the Earth's features and their relationships. They are used for advanced data compilation, cartography, analysis, query, and field data collection. They are similar to the maps what we see; the difference is that we can interact through computer to meet the desired application oriented queries. In addition to maps, other interactive views, such as temporal, globe, and schematic drawings, are used as views into GIS databases. The GIS users attempt to solve a range of simple to most complicated.

The model view: GIS contains a set of information transformation tools. They are used to extract new geographic datasets from existing datasets. An accurate model of the expected location and distribution of geographic features is created to meet the user's query. This is performed by combining datasets and implementing some analytic rules. These three views together are regarded as critical parts of an intelligent GIS and are used at varying levels in all GIS applications.

The efficiency of GIS is in the integration of dataset from various sources to analyse it as a whole and implement it for critical decision-making in planning and management options. The analytic power of a GIS cannot improve unreliable data. If the database is weak, the analysis will be wrong. Putting data into a GIS database without attention to the analytical requirements, then the output will not yield any sensible results. It's a powerful combination of tools which need thorough understanding and careful application. The GIS approach therefore should be very specific and planned. Somers (1990) described the concept of a GIS life cycle to use GIS in a strategic manner. A brief of the concept is outlined below.

Phase 1: Planning

The first stage of the life cycle is planning. It involves:

(i) a systematic review of users, their data, and their information needs.

(ii) to educate decision-makers about the costs and benefits of GIS.

(iii) to include potential users in the planning process.

This phase ends with a thorough understanding of user needs.

Phase 2: System Design

The design phase matches user needs to GIS functionality. It includes:

(i) selection of hardware and software.

(ii) design of the GIS spatial and attribute database including base map specifications for scale, projection, and coordinate systems.

(iii) implementation of prototype or pilot project to learn lessons, which can be used to develop a fully-implemented system.

Phase 3: Implementation

The implementation phase involves:

(i) training and education to all users to utilize and, maintain the system and the database.

(ii) users should be well aware of the capability of GIS that influences them and their data processing tasks including decision-making.

Phase 4: Maintenance

The final stage is the maintenance phase—in which a GIS-application must be maintained and kept update in terms of data and user support.

We live in a dynamic world. We are encountering a sequence of natural hazards that profusely change our natural habitats. This puts us in a problematic situation depriving our safety and security. It is, therefore, growing concern for the planners and decision-makers to estimate possible future situations and deduct necessary adaptation strategies. Although it is beyond the human capability to prevent the natural disasters but the experience shows that the effects of the disasters can be mitigated through planning and preparedness based on risk information. A holistic approach to mitigate disaster has, therefore, become the priority of every nation today. All phases of disaster management depend on data and information from a variety of sources, much of it in the form of maps, images and charts. The value of the right information, at the right time and in the right hands has been immensely realized for an effective response action to be taken. Due to the spatial characteristics of natural disasters (cyclones, earthquakes, floods, etc.) GIS plays a significant role in modeling the disaster events. This provides a set of risk information useful for decision-making and planning for future disasters. Detailed hazard assessment, vulnerability analysis, and financial loss estimation could be possible using a number of GIS techniques. Several probabilistic risk models have been developed in four Indian states viz. Andhra Pradesh, Gujarat, Maharashtra and Orissa.

In recent years, comprehensive geographic datasets of metropolitan areas and individual-level, georeferenced data are becoming more available to social scientists. At the same time, tools for performing spatial analysis in a GIS environment have also become more available. These developments provide many new opportunities for the analysis and theoretical understanding of disaggregate human spatial behavior. If the Geographic Information Systems are to be used more widely, there are still challenges to be addressed. On the other hand, there is still a huge potential for the development of tools to handle geographical data. Heywood et al. (2002) described the visible future challenges ahead in GIS. Those are: data-automatic data collection, data exchange and real time modeling using a range of data types all need to be tackled. Another challenge is the GIS technology. The technological challenges will be easier to address than the organizational and human issues and many new opportunities will come from areas outside the field of GIS. A third challenge they pointed out is how to present geographical data and the results of geographical modeling to a whole new audience. The innovative way of presenting the output of the models and analysis is needed for decision-making and resources management. New interfaces should be developed so that the people can interact in an easier way with the spatial data and decision support systems. The final challenge would be to familiarise the technology by the way of educating and training the users so that a trained audience would emerge who will live with the Geographic Information Systems.

References

Abrams, 1984. Landsat 4, Thematic Mapper and Thematic Mapper Simulator data for a Porphyry Copper Deposit. Photogrammetric Engineering and Remote Sensing, 50: 1171-73.

Ajai, Bhatt, H.P. and Sharma, S.A. 1988. Monitoring drought impact on crop condition by remote sensing. Presented at workshop on agricultural drought, NRSA, Hyderabad.

Antenucci, J.C., Brown, K., Croswell, P.L. and Kevany, M.J. 1991. *Geographic Information Systems: A Guide to the Technology*. New York: Van Nostrand Reinhold.

Arai, K. 1992. A Supervised Thematic Mapper Classification with a Purification of Training Samples. *International Journal of Remote Sensing*, 13 (11): 2039-49.

Arbia, G., Griffith, D. and Haining, R. 1998. Error propagation modeling in raster GIS: Overlay Operations. *International Journal of Geographical Information Science*, 12: 145-67.

Aronoff, S. 1982. Classification accuracy: A user approach. *Photogrammetric Engineering and Remote Sensing*, 48: 1299-1307.

Aronoff, S. 1985. The minimum accuracy value as an index of classification accuracy. *Photogrammetric Engineering and Remote Sensing*, 51: 99-111.

Ashburn, P. 1978. The vegetative index number and crop identification. The LACIE Symposium Proceedings of the Technical Session, pp. 843-50.

Baglio, J.V. and Holroyd, E.W. 1989. Methods for operational snow cover area mapping using the advanced very high resolution radiometer: San Juan Mountains Test Study.

Research Technical Report, U.S. Geological Survey, Sioux Falls and U.S. Bureau of Reclamation, Denver.

Baret, F. and Guyot, G. 1991. Potentials and Limits of Vegetation Indices for LAI and APAR Assessment. *Remote Sensing and the Environment,* 35: 161-73.

Baret, F., Guyot, G. and Major, D.J. 1989. TSAVI: A vegetation index which minimizes soil brightness effects on LAI and APAR estimation. In: Proc. IGARRS '89. 12th Canadian Symposium on Remote Sensing, (3: 1355-58). Vancouver, Canada (10-14 July), 4p.

Barnett, T.L. and Thompson, D.R. 1982. The use of large area spectral data in wheat yield estimation. *Remote Sensing of Environment*, 12: 509-18.

Baxes, G.A. 1994. "Digital Image Processing. Principles & Applications", Wiley & Sons Bernhardsen, T. 1992. Geographic Information Systems. VIAK IT/Norwegian Mapping Authority, Arendal, Norway.

Bernstein, R. 1983. Image geometry and rectification. Chapter 21. In: *Manual of Remote Sensing*, (Eds.) R.N. Colwell. Falls Church, Virginia, American Society of Photogrammetry.

Bernstein, R. and Ferneyhough, D.G. 1975. Digital image processing. *Photogrammetric Engineering and Remote Sensing*, 41: 1465-76.

Bishop, Y., Fienberg, S. and Holland, P. 1975. *Discrete Multivariate Analysis—Theory and Practice* (Cambridge, Massachusetts: MIT Press).

Brooks, F.A. and Miller, W. 1963. Availability of solar energy. In Zarem, A.M., Erway, D., ed., *Introduction to the Utilization of Solar Energy*. McGraw-Hill, New York, NY, USA. pp. 30-58.

Bruniquel, J. and Lopes, A. 1997. Multi-variate optimal speckle reduction in SAR imagery. *International Journal of Remote Sensing*, 18: 603-27.

Buckton, D., O'Mongain, E. and Danaher, S. 1999. The use of neural networks for the estimation of oceanic constituents based on the MERIS instrument. *International Journal of Remote Sensing*, 20: 1841-51.

Burrough, P. 1986. *Principles of Geographical Information Systems for Land Resources Assessment*. Oxford, England: Oxford University Press.

Burrough, P.A. 1990. *Principles of Geographical Information Systems for Land Resources Assessment* (New York, USA: Oxford University Press).

Byrne, G.E., Crapper, P.F. and Mayo. 1980. Monitoring land cover changes by principal component analysis of multitemporal Landsat data. *Remote Sensing of Environment*, 10: 174-84.

Campbell, J.B. 1996. *Introduction to remote sensing*, 2nd Edition, Taylor and Francis, London.

Campbell, J.B. 2002. *Introduction to Remote Sensing*. 3rd Edition, Taylor and Francis, London.

Canas, A.A.D. and Barnett, M.E. 1985. The generation and interpretation of false color composite principal component images. *International Journal of Remote Sensing*, 6: 867-81.

Canters, F. 1997. Evaluating the uncertainty of area estimates derived from fuzzy land-cover classification. *Photogrammetric Engineering and Remote Sensing*, 63: 403-14.

Carlotto, M.J. 1999. Reducing the effects of space-varying, wavelength-dependent scattering in multispectral imagery. *International Journal of Remote Sensing*, 20(17): 3333-44.

Carter, G.A. and Knapp, A.K. 2001. Leaf optical properties in higher plants: linking spectral characteristics to stress and chlorophyll concentration. *American Journal of Botany*, 88: 677-84.

Carter, J.R. 1989. On defining the geographic information system. In *Fundamentals of Geographic Information Systems: A Compendium* (Falls Church, Virginia: ASPRS/ACSM), pp. 3-7.

Carter, J.R. 1989. Relative errors identified in USGS gridded DEMs. Proceedings of Auto-Carto 9, Ninth International Symposium on Computer Assisted Cartography, Baltimore, pp. 255-65.

Carver, K.R. 1988. *SAR Synthetic Aperture Radar—Earth Observing System*, Vol. II, Washington, NASA Instrument Panel Report, 233p.

Caylor, J.A. and Lachowski, H.M. 1988. How to use aerial photographs for natural resources Applications. Washington: USDA Forest Service, 350p.

Chahine, M.T. 1983. Interaction mechanisms within the atmosphere. Chapter 5. In: *The Manual of Remote Sensing*, R.N. Colwell (Ed.), American Society of Photogrammetry and Remote Sensing, Falls Church, Va., USA.

Chandrasekhar, G., Jayaraman, V., Radhakrishnan, K., Manikiam, B. and Honne Gowda, H. 1992. *Remote Sensing Activities in India*, GISdevelopment.net, AARS, ACRS.

Chauhan, P., Nagur, C.R.C., Mohan, M., Nayak, S.R. and Navalgund, R.R. 2001. Surface chlorophyll distribution in Arabian Sea and Bay of Bengal using IRS-P4 Ocean Color Monitor satellite data. *Current Science*, 80: 127-29.

Chavez, G.L. 1982. Statistical method for selecting Landsat MSS ratios. *Journal of Applied Photographic Engineering*, 8(1): 23-30.

Chavez, P.S., Berlin, G.L. and Mitchell, W.B. 1977. Computer enhancement techniques of Landsat MSS digital images for land use/land cover assessments. *Remote Sensing of Earth Resources*, 6: 259.

Chellaiah, S. 2003. Mapping of ultramafics and associated lithological units using satellite data. Map Asia Conference 2003, GISdevelopment.net, pp. 1-4.

Chrisman, N.R. 1997. *Exploring Geographic Information Systems*. Wiley, New York.

Cihlar, J. 2000. Land cover mapping of large areas from satellites: status and research priorities. *International Journal of Remote Sensing*, 21: 1093-1114.

Clarke, K.C. 1990. *Analytical and Computer Cartography*. Prentice-Hall, Englewood Cliffs, New Jersey.

Clevers, J.G.P.W. 1988. The derivation of a simplified reflectance model for the estimation of leaf area index. Remote *Sensing of Environment*, 35: 53-70.

Cohen, W.B. and Justice, C.O. 1999. Validating MODIS terrestrial ecology products: linking in situ and satellite measurements. *Remote Sensing of Environment*, 70: 1-3.

Cohen, W.B., Spies, T.A. and Fiorella, M. 1995. Estimating the age and structure of forests in a multi-ownership landscape of western Oregon, U.S.A. *International Journal of Remote Sensing*, 16: 721-46.

Cohen, W.B., Maiersperger, T.K., Spies, T.A. and Oetter, D.R. 2001. Modelling forest cover attributes as continuous variables in a regional context with Thematic Mapper data. *International Journal of Remote Sensing*, 22: 2279-2310.

Collins, W.E. 1978. Remote sensing of crop types and maturity. *Photogrammetric Engineering and Remote Sensing*, 44(1): 43-55.

Collins, W.E., Raines, G.L. and Canney, F.C. 1977. Airborne spectroradiometer discrimination of Vegetation anomalies over sulfide mineralization: A remote sensing technique. Geology Society of America, 90th Annual Meeting programs and Abstracts, 9(7): 932-33.

Congalton, R.G. 1991. A review of assessing the accuracy of classifications of remotely sensed data. *Remote Sensing of Environment*, 37: 35-46.

Congalton, R.G. 1994. Accuracy assessment of remotely sensed data: future needs and directions. In: Proceedings of Pecora 12 land information from space-based systems. Bethesda: ASPRS, pp. 383-88.

Congalton, R.G. 1996. Accuracy Assessment: A Critical Component of Land Cover Mapping. Gap Analysis, ISBN-1-57083-03603. *American Society for Photogrammetry and Remote Sensing*, pp. 119-31.

Congalton, R.G. and Green, K. 1999. *Assessing the Accuracy of Remotely Sensed Data: Principles and Practices.* Boca Raton: Lewis Publishers.

Congalton, R.G., Balogh, M., Bell, C., Green, K., Milliken, J.A. and Ottman, R. 1998. Mapping and monitoring agricultural crops and other land cover in the Lower Colorado river basin. *Photogrammetric Engineering and Remote Sensing*, 64: 1107-13.

Cracknell, A.P. and Hayes, L.W.B. 1991. *Introduction to Remote Sensing*. London: Taylor and Francis.

Crane, R.G. and Anderson, M.R. 1984. Satellite discrimination of snow/cloud surfaces. *International Journal of Remote Sensing*. 5(1): 213-23.

Crippen, R.E. 1987. The regression intersection method of adjusting image data for band ratioing. *IJRS*, 8(2): 137-55.

Crippen, R.E. 1990. Calculating the vegetation index faster. *Remote Sensing of Environment*, 34: 71-73.

Crippen, R.W. 1989. A simple spatial filtering routine for the cosmetic removal of scan line noise from Landsat TM Tape imagery. *Photogrammetric Engineering and Remote Sensing*, 55: 327-31.

Crist, E.P. 1985. A TM tasseled cap equivalent transformation for reflectance factor data. *Remote Sensing of Environment*, 17: 301-06.

Crist, E.P. and Cicone, R.C. 1984. A physically-based transformation of Thematic Mapper data—the TM Tasseled Cap, IEEE Trans. on *Geosciences and Remote Sensing*, GE-22: 256-63.

Crist, E.P. and Cicone, R.C. 1984b. Application of the tasseled cap concept to simulated Thematic Mapper Data. *Photogrammetric Engineering and Remote Sensing*, 50: 343-52.

Crist, E.P. and Cicone, R.C. 1984c. A physically-based transformation of Thematic Mapper data—the TM Tasseled Cap. IEEE Transactions on Geoscience and Remote Sensing, GE- 22 (3): 256-63.

Crist, E.P. and Kauth, R.J. 1986. The tasseled cap de-mystified. *Photogrammetric Engineering & Remote Sensing*, 52 (1): 81-86.

Crowley, J.K., Williams, D.E., Hammarstrom, Jane M., Piatak, N.M., Chou, I-Ming and Mars, J.C. 2003. Spectral reflectance properties (0.4-2.5 μm) of secondary Fe-oxide, Fe-hydroxide, and Fe-sulphate-hydrate minerals associated with sulphide-bearing mine wastes. *Geochemistry-Exploration, Environment, Analysis*, 3(3): 219-28.

Dale, P.E.R., Chandica, A.L. and Evans, M. 1996. Using image subtraction and classification to evaluate change in sub-tropical inter-tidal wetlands, *International Journal of Remote Sensing*, 17: 703-19.

Dangermond, J. 1988. Introduction and overview of GIS. In Geographic Information Systems Seminar, Data Sharing—Myth or Reality (Ontario, Canada, 3rd-5th October: Ministry of Natural Resources).

Deering, D.W., Rouse, J.W., Haas, R.H. and Schell, J.A. 1975. Measuring "Forage Production" of Grazing Units from Landsat MSS Data. Proceedings of the 10th International Symposium on Remote Sensing of Environment, II: 1169-78.

Dekker, R.J. 1998. Speckle filtering in satellite SAR change detection imagery. *International Journal of Remote Sensing*, 19: 1133-46.

DeMers, M.N. 1997. *Fundamentals of Geographic Information Systems*, Wiley, New York.

Department of Environment, 1987. Handling Geographic Information. Report of the Committee of Enquiry chaired by Lord Chorley, HMSO, London.

Desai, P.S., Honne Gowda, H. and Kasturirangan, K. 2000. Ocean Research in India: Perspective from space. *Current Science*, 78(3): 268-78.

Dias, L.A.V. and Nery, C.E. 1992. Evaluation of interpolators for digital elevation models. Proceedings of the ISPRS and OEEPE Joint Workshop Data by Photogrammetric Methods, Oxford. OEEPE Publication No. 27: 212-57.

Dickinson, H.J. and Calkins, H.W. 1988. The economic evaluation of implementing a GIS. *International Journal of Geographical Information Systems*, 2(4): 307-27.

Dicks, S.E. and Lo, T.H.C. 1990. Evaluation of thematic map accuracy in a land-use and land-cover mapping program. *Photogrammetric Engineering and Remote Sensing*, 56: 1247-52.

Doerffer, R. 1992. Imaging spectroscopy for detection of chlorophyll and suspended matter. GKSS 92/E/54, GKSS-FORSCHUNGSZENTRUM GEESTHACHT GMBH.

Dozier, J. 1989. Remote sensing of snow in visible and near-infrared wavelengths. *Theory and Applications of Optical Remote Sensing*, G. Asrar (Ed.), John Wiley and Sons, New York.

Dubey, R.R. 1983. Manual of image interpretation-II. In Lecture Notes on Orbital Remote Sensing, SAC/RSA/AGPA/TN/03/83, pp. 7.1-7.13.

Duda, R.O. and Hart, P.E. 1973. *Pattern Classification and Scene Analysis*, John Wiley and Sons, New York.

Duggin, M.J. 1987. The Physical Principles Controlling the Remote Sensing Process. In *Remote Sensing Applications in Meteorology and Climatology*, edited by R.A. Vaughan, D. Reidel, Dordrecht, pp. 33-50.

Eckhardt, D.W., Verdin, J.P. and Lyford, G.R. 1990. Automated update of an irrigated lands GIS using SPOT HRV imagery. *Photogrammetric Engineering and Remote Sensing*, 56: 1515-22.

Egenhofer, M., Frank, A. and Jackson, J. 1989. A Topological Data Model for Spatial Databases. Design and Implementation of Large Spatial, Lecture Notes in Computer Science, 409: 271-86.

Elachi, C. 1987. *Introduction to Physics and Techniques of Remote Sensing,* Wiley, New York.

El-Hinnawi, E., Hashmi, M. (ed.). 1982. *Global Environmental Issues*. United Nations Environment Program/Tycooly International Publishing, Dublin, Ireland, 236p.

Eliason, E.M. and McEwen, A.S. 1990. Adaptive box filter for the removal of random noise from digital images. *Photogrammetric Engineering and Remote Sensing*, 56: 453-58.

Elvidge, C.D. and Lyon, R.J.P. 1985. Influence of rock-soil spectral variation on the assessment of green biomass. *Remote Sensing of Environment*, 17: 265-69.

Elvidge, C.D. and Chen, Z. 1995. Comparison of broad-band and narrowband red and near-infrared vegetation indices. *Remote Sensing of the Environment*, 54: 38-48.

Environmental Systems Research Institute, Inc. 1990. Understanding GIS: The ArcInfo Method. Redlands, California: ESRI, Incorporated.

Environmental Systems Research Institute (ESRI). 2001. History of ESRI. Accessed 26 April 2001: http://www.esri.com/company/about/history.html.

ERDAS Inc. 1999. *ERDAS Field Guide*. Fifth Edition (Atlanta, Georgia).

ESRI. 1995. ARC/INFO. Redlands, CA: Environmental Systems Research Institute.

Estes, J.E., Hajic, E.J. and Tinney, L.R. 1983. Fundamental of image analysis: Analysis of visible and thermal infrared data. In *Manual of Remote Sensing*, R.N. Colwell (Editor-in-Chief), 2nd Edition, Falls Church: American Society of Photogrammetry, 1: 987-1124.

Ezra, C.E., Tinney, L.R. and Jackson, R.D. 1984. Considerations for implementing vegetation indices for Agricultural Applications. Technical Papers, 50th Annual Meeting of the American Society of Photogrammetry, pp. 526-36.

Fishwick, A. and Clayson, J. 1995. GIS development project: Final Report to the Countryside Commission. Lake District National Park Authority, Kendal, UK.

Faust, N. 1998. Raster-based GIS. In: T.W. Foresman, ed., *The History of Geographic Information Systems: Perspectives from the Pioneers*. Upper Saddle River, NJ: Prentice Hall, pp. 59-72.

Faust, N.L. 1989. *Image Enhancement. V.20, Supplement 5 of Encyclopedia of Computer Science and Technology*, edited by Allen Kent and James G. Williams, NY, Marcel Dekker, Inc.

Faust, N.L., Sharp, W., Holcomb, D.W., Geladi, P. and K. Esbenson, 1991. Application of Multivariate Image Analysis (MIA) to Analysis of TM and Hyperspectral Image Data for Mineral Exploration. Paper presented at the Eighth Thematic Conference on Geologic Remote Sensing, Denver, Colorado, April/May 1991.

Fischer, W.A. 1983. History of Remote Sensing. *Manual of Remote Sensing*, R.G. Reeves (Ed.), Falls Church, VA: American Society of Photogrammetry, pp. 27-50.

Foody, G. 1992. On the Compensation for Chance Agreement in Image Classification Accuracy Assessment. *Photogrammetric Engineering & Remote Sensing*, 58(10): 1459-60.

Foody, G.M. 2002. Status of land cover classification accuracy assessment. *Remote Sensing of Environment*, 80: 185-201.

Ford, J.P., Blom, M.L., Bryan, M.K., Daily, T.H., Dixon, C.E. and Xenos, E.C. 1980. *Seasat views North America the Caribbean and Western Europe with Imaging Radar*, JPL, Publications #80-67, Los Angeles, CA: NASA Jet Propulsion Laboratory, pp. 1-6.

Forster, B.C. 1984. Deviation of Atmospheric Correction Procedures for Landsat MSS with Particular Reference to Urban Data. *International Journal of Remote Sensing*, 5 (5): 799-817.

Freeman, W.H.F.G., Sadowski, Schowengerdt, N. and Techniques R.A. 1983. *For Image Processing and Classification in Remote Sensing*, Academic Press, NY.

Frulla, L.A., Milovich, J.A. and Gagliardini, D.A. 2000. Automatic computation of speckle standard deviation in SAR images. *International Journal of Remote Sensing, 21*: 2883-99.

Fu, K.S. 1976. Pattern recognition in remote sensing of the Earth resources. IEEE Transaction on Geoscience Electronics, 14: 10-11.

Galbaiti, L.J., Jr. 1990. *Machine Vision and Digital Image Processing Fundamentals*, Prentice Hall, Englewood, New Jersey, Chapter 6: 105-26.

Garbrechet, J., Ogden, F.L., Debarry, P.A. and Maidment, D.R. 2001. GIS and Distributed Models. I: Data Coverage and Sources. *Journal of Hydrologic Engineering*, 6(6): 506-14.

Gauch, H.G. Jr. 1993. Prediction, parsimony, and noise. *American Scientist*, 81(5): 468-78.

Gausmann, H.W. 1974. Leaf reflectance of near infra-red, Photogrammetric Engineering, XL(2): 183-91.

Gausmann, H.W., Escobar, D.E., Everitt, J.H., Richardson, A.J. and Rodriguez, R.R., 1978. Distinguishing succulent plants from crop and woody plants. *Photogrammetric Engineering & Remote Sensing*, 44(4): 487-91.

Gersen, D.J. and Rosenfeld, A. 1975. Automatic sea ice detection in satellite pictures. *Remote Sensing of Environment*, 4: 187-98.

Gillespie, A.R. 1980. Digital techniques of image enhancement. In Siegal, B.S. and Gillispie, A.R. (Eds.), *Remote Sensing in Geology*, John Wiley, New York, pp. 139-226.

Gineste, P. 1999. A simple, efficient filter for multi-temporal SAR images. *International Journal of Remote Sensing*, 20: 2565-76.

Gonzalez, R.C. and Wintz, P. 1977. *Digital Image Processing and Recognition.* Addison-Wesley Publishing company, Reading, Massachusetts, pp. 123-26.

Gonzalez, R.C. and Wintz, P. 1987. *Digital Image Processing*. Addison-Wesley Publishing Company, California, Chapter 4: 140-205.

Gordon, H.R. 1977. Atmospheric correction of ocean colour imagery in the Earth Observing system era. *Journal of Geophysical Research*, 102: 17081-106.

Gordon, H.R. and Clark, D.K. 1980. Remote Sensing optical properties of a stratified ocean. *Applied Optics*, 19: 3428-4330.

Gorden, H.R., Clark, D.K., Muller, J.L. and Hovis, W.A. 1980. Phytoplankton Pigments from the Nimbus-7 Coastal Zone Colour Scanner: Comparison with surface measurements. *Science*, 210: 63-66.

Green, E.P., Mumby, P.J., Edwards, A.J. and Clark, C.D. (Ed. Edwards, A.J.). 2000. *Remote Sensing Handbook for Tropical Coastal Management*. Coastal Management Sourcebooks 3, UNESCO, Paris, 316p.

Haack, B., Jensen, J.R. and Welch, R.A. 1995. *Urban-Suburban Analysis, Manual of Photographic Interpretation*, 2nd ed., W. Phillipson, ed. Bethesda, MD: American Society for Photogrammetry and Remote Sensing.

Hall, E.L. 1978. *Computer Image Processing and Recognition*, Academic Press, NY, CH-4: 159-84.

Hall, E.L. 1979. *Computer Image Processing and Recognition*, Academic Press, NY, CH-4: 158-257.

Hall, E.L. 1983. *Computer Image Processing and Recognition*, Academic Press, New York.

Haralick, R.M. and Fu, K.S. 1983. Pattern recognition and classification: In R.N. Colwell (Ed.) *Manual of Remote Sensing*, American Society of Photogrammetry and Remote Sensing, Falls Church, 1(18): 793-805.

Haralick, R.M. and Shapiro, L.G. 1992. *Computer and Robot Vision*, Vol. 1, Addison-Wesley, Reading, MA.

Haralick, R.M., Shanmugam, K.S. and Dinstein, I. 1973. Textural Features for Image Classification. IEEE

Transactions on Systems, Man and Cybernetics, 3: 610-21.

Harris, R. 1977. Automatic analysis of meteorological satellite imagery In Thomas J.O. and Davey P.G. (Eds.) *Texture Analysis*. British Pattern Recognition Association and Remote Sensing Society, Reading, pp. 45-72.

Harris, T.M. and Elmes, G.A. 1993. The Application of GIS in Urban and Regional Planning: a review of the North American experience.

Henderson, F.M. and Lewis, A.J. 1998. Radar fundamentals: The Geoscience Perspective. In Henderson and Lewis (Eds.) *Principles and Applications of Imaging Radar*, New York: John Wiley & Sons, pp. 131-81.

Heywood, I., Cornelius, S. and Carver, S. 2002. *An introduction to Geographical Information Systems*. 2nd Edition, Pearson Education.

Hillger, D.W. 1996. Meteorological features from principal component image transformation of GOES imagery. International Symposium on Optical Science, Engineering, and Instrumentation (GOES-8 and Beyond Conference), 4-9 August, Denver CO, SPIE, volume 2812: 111-21.

Hoel, E., Menon, S. and Morehouse, S. 2003. Building Robust Topologies. In *Advances in Spatial and Temporal Databases*. Proceedings of the 8th International Symposium on Spatial and Temporal Databases. SSTD 2003. Santorini Island, Greece, Springer-Verlag Lecture Notes in Computer Science, 2750.

Holben, B.N. and Justice, C.O. 1981. An Examination of Spectral Band Ratioing to Reduce the Topographic Effect on Remotely Sensed Data. *International Journal of Remote Sensing*, 2(3): 115-33.

Hord, R.M. 1982. *Digital Image Processing of Remotely Sensed Data*. New York: Academic Press.

Horne, J.H. 2003. A tasseled cap transformation for IKONOS images, ASPRS 2003 Annual Conference Proceedings, Anchorage, Alaska.

Hsu, S. 1978. Texture tone analysis for automated land use mapping. Photogrammetric Engineering and Remote Sensing, 44: 1393-1404.

Hu, C., Carder, K.L. and Mauuler-Karger, F.E. 2000. Atmospheric correction of SeaWiFS Imagery over Turbid Coastal Waters: A Practical Method. *Remote Sensing of Environment*, 74(2): 195-206.

Huang, C., Wylie, B., Yang, L., Homer, C. and Zylstra, G. 2002. Derivation of a tasselled cap transformation based on Landsat 7 at-satellite reflectance. *International Journal of Remote Sensing*, 23(8): 1741-48.

Huete, A.R. 1988. A soil-adjusted vegetation index (SAVI). *Remote Sensing of the Environment*, 25: 295-309.

Huete, A.R. 1989. Soil influences in remotely sensed vegetation-canopy spectra. In: G. Asrar (Ed.), *Theory and Applications of Optical Remote Sensing*. New York: Wiley, pp. 107-41.

Huete, A. and Justice, C. 1999. MODIS Vegetation Index (MOD 13) Algorithm Theoretical Basis Document, Greenbelt: NASA Goddard Space Flight Center, http.// modarch.gsfc.nasa.gov/MODIS/LAND/vegetation-indices, 129p.

Huete, A.R., Jackson, R.D. and Post, D.F. 1985. Spectral response of a plant canopy with different soil backgrounds. *Remote Sensing of the Environment*, 17: 33-37.

Huete, A.R., Post, D.F. and Jackson, R.D. 1984. Soil spectral effects on 4 space vegetation discrimination. *Remote Sensing of the Environment*, 14: 155-65.

Interface, 1996. A bulletin from the NRSA data centre, 7(2).

IOCCG Report. 1999. Data Merging and Validation. In: *Status and Plans for Satellite Ocean-Color Missions: Considerations for Complementary Missions*, Reports of the International Ocean Colour Coordinating Group, Report No. 2, Chapter 5: 33.

Irons, J.R. and Petersen, G.W. 1981. Texture transforms of remote sensing data. *Remote Sensing of the Environment*, 11: 359-70.

ITC, 1997. Ilwis (Integrated Land and Water Information System) 2.2 - Application guide. ITC, Enschede.

Jackson, R.D. 1983. Spectral Indices in N-space. *Remote Sensing of the Environment*, 13: 409-21.

Jain, A.K. 1989. *Fundamentals of Digital Image Processing*, Prentice Hall, Englewood Cliffs, New Jersey, Chapter 7, pp. 233-66.

Janssen, L.L.F. and Van der Wel, F.J.M. 1994. Accuracy assessment of satellite derived land-cover data: a review. *Photogrammetric Engineering and Remote Sensing*, 60: 419-26.

Jensen, J.R. 1981. Urban change detection mapping using Landsat digital data. *The American Cartographer*, 8: 127-47.

Jensen, J.R. 1996. *Introductory Digital Image Processing: A Remote Sensing Perspective*, 2nd edition. Englewood Cliffs, New Jersey: Prentice-Hall.

Jensen, J.R. 2000. *Remote Sensing of the Environment: An Earth Resource Perspective*. Keith C. Clarke (Ed), Pearson Education (Pub.).

Jensen, J.R. 2004. *Remote Sensing of the Environment, An Earth Resource Perspective*. Keith C. Clarke (Ed.), Pearson Education Series in Geographic Information Science, 243p.

Jensen, J.R. 2004. *Remote Sensing of the Environment, An Earth Resource Perspective*, 257p.

Jensen, J.R. 2004. *Remote Sensing of the Environment, an Earth Resource Perspective*, Pearson Education Series in Geographic Information series, 121p.

Jenson, J.R. 1983. Education image processing: An overview. Photogrammetric Engineering and Remote Sensing, 49: 1151-57.

Jenson, J.R. 1986. *Introductory Digital Image Processing: A Remote Sensing Perspective*, Prentice-Hall, Englewood Cliff, New Jersey.

Jordan, C.F. 1969. Derivation of leaf area index from quality of light on the forest floor. *Ecology*, 50: 663-66.

Junge, C.E. 1958. *Atmospheric Chemistry*. Advanced Geophysics, 4: 1-108.

Justice, C., Belward, A., Morisette, J., Lewis, P., Privette, J. and Baret, F. 2000. Developments in the 'validation' of satellite sensor products for the study of the land surface. *International Journal of Remote Sensing*, 21: 3383-90.

Kalkhan, M.A., Reich, R.M. and Czaplewski, R.L. 1995. Statistical properties of five indices in assessing the accuracy of remotely sensed data using simple random sampling. Proceedings ACSM/ASPRS Annual Convention and Exposition, 2: 24-257.

Kaneko, T. 1978. Colour composite pictures from principal axis components of multispectral scanner data. *IBM Journal of Research and Development*, 22: 86-92.

Kasischke, E.S., Meadows, G.A. and Jackson, P.L. 1984. The use of synthetic aperture radar to detect hazards to navigation, Ann Arbor, MI: Environmental Research Institute of Michigan, Publication # 169200-2-F., 194p.

Kauth, R.J. and Thomas, G.S. 1976. The tasseled cap—A graphic description of the spectral-temporal development of agricultural crops as seen by Landsat. Proceeding symposium on machine processing of remotely sensed data, West Lafayette, Ind. Laboratory for the application of remote sensing, pp. 41-51.

Kauth, R.J., Lambeen, P.F., Richardon, W., Thomas, G.S. and Pentland. 1979. Feature extraction applied to agricultural crops as seen by Landsat. Proceedings Technical session, Lacie Symposium, Houston, NASA, pp. 41-51.

Kirsch, R., 1971. Computer determination of the consistent structure of biological images. Computer and Biomedical Research, 4(3): 315-28.

Kloer, B.R. 1994. Hybrid Parametric/Non-Parametric Image Classification. Paper presented at the ACSM-ASPRS Annual Convention, Reno, Nevada.

Kogan, F.N. 1990. Remote Sensing of Weather Impacts on Vegetation in Nonhomogeneous Areas. *International Journal of Remote Sensing*, 11(8): 1405-19.

Koukoulas, S. and Blackburn, G.A. 2001. Introducing new indices for accuracy evaluation of classified images representing semi-natural woodland environments. *Photogrammetric Engineering and Remote Sensing*, 67: 499–510.

Kriegler, F.J., Malila, W.A., Nalepka, R.F. and Richardson, W. 1969. Preprocessing transformations and their effects on multispectral recognition, in Proceedings of the Sixth International Symposium on Remote Sensing of Environment, University of Michigan, Ann Arbor, MI, pp. 97-131.

Kudrat, M., Nag, P. and Bhan, S.K. 1989. Identification of wastelands in Sultanpur district, Uttar Pradesh. *Geographical Review of India*, 50: 71-78.

Kumar, M.S. and Kumar, A.S. 1998. Ocean Color Monitor (OCM) of IRS-P4. *NNRMS Bulletin*-22, pp. 13-19.

Kumar, P. 2004. Integrated resource assessment and sustainable use of coastal wetlands of southern Orissa, India, Ph.D. Thesis, Unpublished.

Kumar, P. and Sangatramani, K. 2003. India in Space—2020. http://www.bharat-rakshak.com/MONITOR/Space%20Essay/entry5.htm.

LeBeau, J.L. 1995. The temporal ecology of calls for police service. In: Block, C.R., Dabdoub, M. and Fregly, S. (Eds.) *Crime Analysis through Computer Mapping*. Washington, DC: Police Executive Research Forum, pp. 111-28.

Leberl, F.W. 1990. Radargrammetric Image Processing, Norwood: Artch House, 595p.

Lee, J. 1981. Speckle analysis and smoothing of synthetic aperture radar images. Computer Graphics and Image Processing, 17: 24-32.

Lee, J. and Ersoy, O. 2006. Consensual and hierarchical classification of remotely sensed multispectral images, Electrical and Computer Engineering, ECE Technical Reports, 4p.

Lewis, H.G. and Brown, M. 2001. A generalised confusion matrix for assessing area estimates from remotely sensed data. *International Journal of Remote Sensing.*

Li, Z. 1988. On the measure of digital terrain model accuracy. Photogrammetric Record, 12(72): 873-77.

Li, Z. 1990. *Sampling strategy and accuracy assessment for digital terrain modeling*, Ph.D Thesis, Glasgow, UK, 299p.

Li, Z. 1992. Variation of the accuracy of digital terrain models with sampling interval, Photogrammetric Record, 14: 113-28.

Liang, S., Vallah-Adl, H., Kalluri, S., Jaja, J., Kaufman, Y.J. and Townshend, J.R.G. 1997. An operational atmospheric correction algorithm for Landsat Thematic Mapper imagery over the land. *Journal of Geophysical Research*, 102: 17173-86.

Lillesand, T.M. and Kiefer, R.W. 1979. *Remote Sensing and Image Interpretation*, John Wiley and Sons, New York.

Lillesand, T.M. Kiefer, R.W. 1987. Remote Sensing and Image Interpretation. John Wiley and Sons, New York.

Lillesand, T.M. and Kiefer, R.W. 1994. *Remote Sensing and Image Interpretation.* John Wileys and Sons, New York.

Lillesand, T.M., Kiefer, R.W. and Chipman, J.W. 2004. *Remote Sensing and Image Interpretation*, 5th ed. Wiley & Sons, Inc., New Jersey, 763p.

Lintz, Jr. J. and Simonett, D. 1976. *Remote Sensing of Environment*, Addison-Wesley, Publication Co.

Liu, W.T. and Kogan, F.N. 1996. Monitoring Regional Drought Using the Vegetation Condition Index. *International Journal of Remote Sensing*, 17(14): 2761-82.

Logan, T.L., Strahler, A.H. and Woodcock, C.E. 1979. Use of a standard deviation based texture channel for Landsat classification of forest strata. Machine Processing of Remotely Sensed Data Symposium. Purdue University, Indiana, pp. 395-404.

Logicon Geodynamics, Inc. 1997. *Multispectral Imagery Reference Guide*. VA: Logicon Geodynamics, Inc.

Loughlin, W.P. 1991. Principal component analysis for alteration mapping. *Photogrammetric Engineering and Remote Sensing,* 57(9): 1163-69.

Ludlam, F.H. 1980. *Clouds and Storms*. The Pensylvania State University Press, University Park and London.

Lytle, D. 1996. Imager Finds 'Hot Spots' in Raging Blazes. *Photonics Spectra*, pp. 16-18.

Ma, Z. and Redmond, R.L. 1995. Tau coefficients for accuracy assessment of classification of remote sensing data. *Photogrammetric Engineering and Remote Sensing*, 61: 435-39.

MacQueen, J.B. 1967. Some Methods for Classification and Analysis of Multivariate Observations. Proceedings of 5th Berkeley Symposium on Mathematical Statistics and Probability, Berkeley, University of California Press, 1: 281-97.

Maguire, D.J. 1991. An overview and definition of GIS. In *Geographical Information Systems Principles and Applications*, edited by Maguire, D.J., Goodchild, M.F. and Rhind, D.W. (New York: Longman Scientific and Technical; John Wiley and Sons, Inc.), pp. 9-20.

Maling, D.H. 1989. *Measurements from Maps*. Principles and Methods of cartometry, New York, Pergamon.

Mark, D.M., Chrisman, N., Frank, A.U., McHaffie, P.H. and Pickles, J. 1997. The GIS History Project. Retrieved 15 January 2001 from the World Wide Web: http://www.geog.buffalo.edu/ncgia/gishist/bar harbor.html.

Marsh, W.M. and Dozier, J. 1981. *Landscape: An Introduction to Physical Geography*, New York: Addison-Wesley, 637p.

Mather, P.M. 1987. *Computer processing of remotely sensed data*. (Chichester): John Wiley & Sons.

McDaid, H. and Oliver, D. 1997. *Smart Weapons: Top Secret History of Remote Controlled Airborne Weapons*, New York: Barnes & Noble, 208p.

Mckenna, F.T. 1997. Object-oriented finite element programming: frameworks for analysis, algorithms and parallel computing. Ph.D. thesis, University of California, Berkeley.

McVeigh, J.C. 1977. *Sun power — an introduction to the application of solar energy*. Pergamon Press, New York, NY, USA. 208p.

Meijerink, A.M.J., de Brouwer, H.A., Mannaerts, C.M. and Valenzuela, C.R. 1994. *Introduction to the use of GIS for Practical Hydrology*. Published by UNESCO.

Merchant, J.W., Yang, L. and Yang, W. 1994. Validation of continental scale land cover data bases developed from AVHRR data. In: Proceedings of Pecora 12 land information from space-based systems. Bethesda: ASPRS, pp. 63-72.

Michels, T. 1979. *Solar Energy Utilization*. Van Nostrand Reinhold, New York, NY, USA. 208p.

Minzheng, Lu D.D. 1999. Atmospheric Correction and Aerosol Remote Sensing Based on Radiative Transfer Model Computation, GIS Development proceedings ACRS, pp. 1-3.

Moore, R.K. 1983. *Imaging Radar Systems. Manual of Remote Sensing*, 2nd Ed., R. Colwell, (Ed.), Fall Church, VA: American Society for Photogrammetry and Remote Sensing, pp. 429-74.

Moore, G.F., Aiken, J., and Lavender, S.J. 1999. The atmospheric correction of water color and the quantitative retrieval of suspended particulate matter in case II waters: application to MERIS. *International Journal of Remote Sensing*, 20(9): 1713-33.

Moore, I.D., Grayson, R.B. and Ladson, A.R. 1991. *Digital terrain modeling: A review of hydrological, geomorphological and biological applications*. Edited by Beven, K.J. and Moore, I.D.

Morrison, D.F. 1976. *Multivariate Statistical Methods*. McGraw-Hill, New York, 415p.

Myneni, R.B. and Williams, D.L. 1994. On the relationship between FAPAR and NDVI. *Remote Sensing of the Environment*, 49: 200-11.

Nag, P. and Kudrat, M. 1998. *Digital Remote Sensing*, Concept Publishing Company, New Delhi.

Navalgund, R.R. 1993. Spectral reflectance of Earth's surface features in different wavelength regions of electromagnetic spectrum. Lecture Note, NNRMS sponsored Training course on remote Sensing—Fundamentals and Applications to Oceanography, Marine Science, Fisheries, June 7-July 16, Space Applications Center, Ahmedabad.

Naveh, Z. and Lieberman, A.S. 1984. *Landscape Ecology Theory and Application*, Springer-Verlag, New York, 356p.

Nayak, S. and Navalgund, R.R. 1998. IRS-P4 OCM applications, International year of the oceans tutorials-cum-workshop on satellite oceanography, Souvenir, 17p.

Nayak, S.R., Sarangi, R.K. and Rajawat, A.S. 2001. Application of IRS-P4 OCM data to study the impact of cyclone on coastal environment of Orissa. *Current Science*, 80(9): 1208-13.

NdiNyoungui, A., Tonye, E. and Akono, A. 2002. Evaluation of speckle filtering and texture analysis methods for land cover classification from SAR images. *International Journal of Remote Sensing*, 23: 1895-1925.

Neumann, A., Walzel, T., Tschentscher, B., Gerasch, B. and Krawczyk, H. 1996. MOS- IRS Data Processing, Software and data products, Pre-Launch Release, 24: 01-96.

NRSA. 1995. Satellite data products source book, NRSA data centre, National Remote Sensing Agency, Hyderabad, 8p.

Oppenheim, A.V. and Ronald, W.S. 1975. *Digital Signal Processing*, Englewood Cliffs, New Jersey: Prentice-Hall, Inc.

Pathan, S.K. 1993. Interpretation of remote sensing data for extraction of thematic information—Urban Studies. Lecture Note, NNRMS sponsored Training Course on Remote Sensing-Fundamentals and applications to Oceanography,

Marine Science, Fisheries, June 7-July 16, Space Applications Center, Ahmedabad.

Peake, W.H. and Oliver, T.L. 1971. The response of terrestrial surfaces of microwave frequencies, Electroscience Laboratory Report Air Force Avionics Laboratory #TR-70-301, Columbus: Ohio State University.

Pearson, R.L. and Miller, L.D. 1972. Remote mapping of standing crop biomass for estimation of the productivity of the short-grass prairie, Pawnee National Grasslands, Colorado Ann Arbor, MI: ERIM, pp. 1357-81.

Peucker, T.K., Fowler, R.J., Little, J.J. and Mark, D.M. 1978. The Triangulated Irregular Network. Proceedings, American Society of Photogrammetry: Digital Terrain Models (DTM) Symposium, St. Louis, Missouri, pp. 516-40.

Peuquet, D.J. 1990. A conceptual framework and comparison of spatial data models. *Introductory Readings in Geographic Information Systems*. Peuquet, D.J. and Marble, D.F. London and Bristol, PA, Taylor & Francis, pp. 250-85.

Philipson, W.R. and Teng, W.L. 1988. Operational Interpretation of AVHRR. Vegetation Indices for World Crop information. *Photogrammetric Engineering and Remote Sensing,* 54: 55-59.

Philipson, W. 1997. *Manual of Photographic Interpretation*, 2nd Edition, Bethesda, MD: American Society for Photogrammetry & Remote Sensing, 555p.

Pickles, J. 1999. Arguments, Debates, and Dialogues: the GIS-Social Theory Debate and the Concern for Alternatives. In Longley, P.A., Goodchild, M.F., MacGuire, D.J. and Rhind, D.W. (Eds.) *Geographical Information Systems: Principles, Techniques, Applications, and Management* (2nd ed.), New York: John Wiley and Sons.

Pinty, S. and Verstrate, M. 1992. GEMI: A non-linear index to monitor global vegetation from satellites. Vegetation, 101: 15-20.

Piper, S.E. 1983. The evaluation of the spatial accuracy of computer classification. In: Proceedings of the 1983 Machine Processing of Remotely Sensed Data Symposium West Lafayette, Purdue University, pp. 303-10.

Pontius, R.G. 2000. Quantification error versus location error in comparison of categorical, 66 maps. *Photogrammetric Engineering and Remote Sensing*: 1011-16.

Potter, J.F. and Mendlowitz, M.A. 1975. On the determination of haze levels from Landsat data. International Symposium on Remote Sensing of Environment, 10th, Ann Arbor, Mich. ; United States; 6-10 Oct., pp. 695-703.

Pratt, W.K. 1978. *Digital Image Processing*, John Wiley and Sons, New York, pp. 201-344.

Pratt, W.K. 1991. *Digital Image Processing*. 2nd Ed. New York: John Wiley & Sons, Inc.

Preisendorfer, R.W. 1988. *Principal Component Analysis in Meteorology and Oceanography*. Elsevier, New York, 425p.

Press, W.H., Flannery, B.P., Teukolsky, S.A. and Vetterling, W.T. 1988. *Numerical Recipes in C. New York*, New York: Cambridge University Press.

Qi, J., Kerr, Y. and Chehbouni, A. 1994. External Factor Consideration in Vegetation Index Development. In Proc. of Physical Measurements and Signatures in Remote Sensing, ISPRS, pp. 723-30.

Qi, J., Chehbouni, A., Huete, A.R. and Kerr, Y.H. 1994. Modified Soil Adjusted Vegetation Index (MSAVI). Remote Sensing of the Environment, 48: 119-26.

Ramage, S. and Woodsford, P. 2002. The Benefits of Topology in the Database, http://spatialnews.geocomm.com/features/laserscan2/.

Ramanathan, V. and Vogelmann, A.M. 1997. Greenhouse effect, atmospheric solar absorption and the earth's solar radiation budget: From the Arrehenius-Langley era to 1990s, Ambio, 26(1): 38-46.

Rao, D.P. 1999. *Remote Sensing for Earth Resources*, 2nd Edition, AEG Publications, Hyderabad, India, 137p.

Ready, P.J. and Wintz, P. 1973. Information extraction, SNR improvement and data compression in multispectral imagery. IEEE Trans. Commu., 21: 1123.

Rees, W. 1990. The Physical Principles of Remote Sensing, Cambridge University Press, Cambridge, UK.

Rees, W.G. 2001. *Physical Principles of Remote Sensing*, 2nd Edition, Cambridge University Press, 296p.

Reeves, R.G., 1979. *Manual of Remote Sensing, American Society of Photogrammetry*. Reeves, R.G. (Ed.), 1975. *Manual of Remote Sensing*, American Society of Photogrammetry, Falls Church, Virginia.

Reginato, R.J., Vedder, J.F., Idso, S.B., Jackson, R.D., Blanchard, M.B. and Goettelman R. 1977. An Evaluation of Total Solar Reflectance and Spectral Band Ratioing Techniques for Estimating Soil Water Content. J. *Geophysical Research*, 82(15): 2101-04.

Richards, J. 1986. *Remote Sensing Digital Image Analysis: An Introduction*. Springer-Verlag. New York, 28p.

Richards, J.A. 1986. *Digital Image Processing*. Springer-Verlag: Berlin.

Richards, J.A. 1993. *Remote Sensing Digital Image Analysis: An Introduction*. 2nd Ed., Springer Verlag, Berlin.

Richards, J.A. and Xiaping, J. 1999. *Remote Sensing Digital Image Analysis: An Introduction*, Springer-Verlag, Berlin.

Richardson, A.J. and Wiegand, C.L. 1977. Distinguishing vegetation from soil background information. *Photogrammetric Engineering and Remote Sensing*, 43(2): 1541-52.

Richter, R. 1990. A fast atmospheric correction algorithm applied to Landsat TM images. *International Journal of Remote Sensing*, 11: 159-66.

Robinson, G.J. 1994. The accuracy of digital elevation models derived from digitized contour data. Photogrammetric Record 14(83): 805-14.

Roujean, J.L. and Breon, F.M. 1995. Estimating PAR absorbed by vegetation from bidirectional reflectance measurements. *Remote Sensing of the Environment*, 51: 375-84.

Rosenfeld, G.H. and Fitzpatrick-Lins, K. 1986. A coefficient of agreement as a measure of thematic classification accuracy. *Photogrammetric Engineering and Remote Sensing*, 52: 223-27.

Rosenfeld, A. and Kak, A.C. 1976. Digital Picture Processing. Academic Press, New York.

Rouse, J.W., Haas, R.H., Schell, J.A., and Deering, D.W. 1973. "Monitoring vegetation systems in the great plains with ERTS," Third ERTS Symposium, NASA SP-351, 1: 309-17.

Rouse, J.W. Jr., Haas, R.H., Deering, D.W., Schell, J.A. and Harlan, J.C. 1974. Monitoring the Vernal Advancement and Retrogradation (Green Wave Effect) of Natural Vegetation. NASA/GSFC Type III Final Report, Greenbelt, MD, 371p.

Running, S.W., Justice, C.O., Solomonson, V., Hall, D., Barker, J., Kaufmann, Y.J., Strahler, A.H., Huete, A.R., Muller, J.P., Vanderbilt, V., Wan, Z.M., Teillet, P. and Carneggie, D. 1994. Terrestrial remote sensing science and algorithms planned for the NASA Strennis Space Center, Washington, DC, NASA, 47p.

Sabins, F.F. 1978. *Remote Sensing: Principles and Interpretation.* Freeman, W.H. San Francisco, CA, USA. 426p.

Sabins, F.F. 1986. *Remote Sensing—Principle and Interpretation*, Freeman and Co., NY, Chapter 7: 246-60.

Sabins, F.F. 1997. *Remote Sensing: Principles and Interpretation.* 3rd Edition. New York: Freeman, W.H. San Francisco, CA, USA.

Saha, S.K., Kudrat, M. and Bhan, S.K. 1988. Evaluation of digitally enhanced Landsat TM data for wasteland mapping in part of Aligarh district, U.P., Proceedings, National Symposium on Remote Sensing in Rural Development, pp. 333-42.

Sahai, B. 1991. Basic Concepts in Remote Sensing, Proceedings, Third Workshop on Remote Sensing Fundamentals and Applications, Chandigarh, September 26-27, Punjab Remote Sensing Center, pp. 1-25.

Sahai, B. 1983. Manual of Image Interpretation-I: In "Lecture Notes on Orbital Remote Sensing", SAC/RSA/AGPA/TN/03/83, pp. 5.1-5.17.

Sahoo, P.K., Soltani, S. and Wong, A.K.C. 1988. A survey of thresholding techniques. Computer Vision, Graphics, and Image Processing, 41: 233-60.

Sarkar, A. 1998. Application of Oceansat-1 MSMR, International year of the oceans tutorials-cum-workshop on satellite oceanography, *Souvenir*, 17p.

Sathyendranath, S., Cota, G., Stuart, V., Maas, S.H. and Platt, T. 2001. Remote sensing of phytoplankton pigments: a comparison of Empirical and Theoretical approaches. *International Journal of Remote Sensing*, 22(2-3): 249-73.

Schott, J.R. 1997. Remote Sensing: the image chain approach, Oxford University Press, NY. Schowengerdt, R.A. 1997. Remote Sensing: Models and Methods for Image Processing, 230p.

Scientific Note, SAC. 1992. Coastal Environment, Scientific Note, RSAM/SAC/COM/SN/11/92, Space Application Center (ISRO), Ahmedabad, pp. 10-16.

Sears, F.W., Zemansky, M.W. and Young, H.D. 1979. *University Physics*. 5th ed. Reading, Mass: Addison-Wesley Publishing.

Sellers, V.D. 1965. *Physical Climatology*. University of Chicago Press, Chicago.

Sharma, S.A., Bhatt, H.P and Ajai. 1990. Generation of brightness and green transformation for IRS-LISS II data, Journal of *Indian Society of Remote Sensing*, 18(3): 25-31.

Shih, E.H.H. and Schowengerdt, R.A. 1983. Classification of arid geomorphic surfaces using spectral and textural features. *Photogrammetric Engineering and Remote Sensing*, 49 (3): 337-47.

Shmutter, B. and Doytsher, Y. 1990. DTM-accuracy estimates. Proceedings of the Fourth International Symposium on spatial data handling, Zurich, pp. 263-72.

Short, N.M., 1982. *The Landsat Tutorial Workbooks: Basics of Satellite Remote Sensing*. NASA Reference Publications, 1078.

Short, N.M. 1999. The Remote Sensing Tutorial, NASA/ Goddard Space Flight Center. On line http:// rst.gcfc.nasa.gov/Front/tofc.html.

Skidmore, A. 1989. Unsupervised Training Area Selection in Forests Using a Non-Parametric Distance Measure and Spatial Information. *International Journal of Remote Sensing*, 10(1): 133-46.

Slater, P.N. 1980. *Remote Sensing: Optics and Optical Systems*, Addison-Wesley, Reading, Mass.

Smith, W.L., Woolf, H.M. and Hayden, C.M. 1979. The TIROS-N Operational Vertical Sounder. *Bull. Amer. Meteor. Soc.*, 60: 1177-87.

Smits, P.C., Dellepiane, S.G. and Schowengerdt, R.A. 1999. Quality assessment of image classification algorithms for land-cover mapping: a review and proposal for a cost-based approach. *International Journal of Remote Sensing*, 20: 1461-86.

Sokal, R.R. and Sneath, P.H.A. 1963. *Principles of Numerical Taxonomy*, Freeman, McGraw-Hill Inc, New York.

Somers, R. 1990. Management topics in GIS development. URISA/GIS '90 Workshop, Vancouver, B.C., Canada. 136p.

Stehmen, S.V. 1997. Selecting and interpreting measures of thematic classification accuracy. *Remote Sensing of the Environment*, 62: 77-89.

Stewart, R. 1985. *Methods of Satellite Oceanography*. University of California Press, Berkeley, California.

Stoner, E.R. and Baumgartner, M.F. 1980. Physico-chemical, Site and Bidirectional Reflectance Factor Characteristics of Uniformly Moist Soils. SR-PO-00431 (Contrast NAS9-

15466), *Purdue* Univ., Feb. (Available as NASA CR-160571).

Story, M. and Congalton, R.G. 1986. Accuracy assessment: A user's perspective. *Photogrammetric Engineering and Remote Sensing*, 52: 397-99.

Sturm, B. 1980. The atmospheric corrections of remotely sensed data and the qualitative determination of suspended matter in marine water surface layers. Edited by A.P. Cracknell, Remote Sensing in Meteorology, Oceanography and Hydrology, Ellis Horwood Ltd., Chichester, pp. 163-97.

Swain, P.H. and Davis, S.M. 1978. *Remote Sensing: Quantitative Approach*. Chapter 3, Mc GrawHill Books Company, New York, pp. 883-96.

Tanre, D., Deschamps, P.Y., Duhaut, P. and Herman, M. 1987. Adjacency effect produced by atmospheric scattering in Thematic Mapper data. *Journal of Geophysical Research*, 92(D10): 12000-06.

Tares, T. and Hallikainen, M. 2003. Electric Field ratio Formalism in Radar Polarimetry, Workshop on applications of SAR polarimetry and polarimetric interferometry, ESA-ESRIN, Italy, pp. 1-4.

Tempfli, K. 1980. Spectral analysis of terrain relief for the accuracy estimation of digital terrain models. *ITC Journal*, 3: 478-510.

Teng, W.L. 1990. AVHRR monitoring of US crops during the 1988 drought. *Photogrammetric Engineering and Remote Sensing*, 56: 1143-46.

Thiam, A.K. 1997. Geographic Information Systems and Remote Sensing Methods for Assessing and Monitoring Land Degradation in the Sahel: The Case of Southern Mauritania. Doctoral Dissertation, Clark University, Worcester Massachusetts.

Thomas, I.L., Howorth, R., Eggers, A.E. and Fowler, A.D.W. 1981. Textural Enhancement of a circular geological feature. *Photogrammetric Engineering and Remote Sensing*, 47: 89-91.

Thomlinson, J.R., Bolstad, P.V. and Cohen, W.B. 1999. Coordinating methodologies for scaling land cover classifications from site-specific to global: steps toward validating global map products. *Remote Sensing of the Environment*, 70: 16-28.

Thompson, D.R. and Wehmanen, O.A. 1980. Using Landsat digital data to detect moisture stress in corn-soyabean growing regions. *Photogrammetric Engineering and Remote Sensing*, 46: 1082-89.

Tou, J.T. and Gonzalez, R.C. 1974. *Pattern Recognition Principles*. Reading, Massachusetts: Addison-Wesley Publishing Company.

Townshend, V.R.G. (Ed.). 1981. *Terrain Analysis and Remote Sensing*, George Allen & Unwin Ltd., London.

Townshend, J.R.G. and Tucker, C.J. 1984. Objective Assessment of Advanced Very High Resolution Radiometer Data for Land Cover Mapping. *International Journal of Remote Sensing*, Vol. 5, No. 2, pp. 497-504.

Townsend, P.A. 2000. A quantitative fuzzy approach to assess mapped vegetation classification for ecological applications. *Remote Sensing of Environment*, 72: 253-67.

Tripathy, G.K., Ghosh, T.K. and Shah, S.D. 1996. Monitoring of Desertification Process in Karnataka State of India Using Multi-Temporal Remote Sensing and Ancillary Information Using GIS. *International Journal of Remote Sensing*, 17(12): 2243-57.

Tsichritzis, T.C. and Lochovsky, F.H. 1977. *Data Base Management Systems*, Academic Press, New York.

Tucker, C.J. 1979. Red and Photographic Infrared Linear Combinations for Monitoring Vegetation. *Remote Sensing of the Environment*, 8: 127-50.

Turk, G. 1979. GT index: a measure of the success of prediction. *Remote Sensing of the Environment*, 8: 75-86.

Turner, M. 1990. Landscape Changes in Nine Rural Counties in Georgia. *Photogrammetric Engineering and Remote Sensing*. 56(3): 379-86.

Tyndal, J. 1861. On the absorption and radiation of heat by gases and vapors, and on the physical concentration of radiation, absorption and conduction, Phil. Trans. Roy. Soc., A151: 1-36.

U.S. Geological Survey. 1999. The Landsat Satellite System Link, USGS on the World Wide Web. URL: http://landsat7.usgs.gov/landsat sat.html. 11/10/99.

Vane, G., Billingsley, F.C. and Dunne, J.A. 1982. Observational parameters for remote sensing in the next decade. Advanced Multispectral Remote Sensing Technology and Applications, Proceedings of SPIE-The International Society for Optical Engineering, Ken J. Ando (ed.), 345: 52-65.

Vanzetti, V. 1972. *Practical Applications of Infrared Techniques*. John Wiley, New York.

Veregin, H. 1994. Integration of simulation modeling and error propagation for the buffer operation in GIS. *Photogrammetric Engineering and Remote Sensing*, 60: 427-35.

VIPS-32, 1986. Software Documents, SEP, France.

Wang, M. 1999. A sensitivity study of the SeaWiFS atmospheric correction algorithm: Effects of spectral band variation. *Remote Sensing of the Environment*, 67: 348-59.

Waring, R.H., Way, E.R., Morrissey, H.L., Ranson, K.J., Weishampel, J.F., Oren, R. and Franlin, S.E. 1995. Imaging radar for ecosystem studies. *BioScience*, 45(10): 715-23.

Warren, S. 1982. Optical properties of snow. Reviews of Geophysics and Space Physics, 20: 67.

Washington-Allen, R.A. 1994. Historical analysis of land use/land cover change on the Bolivian Altiplano: *A remote sensing perspective*. MS Thesis, Utah State University, Logan, USA.

Washington-Allen, R.A., Ramsey, R.D., Norton, B.E. and West, N.E. 1998. Change detection of the effect of severe drought on subsistence agropastoral communities on the Bolivian Altiplano. *International Journal of Remote Sensing*, 19: 1319-33.

Welch, R. 1990. *3-D Terrain Modeling for GIS Applications.* GIS World, 3 (5): 26-30.

Weszka, J.S., Dyer, C.R. and Rosenfeld, A. 1976. A comparative study of texture measures for terrain classification. IEEE Transactions on Systems, Man and Cybernetics, 6: 269-85.

WHO (World Health Organization). 1979. Environmental health criteria 14: Ultraviolet radiation. WHO, Geneva, Switzerland. 110p.

Woodcock, C.E. 1982. *Reducing the Influence of Topography on the Classification of Remotely Sensed Data.* Masters thesis, University of California, Santa Barbara, 69p.

Woodcock, C.E. and Gopal, S. 2000. Fuzzy set theory and thematic maps: accuracy assessment and area estimation. *International Journal of Geographical Information Science,* 14: 153-72.

Worboys, M. 1998. Imprecision in finite resolution spatial data. *GeoInformatica*, 2: 257-79.

Yin, L., Yang, R., Gabbouj, M. and Neuvo, Y. 1996. Weighted Median Filters: A Tutorial. IEEE Trans. on Circuits and Systems, 43(3): 157-92.

Zagolski, F. and Gastellu-Etchegorry, J.P. 1995. Atmospheric Corrections of AIRIS images with a procedure based on the inversion of the 5S model, *International Journal Remote Sensing*, 16(16): 3115-46.

Zebker, H.A. and VanZyl, S.I. 1991. Imaging Radar Polarimetry, Proceedings of IEEE, 79: 1583-1606.

Zhu, Q. and Chen, C. 1998. The fast generation and dynamic updating of TIN. *Journal of Wuhan Technical University of Surveying and Mapping.*

Ziou, D. and Tabbone, S. 1998. Edge Detection Techniques: An Overview. *International Journal of Pattern Recognition and Image Analysis*, 8(4): 537-59.

Index

A

B

C

D

E

O

P

Q

R

T